Lino Guzzella · Antonio Sciarretta

Vehicle Propulsion Systems

Lino Guzzella · Antonio Sciarretta

Vehicle Propulsion Systems

Introduction to Modeling and Optimization

Second Edition

With 202 Figures and 30 Tables

Springer

Prof. Dr. Lino Guzzella
Dr. Antonio Sciarretta
ETH Zürich
Inst. Mess- und Regeltechnik
Sonneggstr. 3
8092 Zürich
Switzerland
lguzzella@ethz.ch
Antonio.Sciarretta@ifp.fr

Library of Congress Control Number: 2007934932

ISBN 978-3-540-74691-1 2nd Edition Springer Berlin Heidelberg New York
ISBN 978-3-540-25195-8 1st Edition Springer Berlin Heidelberg New York

Springer is a part of Springer Science+Business Media

springer.com

© Springer-Verlag Berlin Heidelberg 2005, 2007

Typesetting: Data supplied by the authors
Production: LE-TEX Jelonek, Schmidt & Vöckler GbR, Leipzig
Cover Design: eStudioCalamar S.L., F. Steinen-Broo, Girona, Spain

SPIN 11977568   60/3180/YL   5 4 3 2 1 0   Printed on acid-free paper

Preface

Who should read this text?

This text is intended for persons interested in the analysis and optimization of vehicle propulsion systems. Its focus lies on the control-oriented mathematical description of the physical processes and on the model-based optimization of the system structure and of the supervisory control algorithms.

This text has evolved from a lecture series held during the last years in the mechanical engineering department at the Swiss Federal Institute of Technology (ETH), Zurich. The presumed audience is graduate mechanical or electrical engineering students. The prerequisites are general engineering topics and a first course in optimal control theory. Readers with little preparation in that area are referred to [30]. The most important results of parameter optimization and optimal control theory are summarized in Appendix II.

Why has this text been written?

Individual mobility relies to a large extent on passenger cars. These vehicles are responsible for a large part of the world's consumption of primary energy carriers, mostly fossil liquid hydrocarbons. The specific application profiles of these vehicles, combined with the inexorably increasing demand for mobility, have led to a situation where the reduction of fuel consumption has become a top priority for the society and the economy.

Many approaches that permit to reduce the fuel consumption of passenger cars have been presented so far and new ideas emerge on a regular basis. In most – if not all – cases these new systems are more complex than the traditional approaches. Additional electric motors, storage devices, torque converters, etc. are added with the intention to improve the system behavior. For such complex systems the traditional heuristic design approaches fail.

The only way to deal with such a high complexity is to employ mathematical models of the relevant processes and to use these models in a systematic ("model-based") way. This text focuses on such approaches and provides an introduction to the modeling and optimization problems typically encountered by designers of new propulsion systems for passenger cars.

What can be learned from this text?

This book analyzes the longitudinal behavior of road vehicles only. Its main emphasis is on the analysis and minimization of the energy consumption. Other aspects that are discussed are drivability and performance.

The starting point for all subsequent steps is the derivation of simple yet realistic mathematical models that describe the behavior of vehicles, prime movers, energy converters, and energy storage systems. Typically, these models are used in a subsequent optimization step to synthesize optimal vehicle configurations and energy management strategies.

Examples of modeling and optimization problems are included in Appendix I. These *case studies* are intended to familiarize the reader with the methods and tools used in powertrain optimization projects.

What cannot be learned from this text?

This text does not consider the pollutant emissions of the various powertrain systems because the relevant mechanisms of the pollutant formation are described on much shorter time scales than those of the fuel consumption. Moreover, the pollutant emissions of some prime movers are virtually zero or can be brought to that level with the help of appropriate exhaust gas purification systems. Readers interested in these aspects can find more information in [100].

Comfort issues (noise, harshness, and vibrations) are neglected as well. Only those aspects of the lateral and horizontal vehicle dynamics that influence the energy consumption are briefly mentioned. All other aspects of the horizontal and lateral vehicle dynamics, such as vehicle stability, roll-over dynamics, etc. are not discussed.

Acknowledgments

Many people have implicitly helped us to prepare this manuscript. Specifically our teachers, colleagues, and students have contributed to bring us to the point where we felt ready to write this text. Several people have helped us more explicitly in preparing this manuscript: Hansueli Hörler, who taught us the basic laws of engine thermodynamics, Alois Amstutz and Chris Onder who contributed to the development of the lecture series behind this text, those of our doctoral students whose dissertations have been used as the nucleus of several sections (we reference their work at the appropriate places), and Brigitte Rohrbach, who translated our manuscripts from "Italish" to English.

June 2005   *Lino Guzzella and Antonio Sciarretta*

Preface

Why a second edition?

The discussions about fuel economy of passenger cars have become even more intense since the first edition of this book appeared. Concerns about the limited resources of fossil fuels and the detrimental effects of greenhouse gases have spurred the interest of many people in industry and academia to work towards reduced fuel consumption of automobiles. Not surprisingly, the first edition of this monograph sold out rather rapidly. When the publisher asked us about a second edition, we decided to use this opportunity to revise the text, to correct several errors, and to add new material.

The following list includes the most important changes and additions we made:

- The section describing battery models has been expanded.
- A new section on power split devices has been added.
- A new section on pneumatic hybrid systems has been added.
- The chapter introducing supervisory control algorithm has been rewritten and expanded.
- Two new case studies have been added.
- A new appendix that introduces the main ideas of dynamic programming has been added.

Acknowledgements

We want to express our gratitude to the many colleagues and students who reported to us errors and omissions in the first edition of this text. Several people have helped us improving this monograph, in particular Christopher Onder who actively participated in the revisions.

June 2007   *Lino Guzzella and Antonio Sciarretta*

Contents

1

Introduction

This introductory chapter shows how the problems discussed in this text are embedded in a broader setting. First a motivation for and the objective of the subsequent analysis is introduced. After that the complete energy conversion chain is described, starting from the available primary energy sources and ending with the distance driven. Using average energy conversion efficiency values, some of the available options are compared. This analysis shows the importance of the "upstream" processes. The importance of the selected on-board energy carrier ("fuel") is stressed as well. In particular its energy density and the safety issues connected with the refueling process are emphasized. The last section of this first chapter lists the main options available for reducing the energy consumption of passenger vehicles.

1.1 Motivation

The main motivation to write this book is the inexorably increasing number of passenger cars worldwide. As Fig. 1.1 shows, some 800 million passenger cars are operated today. More interesting than this figure is the trend that is illustrated in this figure for the example of the United States of America (the same trend is observed in Japan and Europe): in wealthy societies the car density saturates at a ratio of approximately 400 to 800 cars per 1000 inhabitants.

It is corroborated empirically that the demand for personal transportation increases with the economic possibilities of a society [220]. Therefore, if the car density mentioned above is taken as the likely future value for other regions of the world, serious problems are to be expected. Countries such as China (1.3 billion inhabitants) or India (1.1 billion inhabitants) in the year 2007 have car densities of around 30 cars per 1000 inhabitants. Accordingly, in the next 20 years the car density in these countries will increase substantially, which will further increase the pressure on fuel prices and cause serious problems to the environment.

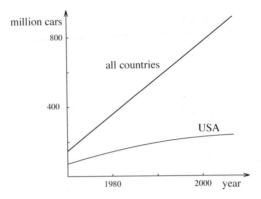

Fig. 1.1. *Schematic* representation of the development of the number of passenger cars operated worldwide.

In the face of these trends, it is clear that new fuel sources must be developed and that the fuel consumption of passenger cars must be reduced substantially. This text focuses on the second approach.

1.2 Objectives

The main objectives of this text are to introduce mathematical models and optimization methods that permit a systematic minimization of the *energy consumption* of vehicle propulsion systems. The objects of this analysis are *passenger cars*, i.e., vehicles that

- are autonomous and do not depend on fixed energy-providing grids;
- have a refueling time that is negligible compared to the driving time between two refueling events;
- can transport two to six persons and some payload; and
- accelerate in approximately 10 to 15 seconds from 0 to 100 km/h, or can drive uphill a 5% ramp at the legal top speed, respectively.[1]

These requirements, which over the last one hundred years have evolved to a quasi-standard profile, substantially reduce the available options. Particularly the first and the second requirement can only be satisfied by few *on-board* energy storage systems, and the performance requirements can only be satisfied by propulsion systems able to produce a maximum power that is substantially larger than the power needed for most driving conditions.

A key element in all considerations is the on-board energy carrier system. This element must:

---

[1] These numerical values are only indicative. It goes without saying that the performance range is very wide.

- provide the highest possible energy density[2];
- allow for the shortest possible refueling time; and
- be safe and cause no environmental hazards in production, operation, and recycling.

The number of components that are necessary to realize modern and in particular future propulsion systems is inexorably increasing. Improved performance and fuel economy can only be obtained with complex devices. Of course, these subsystems influence each other. The best possible results are thus not obtained by an isolated optimization of each single component. Optimizing the entire system, however, is not possible with heuristic methods due to the "curse of exponential growth." The only viable approach to cope with this dilemma is to develop mathematical models of the components and to use model-based numerical methods to optimize the system structure and the necessary control algorithms. These models must be able to *extrapolate* the system behavior. In fact, such an optimization usually takes place before the actual components are available or requires the devices to operate in unexpected conditions. For these reasons, only first-principle models, i.e., models that are based on physical laws, will be used in this text.

Of course, some of the mathematical models and methods introduced in this text may be useful for the design of other classes of vehicles (trains, heavy-duty trucks, etc.). However, there are clear differences[3] that render the passenger car optimization problem particularly interesting.

At least three energy conversion steps are relevant for a comprehensive analysis of the energy consumption of passenger cars. As illustrated in Fig. 1.2, the actual energy source is one of the available primary energy carriers (chemical energy in fossil hydrocarbons, solar radiation used to produce bio mass or electric energy, nuclear energy, etc). In a first step, this energy is converted to an energy carrier that is suitable for on-board storage, i.e., to a "fuel" (examples are gasoline, hydrogen, etc.). This "fuel" is then converted by the propulsion system to mechanical energy that, in part, may be stored as kinetic or potential energy in the vehicle. The third energy transformation is determined by the vehicle parameters and the driving profile. In this step, the mechanical energy produced in the second conversion step is ultimately dissipated to thermal energy that is deposited to the ambient. The terms "well-to-tank," "tank-to-vehicle," and "vehicle-to-miles" are used in this text to refer to these three conversion steps. Unfortunately, all of these conversion processes cause substantial energy losses.

---

[2] The energy density here is defined as the amount of *net* energy available for propulsion purposes divided by the mass of the energy carrier necessary to generate that propulsion energy, including all containment elements but not the on-board energy transformation devices.

[3] For instance, the autonomy requirement and the dominance of part-load operation will be relevant for the optimization problems.

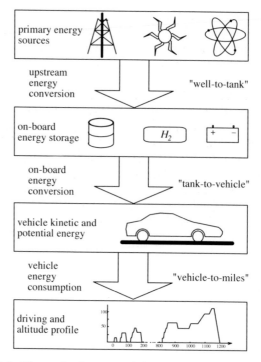

Fig. 1.2. The main elements of the energy conversion scheme.

This text will not address any control problems pertaining to the "well-to-tank" energy conversion. The systems used for that conversion are very large power plants, refineries, or other process engineering systems. Of course, their average efficiency values and pollutant emission have an important impact on the economy and ecology. However, the problems arising in that area and the methods required to solve those problems belong to a different class.

In the next section an overview of the most important energy conversion approaches is presented. With this information, a preliminary estimation of the total energy consumption is possible. Note that a correct comparison is not easy, if at all possible.[4] Readers interested in a broader discussion are referred to [34].

The main physical phenomena influencing the "vehicle-to-miles" energy conversion will be discussed in Chap. 2. That chapter will mainly introduce descriptions that are *quasistatic* (this term will be precisely defined below),

---

[4] For instance, the total "well-to-miles" carbon dioxide emissions are often used to compare two competing approaches. However, such a discussion is not complete unless the "gray" energy invested in the vehicles, refineries and plants is considered. Even more difficult: how to take into account the problems associated with nuclear waste repositories, landscape degradation caused by windmills, or nitric oxide emission of coal-fired power plants?

but also *dynamic* models will be presented. In this context, it is important to understand the impact of the driving profile that the vehicle is assumed to follow. As mentioned above, only those effects are considered that have a substantial influence on the energy consumption.

The main emphasis of this text is on the modeling and optimization of the "tank-to-vehicle" energy conversion systems. For this problem suitable mathematical models of the most important devices will be introduced in Chaps. 3 through 6. Chapter 7 presents methods with which the energy consumption can be minimized. All of these methods are model-based, i.e., they rely on the mathematical models derived in the previous chapters and on systematic optimization procedures to find (local) minima of precisely cast optimization problems. Eight case studies are included in Appendix I. Appendix II then summarizes the most important facts of parameter optimization and optimal control theory and Appendix III introduces the main ideas of dynamic programming.

1.3 Upstream Processes

As mentioned above, a detailed analysis of the "well-to-tank" energy conversion processes is not in the scope of this text. However, the efficiency and the economy of these systems are important aspects of a comprehensive analysis. For this reason a rather preliminary but nevertheless instructive overview of the main energy conversion systems is given in this section.

Figure 1.3 shows a part of that complex network. The efficiency numbers given in that figure are approximate and are valid for available technology. The CO_2 factors relate the amount of carbon dioxide emitted by using one energy unit of natural gas or coal to the amount emitted when using one energy unit of oil.[5] Solar and nuclear primary energy sources are assumed to emit no CO_2, i.e., the gray energy and the associated CO_2 emission are not shown in that figure.

Only three systems are considered in Fig. 1.3 for the conversion of "fuel" to mechanical energy: a spark-ignited (SI) or gasoline internal combustion engine (ICE), a compression-ignited (CI) or Diesel ICE, and an electric motor. Average "tank-to-vehicle" efficiencies of these prime movers are shown in Fig. 1.3 as well.[6] The mechanical energy consumption (the "vehicle-to-miles" efficiency) is approximated by an equation that is valid for the European test cycle (this expression will be introduced in Sect. 2.2). With the information shown in Fig. 1.3 it is easy to make some preliminary, back-of-the-envelope-style calculations that, despite the many uncertainties, are quite instructive.

---

[5] The CO_2 factors reflect the different chemical composition *and* the different heating values. The base line is defined in Table 1.1.

[6] The peak efficiencies of all of these devices are (substantially) higher. However, the relevant data are the cycle-averaged efficiencies, which are close to the values shown in Fig. 1.3.

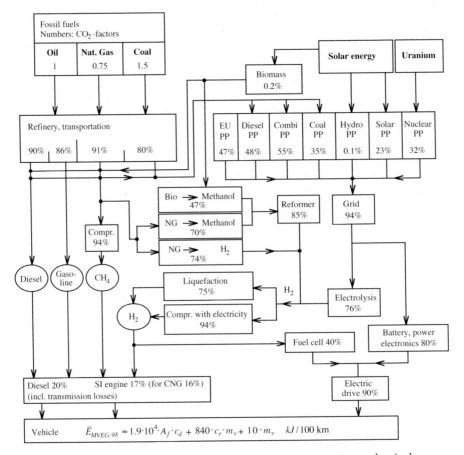

Fig. 1.3. Different paths to convert a primary energy source to mechanical energy needed to drive a car in the MVEG-95 test cycle. Source: [69] and own data.

For that purpose Table 1.1 summarizes some of the most important parameters of the fuels considered below.

Table 1.1. Main parameters of some important energy carriers (lower heating value H_l, hydrogen-to-carbon ration H/C, and mass of CO_2 emitted per mass fuel burned ν); CNG = compressed natural gas.

	$H_l \ (MJ/kg)$	H/C	ν
oil	43	≈ 2	3.2
CNG ($\approx$ methane)	50	4	2.75
coal ($\approx$ carbon)	34	0	3.7
hydrogen	121	∞	0

Figure 1.4 shows the "well-to-miles" carbon dioxide emissions of three ICE-based powertrains. The vehicle assumed in these considerations is a standard mid-size passenger car. The efficiency values of the gasoline and Diesel engines are standard values as well. The efficiency of CNG engines is usually slightly smaller than the one of gasoline SI engines [12].

Of course this analysis neglects several important factors, for instance the greenhouse potential of methane losses in the fueling infrastructure. Nevertheless, the results obtained indicate that increasing the numbers of CNG engines could be one option to reduce CO_2 emissions with relatively small changes in the design of the propulsion system. Unfortunately, as mentioned before, the "well-to-miles" CO_2 emission levels are just one element of the problem space. In this case, the reduced energy density of CNG as on-board energy carrier has, so far, inhibited a broader market penetration of this vehicle class. The next section will show more details on this aspect.

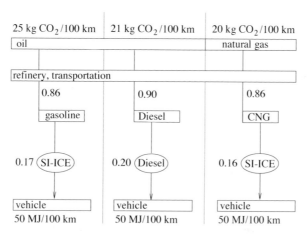

Fig. 1.4. "Well-to-miles" CO_2 emission of three conventional powertrains. The vehicle is described by the parameters $m = 1600$ kg, $c_d \cdot A_f = 0.86\,\mathrm{m}^2$, and $c_r = 0.013$ (see Chap. 2). The fuel properties are defined in Table 1.1.

Figure 1.5 shows what amount of CO_2 emissions can be expected when a battery-electric propulsion system is employed. The base vehicle is assumed to have the same[7] parameters as the one used to compute the values shown in Fig. 1.4. Several primary energy sources are compared in this analysis. The two CO_2-neutral[8] energy sources (solar and nuclear energy) produce no

---

[7] Of course the batteries substantially increase the vehicle mass. Here the (optimistic) assumption is adopted that the recuperation capabilities of the battery electric system compensate for the losses that are caused by this additional mass.

[8] As mentioned, only the CO_2 emission caused by the operation of the power plants are considered.

carbon dioxide emission. However, if the electric energy required to charge the batteries is generated using fossil primary energy sources, surprisingly different CO_2 emission levels result.

In the case of a natural-gas-fired combined-cycle power plant (PP) the CO_2 emission levels are substantially lower than those of traditional ICE-based propulsion systems. However, if the other limit case (coal-fired steam turbines) is taken into consideration, the "well-to-miles" carbon dioxide emission levels of a battery-electric car become even worse than those of the worst ICE-based propulsion system.[9] Moreover, in the next section it will be shown that the energy density of batteries is so small that battery electric vehicles cannot satisfy the specifications of a passenger car as defined in Sect. 1.2.

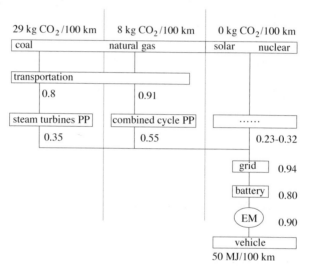

Fig. 1.5. "Well-to-miles" CO_2 emission of a battery electric vehicle. Vehicle parameters as in Fig. 1.4. Battery efficiency includes charging, discharging, and power electronic losses. The fuel properties are defined in Table 1.1.

As a last example, the estimated "well-to-miles" CO_2 emission levels of a fuel cell electric vehicle are shown in Fig. 1.6. Again, the vehicle parameters have been chosen to be the same as in the conventional case. The efficiency of the fuel cell *system* has been assumed to be around 0.40. Despite many more optimistic claims, experimental evidence, as the one published in

---

[9] Of course, low CO_2 primary energy sources should first be used to replace the worst polluting power plants that are part of the corresponding grid. In this sense, each unit of *additional* electric energy used must be considered to have been produced by the power plant in the grid that has the *worst* efficiency. Accordingly, in the example shown in Fig. 1.5 the relevant CO_2 emission number is the one valid for coal-fired power plants.

[212], has shown that the net efficiency of a fuel cell *system* will probably be close to that figure.[10] Even more uncertain are the efficiencies of on-board gasoline-to-hydrogen reformers. Including all auxiliary devices, a net efficiency of approximately 60–70% may be expected.

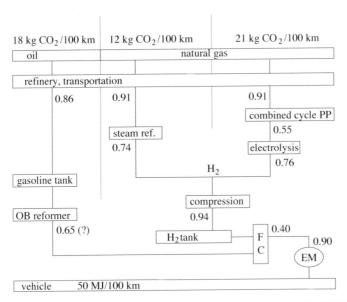

Fig. 1.6. "Well-to-miles" CO_2 emission of a fuel cell electric vehicle. Vehicle parameters as in Fig. 1.4. The efficiency of the on-board gasoline-to-hydrogen reformer is not experimentally verified. The fuel properties are defined in Table 1.1.

The main insight that can be gained from Fig. 1.6 is that as long as fossil primary energy sources are used fuel cell electric vehicles have a potential to reduce the "well-to-miles" CO_2 emission only if the hydrogen is produced in a steam reforming process using natural gas as primary energy source. As shown in Fig. 1.6, fuel-cell-based powertrains have excellent "tank-to-vehicle" but rather poor "well-to-tank" efficiencies. This fact will become very important once *renewable* primary energy sources are available on a large scale. If this comes true, then the "upstream" CO_2 emission levels are zero and the only concern will be to utilize the available on-board energy as efficiently as possible. In this situation fuel-cell-based propulsion systems might prove to be the best choice.

---

[10] Fuel cells must be supercharged to achieve sufficient power densities and to exploit in the best possible way the expensive electrochemical converters. The compressors that are necessary for that consume in the order of 20–25 percent of the electric power produced by the fuel cell [212].

1.4 Energy Density of On-Board Energy Carriers

As mentioned above, the energy density of the on-board energy carrier is one of the most important factors that influence all choices of propulsion systems for individual mobility purposes. Figure 1.7 shows *estimations* of the corresponding figures for some commonly used or often proposed "fuels." All values are approximate and include the average losses caused by the corresponding "tank-to-vehicle" energy conversion system as shown in Figure 1.3.

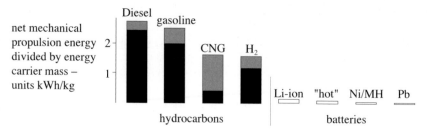

Fig. 1.7. Estimates of the *net* energy density of several on-board energy carriers.

For Diesel and gasoline the lower values (black bars) are obtained using actual average engine efficiencies, the upper values (gray bars) are valid for existing, but not yet standard engine systems (hybrid-electric propulsion, downsized-supercharged gasoline engine systems, etc.).

Compressed natural gas (CNG) is stored in gaseous form. Standard engine systems have 16% efficiency and the corresponding storage systems[11] are rated at 200 bar yielding the energy density indicated in Fig. 1.7 by a black bar. Advanced engine systems can reach more than 20% efficiency and advanced storage systems can go up to 350 bar,[12] yielding the energy density indicated by the gray bar.

Hydrogen may be stored under high pressure (black bar valid for 350 bar and carbon fibre bottles) or liquefied (gray bar).[13] The energy densities shown in Figure 1.7 are based on an cycle-averaged conversion efficiency of fuel cells of 40%. The main reason for the relatively low energy density of gaseous hydrogen is the unfavorable ratio of heating value divided by gas constant.

---

[11] Conventional steel bottles have a mass to volume ratio of approximately 1 kg/l; bottles made of carbon fibres have substantially lower mass to volume ratios of approximately 0.3-0.4 kg/l.

[12] For CNG higher pressures are not foreseeable: the compression losses become too large and outweigh the gains in energy density.

[13] Of course this approach induces an additional substantial penalty of approximately 30% in the "well-to-tank" efficiency and causes problems for long-term on-board storage since no insulation is perfect. The heat transfer to the liquid fuel will therefore either lead to some fuel evaporation and a subsequent venting or require a constant energy supply to avoid such evaporation losses.

The electrochemical on-board energy carriers have the lowest energy density of those "fuels" shown in Fig. 1.7, despite the fact that the electric "tank-to-vehicle" conversion systems have a very high conversion efficiency. State-of-the-art batteries optimized for high energy density achieve approximately 40 Wh/kg for lead–acid, 70 Wh/kg for nickel–metal hydrides, 150 Wh/kg for "hot" batteries such as the sodium–nickel chloride "zebra" battery, and 180 Wh/kg for lithium-ion cells.[14] Taking the average "tank-to-vehicle" conversion efficiency to be around 80%,[15] the values illustrated in Fig. 1.7 result. Not surprisingly, none of the several battery-electric vehicle prototypes that have been developed in the past has ever evolved to become a mass-produced alternative to the existing solutions.

Batteries are, however, very interesting "medium-term" energy storage devices, which can help to improve the efficiency of other propulsion systems. Such *hybrid* powertrains will be analyzed in detail in Chap. 4 and in several case studies in Appendix I.

Note that there are several other ways to store energy on-board, notably supercapacitors (electrostatic energy), hydraulic and pneumatic reservoirs (potential energy), and flywheels (kinetic energy). All of these systems have similar or even smaller net energy densities than lead acid batteries.[16] However, supercapacitors and similar devices have relatively high *power* densities, i.e., they may be charged and discharged with high power.[17] Such "short-term" energy storage devices can be useful for the recuperation of the kinetic energy stored in the vehicle that would otherwise be lost in braking maneuvers. More details on these aspects can be found in Chaps. 2 and 4.

The main point illustrated in Figure 1.7 is that with respect to energy density, liquid hydrocarbons are unquestionably the best fuels for passenger car applications. These fuels have several other advantages:

- the refueling process is fast (several MW of power), safe, and does not require any expensive equipment;
- their long-term storage is possible at relatively low costs; and
- there were and still are large and easily exploitable reserves of crude oil.[18]

---

[14] Unfortunately, lithium-ion cells are not yet available for automotive mass production.

[15] This figure includes the losses in the transmission, in the electric motor, in the power electronics and in the battery discharging process.

[16] Supercapacitors can reach net energy densities in the order of 5 Wh/kg. Systems based on pneumatic air stored at 300 bars in bottles made of carbon fibre can reach net energy densities in the order of 20 Wh/kg.

[17] Batteries can also be optimized for high power density. However, in this case their energy densities are lower than those figures indicated in Fig. 1.7.

[18] Note that liquid hydrocarbons need not originate from crude oil sources. Many approaches that use fossil (natural gas, coal, etc.) and renewable (bio Diesel, ethanol, etc.) primary energy sources are known with which liquid hydrocarbons can be synthesized.

Particularly the last point is, of course, the topic of many vivid debates. This text does not attempt to contribute to these discussions. However, following the paradigm of a "least-regret policy," the standing assumption adopted here is that the improvement of the "tank-to-miles" efficiency, while satisfying the performance and cost requirements, is worth the efforts.

1.5 Pathways to Better Fuel Economy

As illustrated in Fig. 1.2, there are essentially three possible approaches to reducing the total energy consumption of passenger cars:

- Improve the "well-to-tank" efficiency by optimizing the upstream processes and by utilizing alternative primary energy sources.
- Improve the "tank-to-vehicle" efficiency, as discussed below.
- Improve the "vehicle-to-miles" efficiency by reducing the vehicle mass and its aerodynamic and rolling friction losses.

As mentioned above, the optimization of the "well-to-tank" efficiency is an important area in itself. These problems are not addressed in this text. The phenomena which define the "vehicle-to-miles" energy losses will be analyzed in this text, but no attempt is made to suggest concrete approaches to reduce these losses. In fact, the disciplines that are important for that are material science, aerodynamics, etc. which are not in the scope of this monograph.

This text focuses on improving the "tank-to-vehicle" efficiency. Three different approaches are discernible on the component level and two on the system level:

1. Improve the *peak* efficiency of the powertrain components.
2. Improve the *part-load* efficiency of the powertrain components.
3. Add the capability to *recuperate* the kinetic and potential energy stored in the vehicle.
4. Optimize the structure and the parameters of the propulsion system, assuming that the fuel(s) used and the vehicle parameters are fixed.
5. Realize appropriate supervisory control algorithms that take advantage of the opportunities offered by the chosen propulsion system configuration.

Items 2–5 will be discussed in detail in this text, while item 1 is not within its scope. On one hand that optimization requires completely different methods and tools and, on the other hand, the potential for improvements in that area, in most cases, is rather limited. Compared to the peak efficiency of Diesel (≈ 0.40) or gasoline (≈ 0.37) engines, the part-load efficiency, which determines the actual fuel consumption in regular driving conditions, is much smaller (on average ≈ 0.20 for Diesel and ≈ 0.17 for gasoline engines). Therefore, the potential offered by improving these figures is much larger. Optimized powertrain systems and appropriate control algorithms are instrumental to achieve that objective.

2

Vehicle Energy and Fuel Consumption – Basic Concepts

This chapter contains three main sections: First the dynamics of the longitudinal motion of a road vehicle are analyzed. This part contains a discussion of the main energy-consuming effects occurring in the "vehicle-to-miles" part and some elementary models that describe the longitudinal dynamics and, hence, the drivability of the vehicle.

The influence of the driving pattern on the fuel consumption is analyzed in the second section. The main result of this analysis is an approximation of the mechanical energy required to make a road vehicle follow a given driving cycle. The sensitivity of the energy consumption to various vehicle parameters or the potential for the recuperation of kinetic energy when braking is derived from that result.

The third section briefly introduces the most important approaches used to predict the fuel economy of road vehicles, the main optimization problems that are relevant in this context, and the software tools available for the solution of these problems.

2.1 Vehicle Energy Losses and Performance Analysis

2.1.1 Energy Losses

Introduction

The propulsion system produces mechanical energy that is assumed to be momentarily stored in the vehicle. The driving resistances are assumed to drain energy from this reservoir. This separation might seem somewhat awkward at first glance. However, it is rather useful when one has to distinguish between the individual effects taking place.

The energy in the vehicle is stored:

- in the form of kinetic energy when the vehicle is accelerated; and
- in the form of potential energy when the vehicle reaches higher altitudes.

The amount of mechanical energy "consumed" by a vehicle[1] when driving a pre-specified driving pattern mainly depends on three effects:

- the aerodynamic friction losses;
- the rolling friction losses; and
- the energy dissipated in the brakes.

The elementary equation that describes the longitudinal dynamics of a road vehicle has the following form

$$m_v \frac{d}{dt} v(t) = F_t(t) - (F_a(t) + F_r(t) + F_g(t) + F_d(t)) \ , \qquad (2.1)$$

where F_a is the aerodynamic friction, F_r the rolling friction, F_g the force caused by gravity when driving on non-horizontal roads, and F_d the disturbance force that summarizes all other not yet specified effects. The traction force F_t is the force generated by the prime mover[2] minus the force that is used to accelerate the rotating parts inside the vehicle and minus all friction losses in the powertrain. Figure 2.1 shows a schematic representation of this relationship. The following sections contain more information about all of these forces.

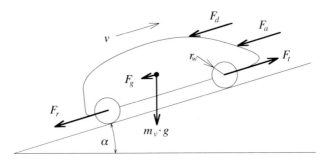

Fig. 2.1. Schematic representation of the forces acting on a vehicle in motion.

Aerodynamic Friction Losses

The aerodynamic resistance F_a acting on a vehicle in motion is caused on one hand by the viscous friction of the surrounding air on the vehicle surface. On the other hand, the losses are caused by the pressure difference between the front and the rear of the vehicle, generated by a separation of the air flow. For idealized vehicle shapes, the calculation of an approximate pressure field and

---

[1] To be thermodynamically correct, the first part of this sentence should read: The amount of *exergy* transformed to useless *anergy* ...
[2] This force can be negative, for instance during braking phases.

the resulting force is possible with the aid of numerical methods. A detailed analysis of particular effects (engine ventilation, turbulence in the wheel housings, cross-wind sensitivity, etc.) is only possible with specific measurements in a wind tunnel.

For a standard passenger car, the car body causes approximately 65% of the aerodynamic resistance. The rest is due to the wheel housings (20%), the exterior mirrors, eave gutters, window housings, antennas, etc. (approximately 10%), and the engine ventilation (approximately 5%) [115].

Usually, the aerodynamic resistance force is approximated by simplifying the vehicle to be a prismatic body with a frontal area A_f. The force caused by the stagnation pressure is multiplied by an aerodynamic drag coefficient $c_d(v, \ldots)$ that models the actual flow conditions

$$F_a(v) = \frac{1}{2} \cdot \rho_a \cdot A_f \cdot c_d(v, \ldots) \cdot v^2 . \qquad (2.2)$$

Here, v is the vehicle speed and ρ_a the density of the ambient air. The parameter $c_d(v, \ldots)$ must be estimated using CFD programs or experiments in wind tunnels. For the estimation of the mechanical energy required to drive a typical test cycle this parameter may be assumed to be constant.

Rolling Friction Losses

The rolling friction force is often modeled as

$$F_r(v, p, \ldots) = c_r(v, p, \ldots) \cdot m_v \cdot g \cdot \cos(\alpha), \quad v > 0 , \qquad (2.3)$$

where m_v is the vehicle mass and g the acceleration due to gravity. The term $\cos(\alpha)$ models the influence of a non-horizontal road. However, the situation in which the angle α will have a substantial influence is not often encountered in practice.

The rolling friction coefficient c_r depends on many variables. The most important influencing quantities are vehicle speed v, tire pressure p, and road surface conditions. The influence of the tire pressure is approximately proportional to $1/\sqrt{p}$. A wet road can increase c_r by 20% and driving in extreme conditions (sand instead of concrete) can easily double that value. The vehicle speed has a small influence at lower values, but its influence substantially increases when it approaches a critical value where resonance phenomena start.

A typical example of these relationships is shown in Fig. 2.2. Note that the tires reach their thermal equilibrium only after a relatively long period (a few ten minutes). Figure 2.2 includes examples of typical equilibrium temperatures for three speed values. For many applications, particularly when the vehicle speed remains moderate, the rolling friction coefficient c_r may be assumed to be constant. This simplification will be adopted in the rest of this text.

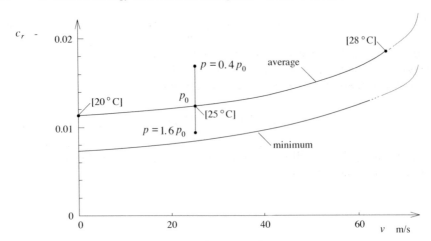

Fig. 2.2. Tire friction coefficient as a function of the vehicle speed v and variations of the tire pressure p.

Uphill Driving Force

The force induced by gravity when driving on a non-horizontal road is conservative and considerably influences the vehicle behavior. In this text this force will be modeled by the relationship

$$F_g(\alpha) = m_v \cdot g \cdot \sin(\alpha) \,, \tag{2.4}$$

which, for small inclinations α, may be approximated by

$$F_g(\alpha) \approx m_v \cdot g \cdot \alpha \tag{2.5}$$

when α is expressed in radians.

Inertial Forces

The inertia of the vehicle and of all rotating parts inside the vehicle causes fictitious (d'Alembert) forces. The inertia force induced by the vehicle mass is included in (2.1) by the term on the left side. The inertia of the rotating masses of the powertrain can be taken into account in the respective submodels. Nevertheless, sometimes for rapid calculation, it may be convenient to add the inertia of the rotating masses to the vehicle mass. Such an analysis usually considers a prime mover and a transmission with a total transmission ratio γ.

The total[3] inertia torque of the wheels is given by

---

[3] The inertia Θ_w includes all wheels and all rotating parts that are present on the wheel side of the gear box. The speed of all wheels ω_w is assumed to be the same.

$$T_{m,w}(t) = \Theta_w \cdot \frac{d}{dt}\omega_w(t) \qquad (2.6)$$

and it acts on the vehicle as an additional inertia force $F_{m,w} = T_{m,w}/r_w$, where r_w is the wheel radius. Usually, the wheel slip is not considered in a first approximation, i.e., $v = r_w \cdot \omega_w$. In this case

$$F_{m,w}(t) = \frac{\Theta_w}{r_w^2} \cdot \frac{d}{dt}v(t) . \qquad (2.7)$$

Consequently, the contribution of the wheels to the vehicle overall inertia is given by the following term

$$m_{r,w} = \frac{\Theta_w}{r_w^2} . \qquad (2.8)$$

Similarly, the inertia torque of the engine is given by

$$T_{m,e}(t) = \Theta_e \cdot \frac{d}{dt}\omega_e(t) = \Theta_e \cdot \frac{d}{dt}(\gamma \cdot \omega_w(t)) = \Theta_e \cdot \frac{\gamma}{r_w} \cdot \frac{d}{dt}v(t) , \qquad (2.9)$$

where Θ_e is the total moment of inertia of the powertrain[4] and ω_m its rotational speed. Again, assuming no wheel slip, this torque is transferred to the wheels as a force

$$F_{m,e}(t) = \frac{\gamma}{r_w} \cdot T_{m,e}(t) = \Theta_e \cdot \frac{\gamma^2}{r_w^2} \cdot \frac{d}{dt}v(t) . \qquad (2.10)$$

Note that this expression is only valid if the gear box has an efficiency of 100%. Since the powertrain inertia is added to the larger vehicle inertia, the errors caused by that simplification are usually small and often may be neglected.

Assuming a constant gear ratio γ, the force (2.10) corresponds to an additional vehicle mass of

$$m_{r,e} = \frac{\gamma^2}{r_w^2} \cdot \Theta_e . \qquad (2.11)$$

In summary, the equivalent mass of the rotating parts is approximated as follows

$$m_r = m_{r,w} + m_{r,e} = \frac{1}{r_w^2} \cdot \Theta_w + \frac{\gamma^2}{r_w^2} \cdot \Theta_e \qquad (2.12)$$

and it should be added to the vehicle mass m_v in (2.1). The *total gear ratio* γ/r_w appears quadratically in this expression. Accordingly, for high gear ratios (lowest gears in a standard manual transmission), the influence of the rotating parts on the vehicle dynamics can be substantial and may, in general, not be neglected.

---

[4] The inertia Θ_e includes the engine inertia and the inertia of all rotating parts that are present on the engine side of the gear box.

2.1.2 Performance and Drivability

General Remarks

Performance and drivability are very important factors that, unfortunately, are not easy to precisely define and measure. For passenger cars, three main quantifiers are often used:

- top speed;
- maximum grade at which the fully loaded vehicle reaches the legal top-speed limit; and
- acceleration time from standstill to a reference speed (100 km/h or 60 mph are often used).

These three quantifiers are discussed below.

Top Speed Performance

For passenger cars, the top speed is often not relevant because that limit is substantially higher than most legal speed limits. However, in some regions and for specific types of cars that information is still provided. This limit is mainly determined by the available power and the aerodynamic resistance.[5] Neglecting all other losses, the maximum speed is obtained by solving the following power balance

$$P_{max} \approx \frac{1}{2} \cdot \rho_a \cdot A_f \cdot c_d \cdot v_{max}^3 \, , \tag{2.13}$$

where P_{max} is the maximum traction power available at the wheels. The relevant information contained in this equation is the fact that the power demand depends on the *cube* of the vehicle speed. In other words, the engine power must be doubled in order to increase the top speed by 25%.

Uphill Driving

For trucks and any other vehicles that carry large loads, a relevant performance metric is the uphill driving capability. The relationship between P_{max} and the maximum gradient angle α_{max} is obtained by neglecting in (2.1) all the resistance forces but F_g

$$P_{max} \approx m_v \cdot v_{min} \cdot g \cdot \sin(\alpha_{max}) \, , \tag{2.14}$$

where v_{min} is the desired uphill speed. For a given rated power and a minimum speed, this equation yields the maximum uphill driving angle. This discussion is not complete without an analysis of the influence of the gear box. However, that analysis deserves to be treated in some detail and this discussion is thus deferred to Chap. 3.

---

[5] At that speed, the power consumed to overcome rolling friction is typically one order of magnitude smaller than the power dissipated by the aerodynamic friction.

Acceleration Performance

The most important drivability quantifier is the acceleration performance. Several metrics are used to quantify that parameter. In this text the time necessary to accelerate the vehicle from 0 to 100 km/h is taken as the only relevant information. This value is not easy to compute exactly because it depends on many uncertain factors and includes highly dynamic effects. An approximation of this parameter can be obtained as shown below.

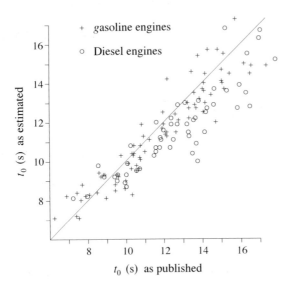

Fig. 2.3. Comparison between acceleration times as published by the manufacturers and values calculated with (2.16) for $v_0 = 100$ km/h.

The energy required to accelerate the vehicle from standstill to any velocity v_0 is given by

$$E_0 = \frac{1}{2} \cdot m_v \cdot v_0^2 . \tag{2.15}$$

If all the resistance forces in (2.1) are neglected, the energy E_0 is also the energy that has to be provided by the powertrain to accomplish the acceleration required. Accordingly, the mean power $\bar{P}$ that has to be provided by the engine is $\bar{P} = E_0/t_0$, where t_0 is the time available for the acceleration. An approximation, which takes into account the varying engine speed and the neglected losses, is obtained by choosing $\bar{P} \approx P_{max}/2$, where P_{max} is the maximum rated power. Consequently, a simple relation between the acceleration time and the maximum power of the engine is given by the following expression

$$t_0 \approx \frac{v_0^2 \cdot m_v}{P_{max}} . \tag{2.16}$$

Figure 2.3 shows the comparison between the acceleration times as published by various manufacturers and as estimated using the approximation (2.16) for $v_0 = 100\,\text{km/h}$. Although this relationship is based on many simplifying assumptions, its predictions of the acceleration times agree well with measured data of gasoline engines. For Diesel engines the agreement is not as good, but the approximation (2.16) still yields acceptable estimates.

2.1.3 Vehicle Operating Modes

From the first-order differential equation (2.1), the vehicle speed v can be calculated as a function of the force F_t. Depending on the value of F_t, the vehicle can operate in three different modes:

- $F_t > 0$, *traction*, i.e., the engine provides a propulsion force to the vehicle;
- $F_t < 0$, *braking*, i.e., the brakes dissipate kinetic energy of the vehicle, the engine can be engaged or disengaged (to consider fuel cut-off); and
- $F_t = 0$, *coasting*, i.e., the engine is disengaged and the resistance losses of the vehicle are exactly matched by the decrease of its kinetic energy.

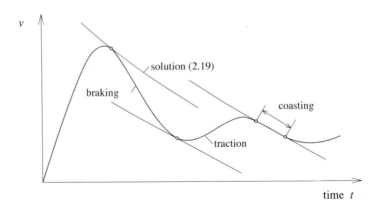

Fig. 2.4. Modes of vehicle motion.

For the limit case $F_t = 0$ on a horizontal road and without disturbances, the coasting velocity $v_c(t)$ of the vehicle can be computed by solving the following ordinary differential equation derived from (2.1)

$$\frac{d}{dt} v_c(t) = \frac{-1}{2 \cdot m_v} \cdot \rho_a \cdot A_f \cdot c_d \cdot v_c^2(t) - g \cdot c_r \qquad (2.17)$$

$$= -\alpha^2 \cdot v_c^2(t) - \beta^2$$

For $v_c > 0$, this equation can be integrated in closed form yielding the result

$$v_c(t) = \frac{\beta}{\alpha} \cdot \tan\left\{\arctan\left(\frac{\alpha}{\beta} \cdot v_c(0)\right) - \alpha \cdot \beta \cdot t\right\} . \qquad (2.18)$$

This solution is important because it can be used to define the three main operating modes of the vehicle. As shown in Fig. 2.4, the vehicle is in:

- *traction mode* if the speed decreases less than the coasting velocity $v_c(t)$ would decrease when starting at the same initial speed;
- *braking mode* if the speed decreases more than the coasting velocity $v_c(t)$ would decrease when starting at the same initial speed; and
- *coasting mode* if the vehicle speed and the coasting speed $v_c(t)$ coincide for a *finite* time interval.

Therefore, once a test cycle (see Sect. 2.2) has been defined, using (2.18) it is possible to decide for each time interval in what mode a vehicle is operated without requiring any information on the traction force F_t. Particularly for the MVEG–95 cycle this analysis is straightforward and yields the result indicated in the lowest bar (index "trac" for traction mode) in Fig. 2.6.

2.2 Mechanical Energy Demand in Driving Cycles

2.2.1 Test Cycles

Test cycles consisting of standardized speed and elevation profiles have been introduced to compare the pollutant emissions of different vehicles on the same basis. After that first application, the same cycles have been found to be useful for the comparison of the fuel economy as well. In practice, these cycles are often used on chassis dynamometers where the force at the wheels is chosen to emulate the vehicle energy losses while driving that specific cycle (see Sect. 2.1.1). These tests are carried out in controlled environments (temperature, humidity, etc.), with strict procedures being followed to reach precisely defined thermal initial conditions for the vehicle (hot soak, cold soak, etc.). The scheduled speed profile is displayed on a monitor[6] while a test driver controls the gas and brake pedals such that the vehicle speed follows these reference values within pre-specified error bands.

There are several commonly used test cycles. In the United States, the federal urban driving cycle (FUDS) represents a typical city driving cycle, while the federal highway driving cyle (FHDS) reflects the extra-urban driving conditions. The federal test procedure (FTP–75) is approximately equal to one and a half FUDS cycle, but it also includes a typical warm-up phase. The first FUDS cycle is driven in cold-soak conditions. The second half of the FUDS cycle (hot soak) is driven after a 10-minute engine-off period. The cold start is important to assess the pollutant emissions (engine and catalyst warming

---

[6] For vehicles with manual transmission, the requested gear shifts are also signalled on the driver's monitor.

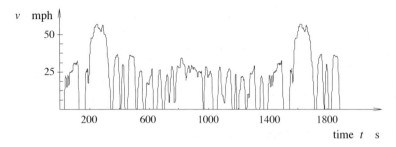

Fig. 2.5. US test cycle FTP–75 (Federal Test Procedure), length: 11.12 miles (17.8 km), duration: 1890 s, average speed: 21 mph (9.43 m/s).

up) but the fuel consumption is also influenced by the cold-start conditions (higher engine friction), though not as dramatically as the pollutant emissions.

In Europe, the urban driving cycle (ECE) consists of three start-and-stop maneuvers. The combined cycle proposed by the motor vehicle expert group in 1995 (MVEG–95) repeats the ECE four times (the first with a cold start) and adds an extra-urban portion referred to as the EUDC. The Japanese combined cycle is the 10–15 mode, consisting of three repetitions of an urban driving cycle plus an extra-urban portion. Figures 2.5 and 2.6 show the speed profiles for the American and the European cycles. Note that for manual transmissions, the MVEG–95 prescribes the gear to be engaged at each time instant.[7]

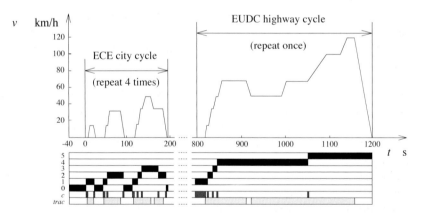

Fig. 2.6. European test cycle MVEG–95, gears 1–5, "c": clutch disengaged, "trac": traction time intervals. Total length: 11.4 km, duration: 1200 s, average speed: urban 5.12 m/s, extra-urban 18.14 m/s, overall 9.5 m/s. The cycle includes a total of 34 gear shifts.

---

[7] This restriction is not particularly reasonable. Using the additional degrees of freedom associated with a flexible gear selection can considerably improve the fuel economy. This will become clearer in the subsequent chapters.

Of course, real driving patterns are often much more complex and demanding (speeds, accelerations, etc.) than these test cycles. All automotive companies have their in-house standard cycles, which better reflect the average real driving patterns. For the sake of simplicity, in this text the FTP and MVEG–95 cycles will be used in most cases. The methods introduced below, however, are applicable to more complex driving cycles as well.

2.2.2 Mechanical Energy Demand

Introduction

In the following subsection the mechanical energy is determined that is required to make a vehicle follow the MVEG–95 cycle. A key role in these considerations is played by the mean tractive force $\bar{F}_{trac}$. The concept of a mean tractive force is particularly useful for the evaluation of a first tentative value for the fuel consumed by a propulsion system and to discuss various aspects of the "vehicle-to-miles" energy conversion step.

The mean tractive force $\bar{F}_{trac}$ is defined as follows

$$\bar{F}_{trac} = \frac{1}{x_{tot}} \cdot \int_{t \in trac} F(t) \cdot v(t) \, dt \,, \qquad (2.19)$$

where $x_{tot} = \int_0^{t_{max}} v(t) \, dt$ is the total distance of the cycle and $trac$ is the set of all time intervals where $F_t(t) > 0$, i.e., those parts of the cycle in which the vehicle drives in $traction\ mode$[8] as defined in Sect. 2.1.3.

Note that in the MVEG–95 cycle the vehicle is never operated in $coasting$ $mode$. As mentioned above, in $braking\ mode$ the vehicle does not require any mechanical energy because the aerodynamic and rolling friction losses are covered by the decreasing kinetic energy of the vehicle. The additional vehicle energy that must be dissipated in these phases is either transformed into heat by the brakes or converted to another energy form if the vehicle is equipped with an $energy\ recuperation$ device. The last mode, $stopped,$ does not cause any mechanical energy losses either. However, if the engine is not shut down in these phases, this part leads to additional fuel consumption referred to as $idling\ losses.$

The evaluation of the integral used to define the mean tractive force is accomplished by a $discretization$ in time of the drive cycle as follows (more on this approach will be said in the next section). The velocity profile of a test cycle is defined for the given time instants $v(t_i) = v_i$, $t_i = i \cdot h$, $i = 0, \ldots, n$. Accordingly, the speed used to compute the mean tractive force is the average speed

$$v(t) = \bar{v}_i = \frac{v_i + v_{i-1}}{2}, \quad \forall \ t \in [t_{i-1}, t_i) \,. \qquad (2.20)$$

---

[8] In the MVEG–95 cycle the vehicle is in traction mode approximately 60% of the total cycle time.

Similarly, the acceleration is approximated by

$$a(t) = \bar{a}_i = \frac{v_i - v_{i-1}}{h}, \quad \forall \ t \in [t_{i-1}, t_i) \ . \tag{2.21}$$

Therefore, an approximation of the tractive force (2.19) can be found using the expression

$$\bar{F}_{trac} \approx \frac{1}{x_{tot}} \cdot \sum_{i \in trac} \bar{F}_{trac,i} \cdot \bar{v}_i \cdot h \ , \tag{2.22}$$

where the partial summation $i \in trac$ only covers the time intervals during which the vehicle is in traction mode.

Case 1: No Recuperation

According to Sect. 2.1.1, the total tractive force $\bar{F}_{trac}$ includes contributions from three different effects

$$\bar{F}_{trac} = \bar{F}_{trac,a} + \bar{F}_{trac,r} + \bar{F}_{trac,m} \ , \tag{2.23}$$

with the three forces caused by aerodynamic and rolling friction and acceleration resistance defined by

$$\bar{F}_{trac,a} \approx \frac{1}{x_{tot}} \cdot \frac{1}{2} \cdot \rho_a \cdot A_f \cdot c_d \cdot \sum_{i \in trac} \bar{v}_i^3 \cdot h \ ,$$

$$\bar{F}_{trac,r} \approx \frac{1}{x_{tot}} \cdot m_v \cdot g \cdot c_r \cdot \sum_{i \in trac} \bar{v}_i \cdot h \ , \tag{2.24}$$

$$\bar{F}_{trac,m} \approx \frac{1}{x_{tot}} \cdot m_v \cdot \sum_{i \in trac} \bar{a}_i \cdot \bar{v}_i \cdot h \ .$$

The sums and the distance x_0 in these equations depend only on the driving cycle and not on the vehicle parameters. For the MVEG–95, the ECE cycle, and the EUDC, the following numerical values can easily be found

$$\frac{1}{x_{tot}} \cdot \sum_{i \in trac} \bar{v}_i^3 \cdot h \approx \{319, 82.9, 455\} \ ,$$

$$\frac{1}{x_{tot}} \cdot \sum_{i \in trac} \bar{v}_i \cdot h \approx \{0.856, 0.81, 0.88\} \ , \tag{2.25}$$

$$\frac{1}{x_{tot}} \cdot \sum_{i \in trac} \bar{a}_i \cdot \bar{v}_i \cdot h \approx \{0.101, 0.126, 0.086\} \ .$$

Once the driving cycle is chosen, (2.22) allows for a simple estimation of the mean tractive force as a function of the vehicle parameters m_v, A_f, c_d, and

c_r. The mean tractive force is equal to $\bar{E}$, the average energy consumed per distance travelled. When the latter is expressed using the units kJ/100 km, the relationship between the two quantities is $\bar{E} = 100 \cdot \bar{F}_{trac}$. In these units, the energy consumed in the MVEG–95 cycle is given by

$$\bar{E}_{MVEG-95} \approx A_f \cdot c_d \cdot 1.9 \cdot 10^4 + m_v \cdot c_r \cdot 8.4 \cdot 10^2 + m_v \cdot 10 \quad (\text{kJ}/100 \,\text{km}) , \quad (2.26)$$

where, to simplify matters, the physical parameters air density ρ_a (at sea level) and acceleration g have been integrated in the constants. Considering typical values for the vehicle parameters, the three contributions in this sum are of the same order of magnitude.

Case 2: Perfect Recuperation

Equation (2.22) is valid for the case that none of the vehicle's kinetic energy is recuperated when braking. In the opposite case, with *perfect recuperation* (recuperation device of zero mass and 100% efficiency), the energy spent to accelerate the vehicle is completely recuperated during the braking phases. As a consequence, the mean force $\bar{F}$ does not have any contributions $\bar{F}_m$ caused by acceleration losses, i.e.,

$$\bar{F} = \bar{F}_a + \bar{F}_r . \quad (2.27)$$

However, in the full recuperation case, the mean force must include the losses caused by aerodynamic and rolling resistances also during the braking phases. In this case, (2.24) must be replaced by

$$\bar{F}_a \approx \frac{1}{x_{tot}} \cdot \frac{1}{2} \cdot \rho_a \cdot A_f \cdot c_d \cdot \sum_{i=1}^{n} \bar{v}_i^3 \cdot h ,$$

$$(2.28)$$

$$\bar{F}_r \approx \frac{1}{x_{tot}} \cdot m_v \cdot g \cdot c_r \cdot \sum_{i=1}^{n} \bar{v}_i \cdot h ,$$

with the following numerical values valid for the MVEG–95, the ECE, and the EUDC cycles

$$\frac{1}{x_{tot}} \cdot \sum_{i=1}^{n} \bar{v}_i^3 \cdot h \approx \{363, 100, 515\},$$

$$(2.29)$$

$$\frac{1}{x_{tot}} \cdot \sum_{i=1}^{n} \bar{v}_i \cdot h = \{1, 1, 1\}.$$

With these results, an approximation can be derived that is similar to (2.26) of the mechanical energy $\bar{E}_{rec,MVEG-95}$ needed to drive 100 km in the MVEG–95 cycle. In the case of full recuperation this quantity is given by

$$\bar{E}_{rec,MVEG-95} \approx A_f \cdot c_d \cdot 2.2 \cdot 10^4 + m_v \cdot c_r \cdot 9.81 \cdot 10^2 \quad (\text{kJ}/100 \,\text{km}) . \quad (2.30)$$

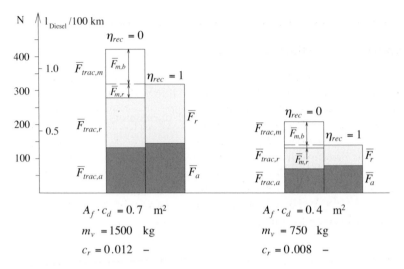

Fig. 2.7. Comparison of the energy demand in the MVEG–95 cycle for a full-size car (left) and a light-weight car (right). A mean force of 1 N is equivalent to 27.78 Wh mechanical energy per 100 km. The lower heating value for Diesel fuel is approximately 10 kWh/l.

Examples

Figure 2.7 shows the mean tractive force and the corresponding specific energy demand in the MVEG–95 cycle for two different vehicles. On the basis of the fuel's lower heating value, a full-size passenger car (without recuperation) requires per 100 km travelled distance an amount of mechanical energy that corresponds to 1.16 l of Diesel fuel.[9] This result is obtained by inserting the vehicle parameters shown in Fig. 2.7 into (2.26). In the case of perfect recuperation (a 100% efficient recuperation device with no additional mass), the corresponding value, obtained by using (2.30), is 0.89 l of Diesel fuel. In the best case, i.e., when equipped with an ideal recuperation device, a hypothetical advanced light car would require the equivalent of about 0.39 l Diesel fuel per 100 km. Without recuperation, the corresponding value would be 0.54 l Diesel fuel per 100 km.

It is worthwhile repeating that even in the case with no additional recuperation device the vehicle utilizes some of its kinetic energy to drive parts of the cycle. As illustrated in Fig. 2.7, out of the mean tractive force necessary for the acceleration, $\bar{F}_{trac,m}$ of (2.24), only the quantity $\bar{F}_{m,b}$ is later dissipated as heat by the brakes. The remaining portion $\bar{F}_{m,r}$ is used to overcome the driving resistances in the non-traction phases.

---

[9] Of course, due to the losses in the engine and in the other components of the powertrain, the actual fuel consumption is much higher.

2.2.3 Some Remarks on the Energy Consumption

Comparison of Different Vehicle Classes

Using the result of (2.26), some typical numerical values of the mechanical power needed to drive the tractive parts of the MVEG–95 cycle can be estimated. As shown in Table 2.1 for a selection of automobile classes, the *average* tractive power $\bar{P}_{MVEG-95}$ necessary to drive the MVEG–95 cycle is much smaller than the power P_{max} necessary to satisfy the acceleration requirements. For instance, a standard full-size car only needs an average of 7.1 kW to follow the MVEG–95 cycle.[10] However, using (2.16), the power necessary to accelerate the same vehicle from 0 to 100 km/h in 10 seconds is found to be around 115 kW. For a light-weight car, the numbers change in magnitude, but the ratio remains approximately the same.

Table 2.1. Numerical values of the average and peak tractive powers for different vehicle classes.

	SUV	full-size	compact	light-weight
$A_f \cdot c_d$	$1.2\,\mathrm{m}^2$	$0.7\,\mathrm{m}^2$	$0.6\,\mathrm{m}^2$	$0.4\,\mathrm{m}^2$
c_r	0.017	0.013	0.012	0.008
m_v	2000 kg	1500 kg	1000 kg	750 kg
$\bar{P}_{MVEG-95}$	11.3 kW	7.1 kW	5.0 kW	3.2 kW
P_{max}	155 kW	115 kW	77 kW	57 kW

This discrepancy between the small average power and the large power necessary to satisfy the drivability requirements is one of the main causes for the relatively low fuel economy of the current propulsion systems. In fact, as shown in the next section in Fig. 2.13, the efficiency of IC engines strongly decreases when these engines are operated at low torque. Several solutions to this *part-load problem* will be discussed in the subsequent chapters.

Sensitivity Analysis

As discussed in the previous sections, the mean tractive force necessary to drive a given cycle depends on the vehicle parameters $A_f \cdot c_d$, c_r, and m_v. Their relative influence can be evaluated with a sensitivity analysis. In the context of this vehicle energy consumption analysis, a meaningful definition of the sensitivity is

$$S_p = \lim_{\delta p \to 0} \frac{[\bar{E}_{MVEG-95}(p + \delta p) - \bar{E}_{MVEG-95}(p)]/\bar{E}_{MVEG-95}(p)}{\delta p/p} , \quad (2.31)$$

---

[10] The last acceleration in the highway part of the MVEG–95 cycle requires the largest power. For this full-size vehicle this power is around 34 kW.

where $\bar{E}_{MVEG-95}$ is the cycle energy as defined by (2.26). The variable p stands for any of the three parameters $A_f \cdot c_d$, c_r, or m_v. Its variation is denoted by δp. The sensitivity (2.31) can also be written as

$$S_p = \frac{\partial \bar{E}_{MVEG-95}}{\partial p}(p) \cdot \frac{p}{\bar{E}_{MVEG-95}(p)} \qquad (2.32)$$

with the three partial derivatives

$$\frac{\partial \bar{E}_{MVEG-95}}{\partial (A_f \cdot c_d)} = 1.9 \cdot 10^4$$

$$\frac{\partial \bar{E}_{MVEG-95}}{\partial c_r} = m_v \cdot 8.4 \cdot 10^2 \qquad (2.33)$$

$$\frac{\partial \bar{E}_{MVEG-95}}{\partial m_v} = (c_r \cdot 8.4 \cdot 10^2 + 10)$$

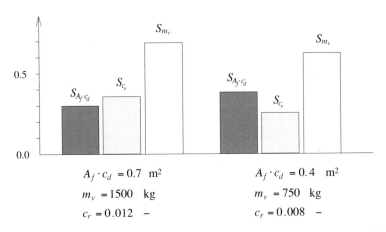

Fig. 2.8. Sensitivities (2.31) of the mechanical energy consumption in the MVEG–95 cycle with respect to the three main vehicle parameters. Two cases: full-size car (left); advanced light-weight vehicle (right).

Figure 2.8 shows the sensitivities of a typical full-size car and of an advanced light-weight vehicle. Two facts are worth mentioning:

- In the case of a standard vehicle, the sensitivity S_{m_v} is by far the most important. Accordingly, the most promising approach to reduce the mechanical energy consumed in a cycle is to reduce the vehicle's mass.[11]

---

[11] Of course this assertion neglects several aspects (economy, safety, ...). The full picture, which unfortunately is very difficult to obtain, would include the cost

- The relative dominance of the vehicle's mass on the energy consumption is not reduced when an advanced vehicle concept is analyzed. For this reason, the idea of recuperating the vehicle's kinetic energy will remain interesting for this vehicle class as well.

As shown in the sensitivity analysis, the vehicle mass m_v plays a very important role for the estimation of the energy demand of a vehicle. Therefore, Fig. 2.9 shows the mean tractive force and the corresponding equivalent of Diesel fuel as a function of the vehicle mass. Contradicting the measured fuel consumption data, the mechanical energy consumption in the test cycle is smaller than the corresponding value for the vehicle driving at constant highway speeds. As will become clear in the subsequent chapters, the reason for this fact is again to be found in the low efficiency of standard IC engines at part-load conditions.

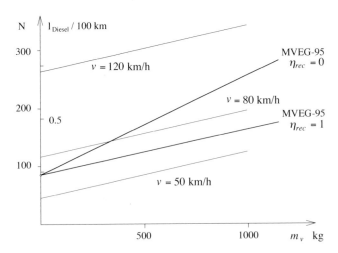

Fig. 2.9. Affine dependency of the mean force on the vehicle mass for the MVEG–95 and three values of constant speed. The example is valid for an advanced vehicle with the parameters $\{A_f \cdot c_d, c_r\} = \{0.4\,\mathrm{m}^2, 0.008\}$.

Figure 2.10 shows the influence of the vehicle mass and the *cycle-averaged* powertrain efficiency on the carbon dioxyde emissions of a passenger car.[12] A limit of 140 g/km can be reached with standard Diesel engines (cycle-averaged efficiency around 0.2) and small to medium-size vehicle (vehicle mass around 1200 kg). If a limit of 120 g/km is to be reached, either a substantial reduction

---

    sensitivity as well. This figure, which indicates how expensive it is to reduce the parameter p, multiplied by the sensitivities (2.31), would show which approach is the most cost effective.

[12] For the sake of simplicity the aerodynamic and rolling friction parameters are kept constant.

of vehicle mass or a substantial increase of powertrain efficiency is needed. As mentioned above, this text focuses on the latter approach. However, it must be emphasized that very-low fuel consumptions and CO_2 emissions can only be reached by combining these two approaches.

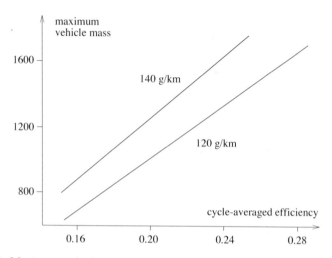

Fig. 2.10. Maximum vehicle mass permitted as a function of cycle-averaged power-train efficiency if the CO_2 emissions are to remain below 140 g/km and 120 g/km, respectively. The other vehicle parameters are $\{A_f \cdot c_d, c_r\} = \{0.7\,\mathrm{m}^2, 0.012\}$.

Realistic Recuperation Devices

So far it has been assumed that the device used to recuperate the vehicle's kinetic energy causes no losses (in both energy conversion directions) and has no mass. Of course, the total recuperation efficiency η_{rec} of all real recuperation devices will be smaller than 100%. Realistic values are around 60%.[13] Moreover, the recuperation device will increase the mass of the vehicle by m_{rec}, which in turn causes increased energy losses.

Therefore, the total energy that can be recuperated $\Delta \bar{E}_{rec}$ is substantially smaller than the theoretic maximum $\Delta \bar{E}_{rec,max}$ given by the difference of (2.26) and (2.30)

$$\Delta \bar{E}_{rec,max} = \bar{E}_{MVEG-95} - \bar{E}_{rec,MVEG-95}(m_{rec} = 0) . \qquad (2.34)$$

The actual value of the energy $\bar{E}(\eta_{rec}, m_{rec})$ necessary to drive the MVEG–95 cycle when a recuperation device with efficiency η_{rec} and mass m_{rec} is

---

[13] Note that this figure represents *two* energy conversions: first from kinetic to, say, electrostatic energy stored in a supercapacitor during braking and then back again into kinetic energy during acceleration.

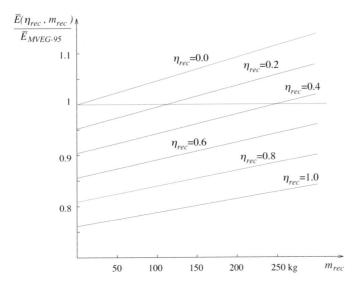

Fig. 2.11. Energy demand in the MVEG–95 cycle of a vehicle equipped with a recuperation device with mass m_{rec} and efficiency η_{rec}, normalized by the energy $\bar{E}_{MVEG-95}$ consumed by a vehicle without recuperation device. The example is valid for a vehicle with the parameters $\{A_f \cdot c_d, c_r, m_v\} = \{0.7\,\mathrm{m}^2, 0.012, 1500\,\mathrm{kg}\}$.

installed is obtained by linear interpolation between the two extreme cases (2.26) and (2.30) and by considering the additional mass of the recuperation device

$$\bar{E}(\eta_{rec}, m_{rec}) = \bar{E}_{rec,MVEG-95}(m_{rec})$$

$$+ (1 - \eta_{rec}) \cdot \left[\bar{E}_{MVEG-95}(m_{rec}) - \bar{E}_{rec,MVEG-95}(m_{rec}) \right]$$

$$= A_f \cdot c_d \cdot \left[2.2 \cdot 10^4 - (1 - \eta_{rec}) \cdot 3 \cdot 10^3 \right]$$

$$+ c_r \cdot (m_v + m_{rec}) \cdot \left[9.8 \cdot 10^2 - (1 - \eta_{rec}) \cdot 1.4 \cdot 10^2 \right]$$

$$+ (1 - \eta_{rec}) \cdot 10 \cdot (m_v + m_{rec}) .$$

$$(2.35)$$

For a typical mid-size car Fig. 2.11 shows the normalized energy recuperation potential as a function of the mass and the efficiency of the recuperation device. The maximum recuperation potential is approximately 25% in the hypothetical case $\eta_{rec} = 1$ and $m_{rec} = 0$. With the realistic values $\eta_{rec} \approx 0.6$ and $m_{rec} \approx 100\,\mathrm{kg}$, only slightly more than 10% of the energy necessary to drive the MVEG–95 cycle can be recuperated.

2.3 Methods and Tools for the Prediction of Fuel Consumption

Once the mechanical energy required to drive a chosen test cycle is known, the next step is to analyze the efficiency of the propulsion system. Three possible approaches are *introduced* in this section. More details on each of the three methods are discussed in the subsequent chapters and, in particular, in the case studies included in Appendix I.

2.3.1 Average Operating Point Approach

This method is often used for a first preliminary estimation of the fuel consumption of a road vehicle. The key point is to lump the full envelope of all engine operating points into one single representative average operating point and to compute the fuel consumption of the propulsion system at that regime.

This approach requires a test cycle to be specified a priori. Once the driving pattern is chosen, the mean mechanical power at the wheel $\bar{P}_v$ can be estimated with the methods discussed in the last section. This information is then used to "work backwards" through the powertrain.

Note that only the tractive part of the drive cycle is relevant for that calculation. In the coasting or braking parts of the cycle the propulsion system can operate at idle, with a concomitant idling fuel consumption, or at "fuel cut-off," with no fuel consumed at all. The example of Sect. 3.3.2 includes a discussion of these points for the case of a standard IC engine powertrain.

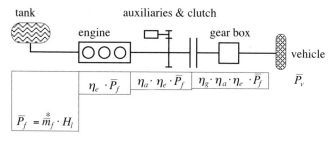

Fig. 2.12. Illustration of the components considered in the fixed operating point approach for the case of a conventional IC engine powertrain.

Figure 2.12 shows the components relevant for such an analysis for the case of a conventional IC engine powertrain. This figure illustrates:

- the losses caused by the gear box and differential (lumped into one device for the sake of simplicity) represented by the efficiency η_g;
- the losses caused by the auxiliary devices (generation of electric energy, power steering, etc.) and the clutch represented by the efficiency η_a; and

- the losses caused by the IC engine represented by the efficiency η_e.

With $\bar{P}_f$ being the fuel power consumed by the engine, the overall power balance

$$\bar{P}_v = \eta_g \cdot \eta_a \cdot \eta_e \cdot \bar{P}_f \qquad (2.36)$$

permits the computation of the mean fuel mass flow

$$\overset{*}{m}_f = \bar{P}_f / H_l , \qquad (2.37)$$

where H_l is the fuel's lower heating value.

The engine efficiency η_e can be estimated using approximations, such as the Willans rule introduced in the next chapter, or using measured *engine maps*. Figure 2.13 shows such a map. Note that η_e strongly depends on engine torque, whereas the engine speed has less of an influence. In the mean operating point approach, one characteristic engine torque and one characteristic engine speed must be found. These data points follow from the power $\bar{P}_v$ consumed at the wheels, the mean vehicle velocity $\bar{v}$, and the mean gear ratio $\bar{\gamma}$. Section 3.3.2 contains an example that shows in detail how to apply this approach to a standard ICE powertrain.

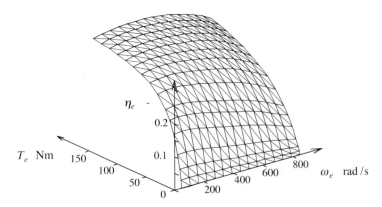

Fig. 2.13. IC engine efficiency as a function of engine torque and speed.

The average operating point method is able to yield reasonable estimates of the fuel consumption of *simple* powertrains (IC engine or battery electric propulsion systems). It is not well suited to problems in which complex propulsion systems must be optimized. In particular it does not offer the option of including the effect of energy management strategies in these computations.

2.3.2 Quasistatic Approach

In quasistatic simulations, the input variables are the speed v and the acceleration a of the vehicle as well as the grade angle α of the road.[14] With this

---

[14] Regulatory test cycles or driving patterns recorded on real vehicles can be used.

information, the force F_t is computed that has to be acting on the wheels to drive the chosen profile for a vehicle described by its main parameters $\{A_f \cdot c_d, c_r, m_v\}$. In this step the vehicle is assumed to run at constant speed v, acceleration a, and grade α for a (short) time period h. Once F_t is known, the losses of the powertrain are considered by a power balance similar to the one expressed in (2.36) that was introduced in the last section. The fuel consumption during the time interval $t \in [(i-1) \cdot h, i \cdot h)$ is calculated using an equation similar to (2.37).

Evidently, the average operating point method and the quasistatic method are very similar. The only difference is that in the latter approach the test cycle is divided into (many) intervals in which the average operating point method is applied. For each time interval, the *constant* speed and acceleration that the vehicle is required to follow are given by (2.20) and (2.21), respectively.

The basic equation used to evaluate the force required to drive the chosen profile is Newton's second law (2.1) as derived in the previous section. Using this equation, the force is calculated as follows

$$\bar{F}_{t,i} = m_v \cdot \bar{a}_i + F_{r,i} + F_{a,i} + F_{g,i} \tag{2.38}$$

$$= m_v \cdot \bar{a}_i + \frac{1}{2} \cdot \rho_a \cdot A_f \cdot c_d \cdot \bar{v}_i^2 + c_r \cdot m_v \cdot g \cos(\alpha_i) + m_v \cdot g \cdot \sin(\alpha_i) \ .$$

The rotating parts can be included by adding the mass m_r (2.12) to the vehicle mass.

In the quasistatic approach the velocity and accelerations are assumed to be constant in a time interval h chosen small enough to satisfy this assumption. Usually this time interval is constant (in the MVEG–95 and FTP–75 cycles h is equal to 1 s) but the quasistatic approach can be extended to a non-constant value of h.

If the driving profile contains idling or slow-speed phases, the propulsion system can be operated at very low loads in the corresponding time intervals. Therefore, the efficiency of the energy converters cannot be mapped as shown in Fig. 2.13. In fact, at very low loads (or torques, in the case of an IC engine), the efficiency approaches zero, with the consequence that small measurement uncertainties can cause substantial errors in the estimation of the fuel consumption. Figure 2.14 illustrates a more suitable approach for the representation of the efficiency of an IC engine. Similar maps can be obtained for other energy converters.

In this approach, the fuel consumption necessary to sustain a pre-defined torque and speed combination is directly mapped to the torque-speed plane. This representation allows a visualization of the *idling losses* and the *fuel cut-off* limits. If the deceleration of the vehicle provides enough power to cover, in addition to the aerodynamic and rolling friction losses of the vehicle itself, the losses of the complete propulsion system (friction losses of the powertrain, power consumed by the auxiliaries, engine friction and pumping losses, etc.), the fuel may be completely cut off.

T_e Nm
150
100
50
idling
1.2
2.5
5
7.5
12
18
25 l/h
200
400
600
fuel cut-off limit
ω_e rad/s

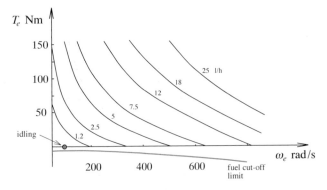

Fig. 2.14. IC engine fuel consumption in liters per hour as a function of engine torque and speed.

For the example of a standard IC engine powertrain, Fig. 2.15 shows the simplified structure of an algorithm that computes on a digital computer the vehicle's fuel consumption using the quasistatic method. For relatively simple powertrain structures the algorithm illustrated in Fig. 2.15 can be compactly formulated using a vector notation for all variables (speed, tractive force, etc.). This form is particularly efficient when using Matlab.[15]

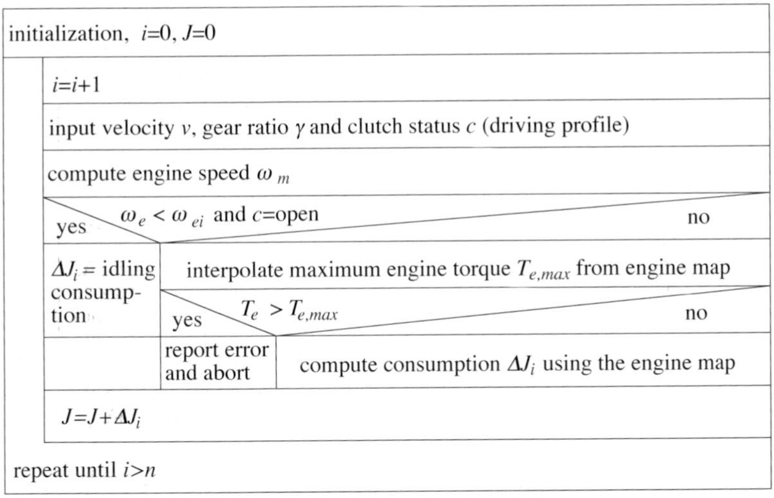

initialization, i=0, J=0
i=i+1
input velocity v, gear ratio γ and clutch status c (driving profile)
compute engine speed ω_m
$\omega_e < \omega_{ei}$ and c=open
yes
no
ΔJ_i = idling consump-tion
interpolate maximum engine torque $T_{e,max}$ from engine map
$T_e > T_{e,max}$
yes
no
report error and abort
compute consumption ΔJ_i using the engine map
J=J+ΔJ_i
repeat until i>n

Fig. 2.15. Structure of an algorithm for quasistatic fuel consumption.

The quasistatic method is well suited to the minimization of the fuel consumption of complex powertrain structures. With this approach it is possible

[15] Matlab/Simulink is a registered trademark of TheMathWorks, Inc., Natick, MA.

to design *supervisory* control systems that optimize the power flows in the propulsion system. The influence of the driving pattern can be included in these calculations. Despite these capabilities, the numerical effort remains relatively low. The main drawback of the quasistatic method is its "backward" formulation, i.e., the physical causality is not respected and the driving profile that has to be followed has to be known a priori. Therefore, this method is not able to handle feedback control problems or to correctly deal with state events.

Experiments on engine dynamometers are one option to verify the quality of the predictions obtained by quasistatic simulations. Figure 2.16 shows a picture of a typical engine dynamometer system. The electric motor shown on the left side in that picture is computer-controlled such that it produces the braking torque that the engine, shown on the right side, would experience when installed in the simulated vehicle while driving the pre-specified driving cycle.

Fig. 2.16. Photograph of an engine test bench.

Modeling the propulsion system with the methods introduced in this text and simulating the behavior of the vehicle while following the MVEG-95 test cycle yields results similar to the ones shown in Fig. 2.17. While substantial deviations do occur for short time intervals, the overall fuel economy predictions are surprisingly accurate.

2.3.3 Dynamic Approach

The *dynamic approach* is based on a "correct" mathematical description of the system.[16] Usually, the model of the powertrain is formulated using sets of ordinary differential equations in the state-space form

---

[16] Of course, no model will exactly describe the system behavior. Modeling errors must be taken into account by appropriate robustness guarantees in the later design steps.

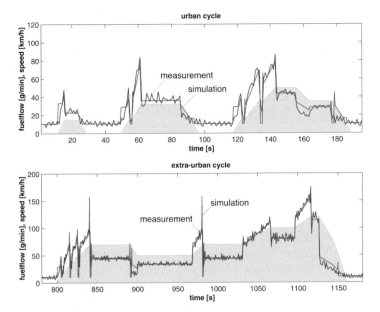

Fig. 2.17. Comparison of measured and simulated fuel consumption.

$$\frac{d}{dt}x(t) = f(x(t), u(t)), \qquad x(t) \in \Re^n, \ u \in \Re^m, \tag{2.39}$$

but other descriptions are possible, e.g., partial differential equations or equations with differential *and* algebraic parts. The formulation (2.39) may be used to describe many dynamic effects in powertrains. While some of these effects are relevant for the estimation of the fuel consumption (engine temperature dynamics, etc.) certain others are not (inlet manifold dynamics of SI engines, EGR rate dynamics in Diesel engines, etc). Most of the relevant effects are relatively slow, while the fast effects are significant for the optimization of comfort, drivability, and pollutant emission. Readers interested in these effects are referred to [137] and [100]. The slower effects that are important for the optimization of the fuel economy will be discussed in the subsequent chapters.

Equation (2.39) is the starting point for a plethora of feedforward and feedback system analysis and synthesis approaches. Readers interested in these approaches are referred to the standard textbooks, e.g. [131], [261], [120]. Note that in the dynamic method the inputs to the powertrain model are the same signals as those that are present in the real propulsion system. Accordingly, a module that emulates the behavior of the driver has to be included in these simulations.

The description (2.39) is very versatile and many optimization problems can only be solved using the dynamic method. Notable examples of such problems are the design of feedback control systems or the detection and correct handling of state events in optimization problems. The drawback of using (2.39) and the associated analysis and synthesis approaches is the relatively high computational burden of these methods. For this reason, this text emphasizes the quasistatic methods. Dynamic methods are only chosen when no other option is available.

2.3.4 Optimization Problems

At least three different layers of optimization problems are present in most vehicle propulsion design problems:

- *structural optimization* where the objective is to find the best possible powertrain structure;
- *parametric optimization* where the objective is to find the best possible parameters for a fixed powertrain structure; and
- *control system optimization* where the objective is to find the best possible supervisory[17] control algorithms.

Of course, these three tasks are not independent. Unfortunately, a complete and systematic optimization methodology that *simultaneously* considers all three problem layers is still missing. Moreover, the notion of an optimal solution is somewhat elusive in the sense that to find an optimum very restricting assumptions often must be adopted.

The chosen driving profile will also substantially influence the results and in this context the distinction between *causal* and *non-causal* solutions is relevant in the design of supervisory control algorithms. The first set of solutions can be directly used in real driving situations because the output of the control system only depends on actual and past driving profile data. Non-causal control algorithms also utilize future driving profile data to produce actual control system outputs. Clearly, such an approach is only possible in situations where the complete driving profile is known at the outset.

This situation can arise when trip planning instruments become generally available (GPS-based navigation, on-line traffic situation information, etc.). Even more important is the role of non-causal optimal solutions as benchmarks for causal solutions. A causal solution only can approach this result. Its "distance" from the non-causal optimum is a good indicator of the quality of a causal solution. In fact, if a causal solution achieves 95% of the benefits of the non-causal approach, there is little room for further improvements, such that refining the causal control system might not be worthwhile.

In summary, the minimization of the fuel consumption of a powertrain, in general, is not a simple and straightforward problem that can be completely solved with systematic procedures. Many iteration loops and intuitive

---

[17] In the context of this book the emphasis is on those control algorithms that produce the set points for all low-level control loops. Such systems will be denoted by the term *supervisory controllers*.

shortcuts are necessary in all but the simplest cases. Only well-defined partial problems can be solved using systematic optimization problems. The case studies shown in Appendix I exemplify this part of the design procedure, and the main mathematical tools used in this analysis are introduced in Appendix II and Appendix III.

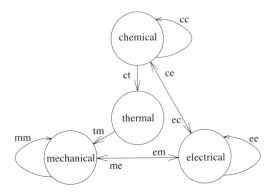

Fig. 2.18. Energy domains relevant for the modeling and optimization of vehicle propulsion systems. Also shown are the feasible energy conversion paths (reversible with double arrows, non-reversible with single arrows) [210].

2.3.5 Software Tools

General Remarks

Most non-trivial system analysis and synthesis problems can only be solved using numerical approaches. For that purpose efficient and reliable numerical computer tools must be available. The basic functions of such tools usually are provided by general-purpose software packages. Using these functions, tools specifically developed for the systems analyzed in this text can then be developed.

Such a software package must fulfill at least the following requirements:

- it allows for the interconnection of all relevant powertrain elements even if they operate in different energy domains (see Fig. 2.18);
- it allows for the scaling of all powertrain elements such that parametric optimizations can be accomplished; and
- it integrates well with the visualization and numerical optimization tools provided by the underlying general-purpose software package.

Quasistatic Simulation Tools

For quasistatic simulations, the ADVISOR[18] software package is often used. More information is available at the URL `http://www.avl.com/advisor`. Originally, this program was developed by the National Renewable Energy Laboratory (NREL). For more information on the previous versions of that software see `http://www.ctts.nrel.gov/analysis/advisor.html`.

The examples shown in this text have been calculated with the QSS Toolbox. The QSS Toolbox is a collection of Matlab/Simulink blocks and the appropriate parameter files that can be run in any Matlab/Simulink environment. This package is available for academic purposes and can be downloaded at the URL `http://www.imrt.ethz.ch/research/qss/`.

The QSS Toolbox fulfills the three requirements stated at the beginning of this section. The issue of scalability is discussed in the subsequent chapters. The interconnectability is guaranteed by interface structures that are compatible with the quasistatic method. Figure 2.19 shows some of the QSS Toolbox elements. Two examples that use the QSS Toolbox have been included in this text (see the example in Sect. 3.3.3 and the case study 8.1).

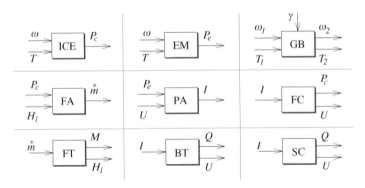

Fig. 2.19. Interfaces of main blocks of a QSS simulation environment. ICE = IC engine, FA = fuel amplifier (fuel control), FT = fuel tank, EM = electric motor, PA = power amplifier (current control), BT = battery, GB = gear box, FC = fuel cell, SC = supercapacitor.

Complex powertrain structures can be built with the basic blocks shown in Fig. 2.19. Figure 2.20 shows the example of a fuel-cell electric powertrain that includes a supercapacitor element for load-leveling purposes. A similar setup was used in [6] to analyze the potential for CO_2 reduction of fuel cell systems. The block "SEC" in Fig. 2.20 represents the supervisory energy control algorithm. Based on its two inputs P_e (the total required electric power) and U_2 (the supercapacitor voltage that indicates the amount of electrostatic

---

[18] ADVISOR is a registered trademark of AVL, Graz, Austria

energy stored in that device), it controls the amount of electric power that the fuel cell FC and the supercapacitor SC have to produce.[19]

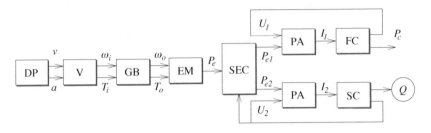

Fig. 2.20. Top level of a QSS Toolbox model of a fuel-cell electric powertrain (detailed model description in [6]). Blocks not yet introduced so far: DP = driving profile, V = vehicle, SEC = supervisory energy control system.

Dynamic Simulation Tools

Many packages are available for the dynamic simulation of powertrain systems. The main problem encountered with the usual tools like Matlab/Simulink is the missing flexibility when the system topology is changed. Such a change usually requires a complete redesign of the mathematical model. Attempts have been made to improve this situation. Notably the approach proposed by the Dymola/Modelica software tools is cited by several authors as a very promising step in that direction (see for instance [252]).

Since in this text the emphasis is on the quasistatic approach, that set of tools is not used in the subsequent parts. The few problems of dynamic system simulation and optimization are solved using classical software tools.

---

[19] Of course the power balance $P_e = P_{e1} + P_{e2}$ has to be satisfied.

3

IC-Engine-Based Propulsion Systems

In this chapter, standard IC-engine-based propulsion systems are analyzed using the tools that will be later applied for the optimization of more complex powertrains. The main components of IC-engine-based propulsion systems are the engine and the gear box. Clutches or torque converters, which also are part of such a powertrain, are needed during the relatively short phases in which the engine must be kinematically decoupled from the vehicle. As shown in the first section of this chapter, in the context of this book the engine may be described by an engine map and two normalized engine variables. The second section shows how the gear ratios must be chosen to satisfy drivability requirements. Once these two components and the main vehicle parameters are specified, estimations of the fuel consumption can be made using any of the methods introduced in the previous chapter. The third section of this chapter includes two examples of such an analysis.

3.1 IC Engine Models

3.1.1 Introduction

As mentioned in the previous chapter, two different descriptions of the powertrain elements are used in this text: the quasistatic and the dynamic formulation. For IC engines, the corresponding input and output variables are shown in Fig. 3.1. The thermodynamic efficiency of such a device is defined by

$$\eta_e = \frac{\omega_e \cdot T_e}{P_c} , \tag{3.1}$$

where ω_e is the engine angular speed, T_e the engine torque, and P_c the enthalpy flow[1] associated with the fuel mass flow

---

[1] The index c stands for "chemical" because the fuel carries the energy in the form of chemical energy.

$$\overset{*}{m}_f = P_c/H_l \, , \tag{3.2}$$

where H_l is the fuel's lower heating value.

(a) Quasistatic approach

(b) Dynamic approach

Fig. 3.1. Engine input and output variables in the quasistatic and in the dynamic system description.

The thermodynamic efficiency η_e of IC engines mainly depends on the engine speed and torque. The modeling of all relevant phenomena is a vast and well-documented area. A rich literature exists that describes the important points of this topic. The standard text [110] summarizes the main ideas and contains references to many other publications. In Sect. 3.1.3 below, simplified formulations of the dependency $\eta_e(\omega_e, T_e)$ are shown that are suitable for the purposes of this text.

The variables T_e and ω_e have a clear physical interpretation. Unfortunately, their range depends on the specific engine that is modeled (size, geometry, etc.). For this reason, normalized variables are introduced in the next section. Using these variables, the engine size can be used as an optimization parameter.

3.1.2 Normalized Engine Variables

When the engine runs in steady-state conditions, two normalized variables describe its operating point. These two quantities are the *mean piston speed*

$$c_m = \frac{\omega_e \cdot S}{\pi} \tag{3.3}$$

and the *mean effective pressure*

$$p_{me} = \frac{N \cdot \pi \cdot T_e}{V_d} \, , \tag{3.4}$$

where ω_e is the engine speed, T_e the engine torque, V_d the engine's displacement, and S its stroke. The parameter N depends on the engine type: for a four-stroke engine $N = 4$ and for a two-stroke engine $N = 2$ must be inserted in (3.4).

Obviously, the mean piston speed is the piston speed averaged over one engine revolution. It is limited at the lower end by the idling speed limit and at the upper end by aerodynamic friction in the intake part and by mechanical stresses in the valve train. Typical maximum values of c_m are below 20 m/s.

The mean effective pressure is that amount of constant pressure that must act on the piston during one full expansion stroke to produce that amount of mechanical work that a constant engine torque T_e produces during one engine cycle. For naturally aspirated engines the maximum value of p_{me} is around 10^6 Pa (10 bar). Typical turbocharged Diesel engines reach maximum mean effective pressures close to 20 bar. Even higher values are possible with special supercharging devices (twin turbochargers, pressure-wave superchargers, etc.).

The key advantages of using the normalized engine variables c_m and p_{me} are that their range is approximately the same for all[2] engines and that they are not a function of the engine size. Since for engines of similar type the speed boundaries vary less than the torque limits, engine maps are often shown in practice with c_m replaced by n_e, i.e., the engine speed in *rpm*.

For a fixed mean effective pressure and a mean piston speed, the equation

$$P_e = z \cdot \frac{\pi}{16} \cdot B^2 \cdot p_{me} \cdot c_m \qquad (3.5)$$

describes how the mechanical power P_e produced by the engine correlates with the number of cylinders z and the cylinder bore B.

With (3.5) it is possible to estimate the necessary engine size once the desired rated engine power P_{max} has been chosen. For instance, if a light-weight vehicle with $m_v = 750$ kg plus a payload of 100 kg is designed to reach 100 km/h in $t_0 = 15$ s, an estimation of the necessary rated power of 45 kW is obtained using the the approximation (2.16). Assuming the engine to be a naturally aspirated one, (3.5) indicates that the choice $z = 3$ and $B = 0.067$ m are reasonable values.[3]

3.1.3 Engine Efficiency Representation

The engine efficiency (3.1) is often plotted in the form of an *engine map*. Figure 3.2 shows such a map of the engine specified in the last section. On top of the $p_{me} = 0$ line, this map shows the engine efficiency as calculated with a standard thermodynamic engine process simulation program. No mixture enrichment at high loads is considered in these calculations. For that reason the best efficiencies are reached at full load conditions. Also shown are the constant power curves and the estimated maximum mean effective pressure limits.

---

[2] Of course the engines have to be of the same type, e.g., naturally aspirated SI engines.

[3] A four-cylinder configuration would require a bore B which is too small to yield a satisfactory thermodynamic efficiency.

As mentioned in Sect. 2.3, engine maps similar to the one shown in Fig. 3.2 are not the only way to describe the engine efficiency. It is easy to convert this form to the "fuel-flow" description that is used in Fig. 2.14.

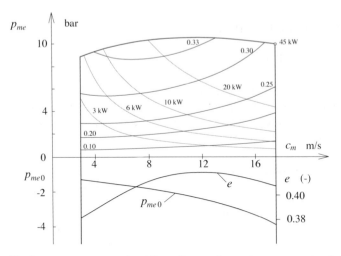

Fig. 3.2. Engine map computed with a thermodynamic process simulation program. Naturally aspirated SI engine, strictly stoichiometric air/fuel mixture. Engine parameters: displaced volume $V_d = 710 \cdot 10^{-6}\,\mathrm{m}^3$, bore and stroke $B = S = 0.067\,\mathrm{m}$, compression ratio $\epsilon = 12$.

Quite often it is possible to further simplify the engine model. A very simple, but nevertheless rather useful approximation is the Willans description [210], [274]. In this approach the engine mean effective pressure is approximated by

$$p_{me} \approx e(\omega_e) \cdot p_{mf} - p_{me0}(\omega_e) , \tag{3.6}$$

where the input variable p_{mf} is the *fuel mean pressure*. This variable is the mean effective pressure that an engine with an efficiency of 100% would produce by burning a mass m_f of fuel with a (lower) heating value H_l

$$p_{mf} = \frac{H_l \cdot m_f}{V_d} . \tag{3.7}$$

The parameter $e(\omega_e)$ stands for the indicated engine efficiency, i.e., the efficiency of the thermodynamic energy conversion from chemical energy to pressure inside the cylinder. The parameter p_{me0} summarizes all mechanical friction and pumping losses in the engine. For a specific engine system, these two parameters significantly depend mainly on the engine speed. Figure 3.2 contains two curves in the lower part that exemplify these dependencies.[4]

---

[4] In this example, the mean friction pressure is larger than in most modern engines. This fact is a consequence of the small engine size chosen in this example.

A parametrization using low-order polynomials or spline functions is usually sufficient. For preliminary computations the variables e and p_{me0} are often assumed to be constant parameters.

3.2 Gear-Box Models

3.2.1 Introduction

Gear boxes are elements that transform the mechanical power provided by a power source at a certain speed ω_1 and torque T_1 to a different speed ω_2 and torque T_2 level. Neglecting all losses that are caused by such a device, the following relations are valid

$$\omega_1 = \gamma \cdot \omega_2, \qquad T_2 = \gamma \cdot T_1 , \qquad (3.8)$$

where γ is the gear ratio.

As usual, a quasistatic and a dynamic formulation may be used to describe this element. The corresponding inputs and outputs are shown in Fig. 3.3. Mathematical models of these devices are introduced in this section and later in Chap. 4.

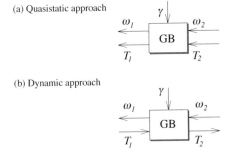

Fig. 3.3. Gear box input and output variables in the quasistatic and in the dynamic system description.

In addition to gear boxes most powertrains include devices that kinematically decouple the prime mover from the vehicle. Two types are commonly used: friction clutches (dry or wet) and hydraulic torque converters. These devices will be described in Sect. 3.2.4.

3.2.2 Selection of Gear Ratios

The ratio of the maximum to the minimum speed of a standard IC engine is limited to a value of approximately 10 for SI engines and to a value of 6 for

Diesel engines. Clearly, this is not sufficient for most practical applications. Excluding the very small speed values when the clutch or the torque converter are slipping, the ratio of the maximum to the minimum speed of a standard vehicle has a value of around 30. Accordingly, a gear box must be added to the powertrain. This device must realize a minimum gear ratio of five to meet the standard driving requirements.

Three types of gear boxes are encountered in practical applications:

- manual gear boxes, which have a finite number of fixed gear ratios and are manually operated by the driver;
- automatic transmissions, which combine a fixed number of gear ratios with a gear shift mechanism and a hydrodynamic torque converter or an automated standard clutch; and
- continuously variable transmissions (CVTs), which are able to realize any desired gear ratio within the limits of this device.

Choosing the gear ratios requires the solution of a complex optimization problem [78]. In a first iteration, the gear box efficiency and its dynamic properties may be neglected. The largest gear ratio (the smallest gear in a manual gear box) is often chosen to meet the towing requirements. Using (2.4), this gear ratio is found to be

$$\gamma_1 = \frac{m_v \cdot g \cdot r_w \cdot \sin(\alpha_{max})}{T_{e,max}(\omega_e)} . \tag{3.9}$$

The maximum engine torque $T_{e,max}$ depends on the engine speed ω_e. For this reason iterations may become necessary.

The smallest gear ratio (the highest gear in a manual gear box) can be chosen either to reach the top-speed limit or to maximize the fuel economy. These two approaches can be combined by choosing the second-smallest gear to satisfy the maximum speed requirements and the smallest gear to maximize the fuel economy ("overdrive" configuration). This approach is chosen in the example illustrated in Fig. 3.4. In this case, to determine the ratio of the fourth gear, first the following equation[5] must be solved for v_{max}

$$P_{e,max} = v_{max} \cdot F_{max} = v_{max} \cdot \left(m_v g c_r(v_{max}) + \frac{1}{2}\rho_a A_f c_d v_{max}^2 \right) . \tag{3.10}$$

In general, this equation must be solved numerically. If c_r may be assumed to be constant, a third-order polynomial equation results. Using Cartan's formula, it can be proven that in this case only one real solution of (3.10) exists such that no ambiguities arise.

Once the achievable top vehicle speed v_{max} and the top engine speed $c_{m,max}$ are known, the gear ratio γ_4 is found using the following equation

---

[5] This equation is similar to (2.13). However, it includes the rolling friction losses that become relevant in this context.

$$\gamma_4 = \frac{r \cdot c_{m,max} \cdot \pi}{v_{max} \cdot S} \,. \tag{3.11}$$

With this information, the vehicle resistance curves can be plotted in the engine map. Figure 3.4 shows the resulting resistance curves for the light-weight vehicle and the three-cylinder engine used as an example in this section.

The fifth gear can be chosen according to several fuel-economy optimization criteria. One possibility is to choose a value with which the vehicle can run at the most frequently used city speed without violating the smooth-running limits. In the case illustrated in Fig. 3.4 that speed is assumed to be 50 km/h.

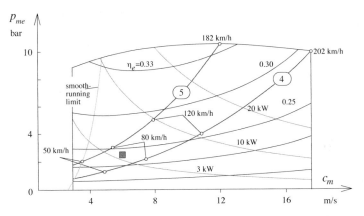

Fig. 3.4. Engine map 3.2 combined with two vehicle resistance curves (fourth and fifth gears on horizontal road) for the vehicle specified in Sect. 3.1.2. The grey square indicates the MVEG–95 average operating point.

Of course, there are many other criteria that must be observed when choosing the gear ratios. In all cases, the *gear spread*, i.e., the ratio of two neighboring gear ratios, must remain within certain boundaries. A geometric law is often chosen[6]

$$\gamma_k = \kappa \cdot \gamma_{k-1}, \quad k = 2, \ldots, k_{max}, \quad \kappa \approx \frac{2}{3} \,. \tag{3.12}$$

Moreover, when using gear boxes that have discrete values of gear ratios, *gear-box gaps* cannot be avoided. As illustrated by the shaded areas in Fig. 3.5, such gear boxes cannot exploit the full traction potential of the engine. The condition (3.12) ensures that these regions do not become too large.

An example of a model-based numeric optimization of the gear ratios is shown in Appendix I in Case Study 8.1.

---

[6] Of course, only rational gear ratios can be realized in practice.

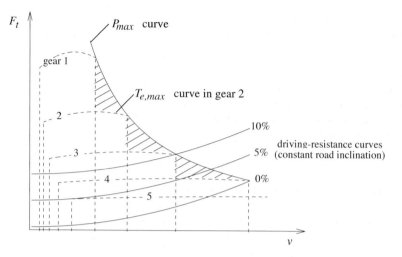

Fig. 3.5. Traction force F_t as a function of vehicle speed v at maximum engine power P_{max}. The hyperbola is realizable using a CVT and the five dome-shaped regions using a manual gear box with five different gear ratios. Also shown are the total vehicle driving-resistance curves for three different values of constant road inclinations. Vehicle parameters $\{A_f \cdot c_d, c_r, m_v\} = \{0.4\,\mathrm{m}^2, 0.008, 850\,\mathrm{kg}\}$; engine map as illustrated in Fig. 3.2.

3.2.3 Gear-Box Efficiency

The main dynamic effects caused by gear boxes have been discussed in Sect. 2.1.1. However, the result summarized in (2.11) is only valid if the efficiency η_{gb} of the gear box is 100%. Of course, this is not realistic. The losses caused by gear boxes and similar powertrain components depend on many influencing factors: speed, load, temperature, etc., just to name the most important ones. Figure 3.6 displays the structure of the system analyzed in this section and illustrates the variables that will be important in this analysis.

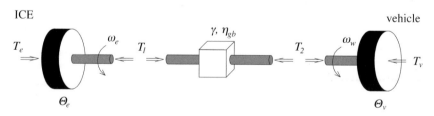

Fig. 3.6. Illustration of the definitions relevant for the modeling of the gear-box efficiency.

An approximation of the losses in gear boxes can be formulated using an affine dependency between the gear box input and output power

$$T_2 \cdot \omega_w = e_{gb} \cdot T_1 \cdot \omega_e - P_{0,gb}(\omega_e), \quad T_1 \cdot \omega_e > 0 , \tag{3.13}$$

where $P_{0,gb}$ is the power that the gear box needs to idle at an engine speed ω_e. Equation (3.13) is valid when the vehicle is in traction mode. If $T_1 \cdot \omega_e < 0$ a similar equation can be formulated to describe the losses in the gear box that affect the fuel cut-off torque

$$T_1 \cdot \omega_e = e_{gb} \cdot T_2 \cdot \omega_w - P_{1,gb}(\omega_e), \quad T_1 \cdot \omega_e < 0 . \tag{3.14}$$

For automotive cog-wheel gear boxes, typical values for e_{gb} are between 0.95 and 0.97. Depending on the size of the gear box and on its lubrication system, the idling losses $P_{0,gb}$ can reach up to 3% of the rated power of the gear box.

The evaluation of the efficiency of CVTs is discussed in Chap. 5, where hybrid-inertial vehicles are treated.

3.2.4 Losses in Friction Clutches and Torque Converters

During those phases in which the vehicle and the engine speed are not matched, the powertrain has to be kinematically decoupled. For that purpose either friction clutches or hydrodynamic torque converters are used in practice. Figure 3.7 illustrates the structure of the powertrain and the corresponding main system variables.

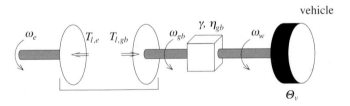

Fig. 3.7. Illustration of the definitions relevant for the modeling of the clutch and torque converter efficiency.

Friction Clutches

Dry- or wet-friction clutches have no torque amplification capability, i.e., their input and output torques are identical

$$T_{1,e}(t) = T_{1,gb}(t) = T_1(t) \quad \forall \, t . \tag{3.15}$$

Friction clutches produce substantial losses only during the first acceleration phase when the vehicle starts at zero velocity. If the engine speed ω_e

is assumed to be constant during this start phase, the clutch dissipates the following amount of mechanical energy

$$E_c = \frac{1}{2} \cdot \Theta_v \cdot \omega_{w,0}^2 \, , \tag{3.16}$$

where $\omega_{w,0}$ is that wheel velocity at which the clutch input speed ω_e and the output speed ω_{gb} coincide for the first time. The inertia Θ_v includes the vehicle inertia and all inertias due to the rotating parts located after the clutch. The amount of energy dissipated does not depend on the clutch torque profile during the clutch-closing process.

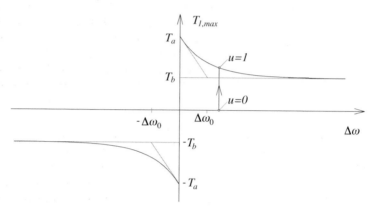

Fig. 3.8. Approximation of the maximum torque of a friction clutch.

Note that during all phases in which the clutch is slipping, the torque $T_1(t)$ at the gear box input is not limited by the engine but is defined by the clutch characteristics and its actuation system. As illustrated in Fig. 3.8, the clutch torque $T_1(t)$ depends on the speed difference $\Delta\omega(t) = \omega_{1,e}(t) - \omega_{1,gb}(t)$ and on the actuation input $u(t)$

$$T_1(t) = T_{1,max}(\Delta\omega(t)) \cdot u(t), \quad 0 \leq u(t) \leq 1 \, . \tag{3.17}$$

The maximum clutch torque can be approximated by

$$T_{1,max}(t) = \text{sign}(\Delta\omega(t)) \cdot \left[T_b - (T_b - T_a) \cdot e^{-|\Delta\omega(t)|/\Delta\omega_0} \right] \, . \tag{3.18}$$

The parameters $\Delta\omega_0$, T_a and T_b must be determined experimentally. In general, they depend on the temperature and wear of the clutch.

Torque Converters

Most automatic transmissions consist of an automated cog-wheel gear system and a hydraulic torque converter. The latter device produces additional losses in those operating phases in which it is not locked up. The losses incurred in these phases can be modeled as shown below.

The torque at the input of the converter may be modeled as follows

$$T_{1,e}(t) = \xi(\phi(t)) \cdot \rho_h \cdot d_p^5 \cdot \omega_e^2(t) . \tag{3.19}$$

The converter input speed $\omega_e(t)$ (the "pump speed") has a strong influence on the converter input torque, but the speed ratio

$$\phi(t) = \frac{\omega_{gb}(t)}{\omega_e(t)} \tag{3.20}$$

is important as well. The parameters ρ_h and d_p stand for the density of the converter fluid and for the pump diameter, respectively. The function $\xi(\phi)$ must be determined using experiments. Its qualitative form is illustrated in Fig. 3.9.

The converter output torque $T_{1,gb}$, which in this case may be larger than the input torque, is determined by the pump–turbine interaction

$$T_{1,gb} = \psi(\phi(t)) \cdot T_{1,e}(t) . \tag{3.21}$$

The function $\psi(\phi)$ must be experimentally determined as well. Qualitatively, it will have a form similar to the one shown in Fig. 3.9. Equations (3.19) and (3.21) are valid in steady-state conditions. However, since the fluid dynamic processes inside the torque converter are substantially faster than the typical time constants of the vehicle longitudinal dynamics, the fluid dynamic effects may often be neglected.

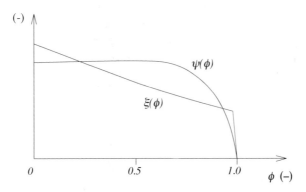

Fig. 3.9. Qualitative representation of the main parameters of a torque converter.

With these preparations, the efficiency of the torque converter in traction mode is easily found to be

$$\eta_{tc} = \frac{\omega_{gb} \cdot T_{1,gb}}{\omega_e \cdot T_{1,e}} = \psi(\phi) \cdot \phi . \tag{3.22}$$

3.3 Fuel Consumption of IC Engine Powertrains

3.3.1 Introduction

In the following two sections the two methods introduced in the last chapter in Sects. 2.3.1 and 2.3.2 are used to predict the fuel consumption of an experimental light-weight vehicle. The powertrain consists of a downsized high-speed SI engine and a manual cog-wheel gear box. The vehicle is a standard small four-seat passenger car with improved aerodynamics and rolling friction parameters.

The data of this vehicle are similar to those of the experimental SmILE prototype that was developed and realized by Swissauto/WENKO AG, Burgdorf Switzerland.[7] A picture of that vehicle is shown in Fig. 3.10. A detailed description of that project and its results can be found in [103].

Fig. 3.10. The SmILE vehicle, courtesy Swissauto/WENKO AG, Burgdorf.

3.3.2 Average Operating Point Method

At this point all the elements have been introduced that are needed to estimate the fuel consumption of a conventional powertrain using the approach introduced in Sect. 2.3.1. The lightweight vehicle analyzed in this section is characterized by its main parameters $\{A_f \cdot c_d, c_r, m_v\} = \{0.4\,\text{m}^2, 0.008, 750\,\text{kg}\}$. The engine analyzed below is the one whose main characteristics are illustrated in Fig. 3.2.

Using (2.23) and (2.24), and adding a payload mass of $100\,kg$ to the vehicle mass, the mean traction force at the wheel $\bar{F}_{trac}$ necessary to drive the

---

[7] The engine developed in that project was a downsized supercharged SI engine. That approach proved to be very effective to improve the part-load efficiency of a stoichiometrically operated SI engine.

MVEG–95 cycle is found to be approximately 210 N. This yields an average traction power at the wheels of approximately

$$\bar{P}_{trac} = \frac{\bar{F}_{trac} \cdot \bar{v}}{trac} = \frac{210\,\mathrm{N} \cdot 9.5\,\mathrm{m/s}}{0.6} \approx 3.3\,\mathrm{kW} . \qquad (3.23)$$

The parameter $trac$ denotes the time fraction in which the vehicle is in traction mode (see Fig. 2.6) and the mean power $\bar{P}_{trac}$ is the relevant information needed to compute the engine load.[8]

The powertrain is assumed to include a conventional cog-wheel gear box and friction clutch. Using (3.13) the power at the input of the gear box can be estimated to be

$$P_1 = \frac{1}{e_{gb}} \cdot \left(\bar{P}_{trac} + P_{0,gb}\right) = \frac{1}{0.97} \cdot (3.3\,\mathrm{kW} + 0.3\,\mathrm{kW}) \approx 3.7\,\mathrm{kW} . \qquad (3.24)$$

The auxiliaries, including the electric power generator, and the friction clutch consume some of the power produced by the engine. In this approach the losses caused by the auxiliaries are taken into account by an additional average mechanical power $\bar{P}_{aux}$ of 0.25 kW. This value is rather low, i.e., the vehicle is assumed to have no power steering and no air conditioning.

According to (3.16), each start causes an energy loss of

$$E_c = \frac{1}{2} \cdot \Theta_v \cdot \omega_{w,0}^2 = \frac{1}{2} \cdot m_v \cdot v_0^2 = \frac{1}{2} \cdot 850\,\mathrm{kg} \cdot (3\,\mathrm{m/s})^2 \approx 3.8\,\mathrm{kJ} . \qquad (3.25)$$

In the MVEG–95 on average one start from rest occurs every kilometer. Since in this cycle the average velocity is 9.5 m/s, such an event takes place every 105 s. Accordingly, the average power consumed by the starts is around 3.8 kJ/105 s ≈ 35 W.

In summary, during the traction phases the engine has to produce the average power $\bar{P}_e = (3.7 + 0.25 + 0.035)\,\mathrm{kW} \approx 4\,\mathrm{kW}$. Therefore, assuming a mean piston speed of $\bar{c}_m = 6\,\mathrm{m/s}$,[9] the engine is operated with an average mean effective pressure $\bar{p}_{me}$ of approximately 2.5 bar. This value is obtained by inserting the engine parameters into (3.5).

The efficiency of the engine at that operating point is found using (3.6), whereas the numerical values for $e(c_m)$ and $p_{me0}(c_m)$ are taken from Fig. 3.2

$$\eta_e = \frac{p_{me}}{p_{mf}} = \frac{e(c_m) \cdot p_{me}}{p_{me} + p_{me0}(c_m)} \approx \frac{0.4 \cdot 2.5\,\mathrm{bar}}{2.5\,\mathrm{bar} + 1.6\,\mathrm{bar}} \approx 0.24 . \qquad (3.26)$$

Therefore, the average fuel power consumed by the engine in the MVEG–95 cycle is approximately

$$\bar{P}_f = trac \cdot \bar{P}_e/\eta_e = 0.6 \cdot 4\,\mathrm{kW}/0.24 \approx 10\,\mathrm{kW} . \qquad (3.27)$$

---

[8] To obtain the correct average fuel consumption, the factor $1/trac$ will be compensated later in (3.27).

[9] For the engine specified this corresponds to approximately 2700 rpm.

This corresponds to a fuel flow of

$$\overset{*}{V}_f = \bar{P}_f / (H_l \cdot \rho_f) \ . \tag{3.28}$$

Assuming standard RON–95 gasoline and inserting the corresponding numerical values of $H_l = 43.5 \cdot 10^6 \, \text{J/kg}$ for the fuel's lower heating value and $\rho_f = 0.75 \, \text{kg/l}$ for its density yields a fuel consumption of approximately $3.1 \cdot 10^{-4} \, \text{l/s}$ or, with the value of $9.5 \, \text{m/s}$ for the average speed in the MVEG–95 cycle, of approximately $3.3 \, \text{l/100 km}$.

So far it has been assumed that in all braking *and* idling phases the engine is shut down. While it is easy to cut off fuel in the braking phases, automatic starters that avoid idling losses are more expensive and, thus, most engines have non-zero idling losses.

The idling fuel mean pressure can be estimated from (3.6) by setting $p_{me} = 0$

$$p_{mf,0} = p_{me0}(\omega_{e,idle})/e(\omega_{e,idle}) \tag{3.29}$$

from which the fuel flow follows to be

$$\overset{*}{V}_{f,idle} = p_{mf,0} \cdot \frac{V_d}{H_l \cdot \rho_f} \cdot \frac{c_{m,idle}}{N \cdot S} \ . \tag{3.30}$$

Choosing $c_{m,idle} = 2.5 \, m/s$ as the idling mean piston speed[10] and assuming a four-stroke engine yields the following numerical values

$$\overset{*}{V}_{f,idle} = 4 \cdot 10^5 \, \text{Pa} \cdot \frac{710 \cdot 10^{-6} \, \text{m}^3}{43.5 \cdot 10^6 \, \text{J/kg} \cdot 0.75 \, \text{kg/l}} \cdot \frac{2.5 \, \text{m/s}}{4 \cdot 0.067 \, \text{m}} \approx 8.3 \cdot 10^{-5} \, \text{l/s} \ . \tag{3.31}$$

In the MVEG–95 cycle the engine is idling for approximately $300 \, \text{s}$. Accordingly, in that cycle the fuel spent for idling is approximately

$$300 \, \text{s} \cdot 8.3 \cdot 10^{-5} \, \text{l/s} \cdot 100/11.4 \, \text{km} \approx 0.2 \, \text{l/100 km} \ . \tag{3.32}$$

This figure has to be added to the $3.3 \, \text{l/100 km}$ used for vehicle propulsion. The sum of $3.5 \, \text{l/100 km}$ is the estimated total fuel consumption for the chosen example of a lightweight vehicle and downsized engine system. Cold-start losses and other detrimental effects not considered so far are likely to increase that figure somewhat.

3.3.3 Quasistatic Method

In this section a quasistatic approach is used to predict the fuel consumption of the vehicle and powertrain described in the previous section. The QSS toolbox serves as the computational platform for the powertrain modeling and simulation. Figure 3.11 shows the top layer of the resulting model description. Readers familiar with Matlab/Simulink will immediately recognize the characteristic elements of that software tool.

---

[10] For the engine chosen in this example this corresponds to approximately 1100 rpm. This is a reasonable value for such a small engine.

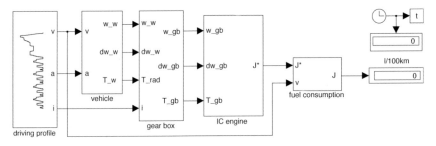

Fig. 3.11. Top layer of the QSS model of the powertrain analyzed in this section.

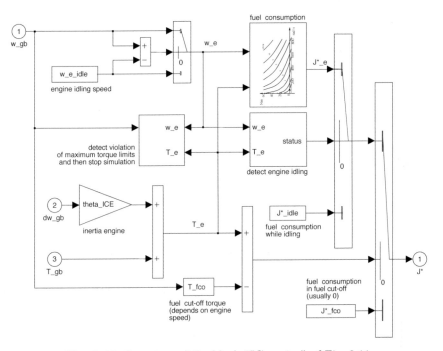

Fig. 3.12. Structure of the block "IC engine" of Fig. 3.11.

As an example, the contents of the block "IC engine" are shown in Fig. 3.12. The complete model, including all necessary system parameters, is part of the QSS toolbox package that may be downloaded at the URL http://www.imrt.ethz.ch/research/qss/. The interested reader is referred to that source for a detailed description of all elements of that module.

Simulating the behavior of this powertrain yields a total fuel consumption of 3.6 l/100 km in the MVEG–95 cycle. This value correlates well with the value of 3.5 l/100 km obtained in the last section.

In fact, as Fig. 3.13 shows, the engine is operated with many different load/speed combinations. Such a variability offers many opportunities for energy optimization, particularly if more than one mechanical power source

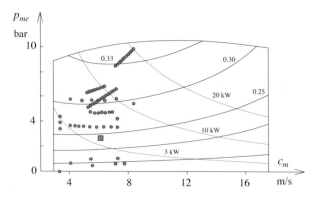

Fig. 3.13. Engine operating points of the example of this section for the MVEG–95 cycle. Also shown is the average operating point used in the previous section.

and energy storage device are available. Accordingly, for hybrid vehicles the average-point method may not be applied. In these cases, reliable fuel consumption estimations must be based on quasistatic simulations that include the supervisory control loops. Figure 3.13 must be analyzed with some care. In fact, the distribution of the load/speed points must be complemented by information on the frequency of these points. A representation similar to the one shown in Fig. 3.14 helps to understand which engine operation points are relevant.

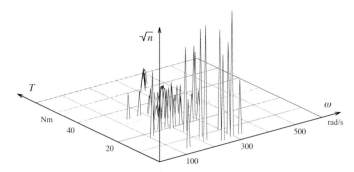

Fig. 3.14. Distribution and frequency of the engine operating points (*square root of the number of counts for each non-zero engine load/speed point*).

4

Electric and Hybrid-Electric Propulsion Systems

While in conventional ICE-based vehicles the energy carrier is a fossil fuel, electric and hybrid-electric propulsion systems are characterized by the presence of an electrochemical or electrostatic energy storage system. Moreover, at least one electric motor is responsible — totally or partially — for the vehicle propulsion.

In this chapter, first purely electric vehicles will be briefly discussed. Then various types of hybrid-electric vehicles will be introduced. The subsequent sections describe the quasi-stationary and the dynamic models of typical electric components of such vehicles, including electric motors/generators, electrochemical batteries, and supercapacitors. The modeling representations of an electric power bus, a torque coupler, and a planetary gear set are added as separate sections due to the importance of the mentioned components in most hybrid-electric powertrains. A mean-value analysis of the energy consumption of various powertrain configurations concludes the chapter.

4.1 Electric Propulsion Systems

Purely electric propulsion systems (*electric vehicles*, EVs, or *battery-electric vehicles*, BEVs) are characterized by an electric energy conversion chain upstream of the drive train, roughly consisting of a battery (or another electricity storage system) and an electric motor with its controller. The resulting vehicle is not autonomous (see the definition in the Introduction), since the energy density of batteries does not permit sufficient driving autonomy. Moreover, the time required for recharging is usually not negligible, and surely larger than the typical refueling time of ICE-based vehicles. For these reasons, only some brief considerations on the energy performance of such vehicles will be presented here.

Some passenger cars appearing on the market in the last years have been equipped with electric propulsion systems. Asynchronous AC, permanent-magnet AC, or DC machines are all used as traction motors, with a peak power in the range from 20 to 50 kW. The key component is the battery, which is usually of the nickel–metalhydrate or of the nickel–cadmium type, even though newer technologies (e.g., lithium-ion) have been demonstrated already. The energy density typically ranges from 30 to 65 Wh/kg. Top speed normally does not exceed 100 km/h, while the range is typically between 75 and 95 km per charge, but it may reach 100 km in the newer models. Typical values for the electric energy consumption are 15–30 kWh/100 km, which means that the overall efficiency ranges from 40 to 60% [157, 46, 136].

The arguments listed in Sects. 1.3 and 1.4 seem to suggest that the most suitable application of EVs is in micro-cars for use in urban contexts, especially within car-sharing organizations. However, whether the technology of battery-electric vehicles will have a future in the 21st century or not, and under which conditions, is still a matter of debate [105, 267, 170, 72].

4.2 Hybrid-Electric Propulsion Systems

In contrast to ICE-based and battery-electric vehicles, hybrid vehicles are characterized by two or more prime movers and power sources. Usually, the term "hybrid vehicle" is used for a vehicle combining an engine and an electric motor. More appropriately, such a combination should be called a hybrid-electric vehicle (HEV), since other, different "hybrid" configurations have been proposed (see Chap. 5 for mechanical, pneumatic, and hydraulic hybrids and Chap. 6 for fuel cell vehicles).

In general, an HEV includes an engine (see Chap. 3) as a fuel converter or irreversible prime mover (fuel cells are treated in Chap. 6; gas turbines or Stirling engines are not considered here[1]). As electric prime movers, different types of motors (standard DC, induction AC, brushless DC, etc.) are used. In some configurations, a second electric machine is required, which acts primarily as a generator. The electric energy storage system is usually an electrochemical battery, though supercapacitors may be used in some prototypes. Sections 4.3 – 4.5 describe motor, battery, and supercapacitor models, respectively.

One of the main motivations for developing HEVs is the possibility to combine the advantages of the purely electric vehicles, in particular zero local emissions, with the advantages of the ICE-based vehicles, namely high energy and power density. HEVs can profit from various possibilities for improving the fuel economy with respect to ICE-based vehicles. In principle, it is possible to:

1. downsize the engine and still fulfill the maximum power requirements of the vehicle;

---

[1] Gas turbines and Stirling engines have been proposed in series hybrid configurations, where they are operated in their preferred steady-state conditions. However, the low efficiencies of these two thermal engines have kept them from gaining general acceptance.

2. recover some energy during deceleration instead of dissipating it in friction braking;
3. optimize the energy distribution between the prime movers;
4. eliminate the idle fuel consumption by turning off the engine when no power is required (stop-and-start); and
5. eliminate the clutching losses by engaging the engine only when the speeds match.

These possible improvements are partially counteracted by the fact that HEVs are about 10–30% heavier than ICE-based vehicles [122, 116].

Generally, not all the possibilities (1)–(5) are used simultaneously. The following section describes the different types of hybrid-electric vehicles developed or proposed and their modes of operation.

4.2.1 System Configurations

Hybrid-electric vehicles are classified into three main types:

- **Parallel hybrid**: both prime movers operate on the same drive shaft, thus they can power the vehicle individually or simultaneously.
- **Series hybrid**: the electric motor alone drives the vehicle. The electricity can be supplied either by a battery or by an engine-driven generator.
- **Series-parallel, or combined hybrid**: This configuration has both a mechanical and an electrical link.

Additionally, certain new concepts have been introduced that cannot be adequately classified into either of the three basic types. For these concepts, the term "complex" hybrid is sometimes used [45, 116].

Series HEVs

Series hybrid propulsion systems utilize the internal combustion engine as an auxiliary power unit (APU) to extend the driving range of a purely electric vehicle. Using a generator, the engine output is converted into electricity that can either directly feed the motor or charge the battery (Fig. 4.1). Regenerative braking is possible using the traction motor as a generator and storing the electricity in the battery (point 2 in Sect. 4.2). The engine operation is not related to the power requirements of the vehicle (4), thus the engine can be operated at a point with optimal efficiency and emissions (3). An added advantage may be the fact that the transmission does not require a clutch, i.e., the engine is never disengaged since it is mechanically decoupled from the drive axle (5). However, a series hybrid configuration needs three machines: one engine, one electric generator, and one electric traction motor. At least the traction motor[2] has to be sized for the maximum power requirements of

---

[2] Possibly, the generator also in configurations with larger APUs.

the vehicle. Thus a series hybrid in principle offers the possibilities of reducing fuel consumption following the approaches (2) to (5) listed above. The overall tank-to-wheel efficiency for series hybrid vehicles is on a par with the values of vehicles powered by modern, fuel-efficient IC engines.[3] The additional weight due to car body reinforcement, electric machines, battery, etc. may push the fuel consumption above the value of good ICE-based vehicles, however.

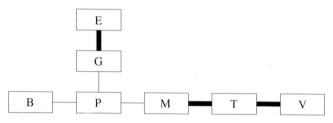

Fig. 4.1. Basic series hybrid configuration. B: battery, E: engine, G: generator, M: motor, P: power converter, T: transmission (including clutch and gears), V: axles and vehicle. Bold lines: mechanical link, solid lines: electrical link.

Parallel HEVs

While series hybrid vehicles may be considered as purely electric vehicles with an additional ICE-based energy path, parallel hybrid vehicles are rather ICE-based vehicles with an additional electrical path (Fig. 4.2). In parallel HEVs both engine and electric motor can supply the traction power either alone or in combination. This leaves an additional degree of freedom in fulfilling the power requirements of the vehicle, which can be used to optimize the power distribution between the two parallel paths (point 3 in the previous section). Typically, the engine can be turned off at idle (4) and the electric motor can be used to assist accelerations and, in general, high power demands. Both machines can therefore be sized for a fraction of the maximum power (1). Together with the fact that only two machines are needed, this is an advantage with respect to series hybrid vehicles. A disadvantage is the need for a clutch, since the engine is mechanically linked to the drive train (5). The electric motor can be utilized as a generator to charge the battery, being fed by regenerative braking (2) or by the engine. Even though the additional weight still plays an important role, all the possibilities (1) to (4) listed above increase in principle the system efficiency as compared to an ICE-based vehicle.

The most simple parallel hybridization is the so-called *mild hybrid* concept, which is essentially an ICE-based powertrain with a small electric motor. This motor is typically belt-driven and mounted on the front of the engine

---

[3] Although not a topic of this text, it is worth mentioning that series HEVs have extremely low pollutant emissions.

Fig. 4.2. Full parallel hybrid configuration. B: battery, E: engine, M: motor, P: power converter, T: transmission (including clutch and gears), V: axles and vehicle. Bold lines: mechanical link, solid lines: electrical link.

(Fig. 4.3). It does not require any high battery capacity or complex power electronics, since its role consists of the automatic engine stop-and-start (4), providing a limited power boosting capability useful for engine downsizing (1), and offering a limited capability for energy recuperation (2). The motor may also act as an alternator for the electrical loads, and this operation is particularly useful with regard to 42 V on-board networks (see Sect. 4.6), which will require higher power levels than conventional networks [121].

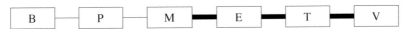

Fig. 4.3. Mild parallel hybrid configuration. B: battery, E: engine, M: motor, P: power converter, T: transmission (including clutch and gears), V: axles and vehicle. Bold lines: mechanical link, solid lines: electrical link.

Combined HEVs

Somehow intermediate between series and parallel hybrids is the combined hybrid configuration. This is mostly a parallel hybrid, but it contains some features of a series hybrid. Actually, both a mechanical and an electric link are present, together with two distinct electric machines. As in a parallel hybrid configuration, one is used as a prime mover or for regenerative braking. The other machine acts like a generator in a series hybrid system. It is used to charge the battery via the engine or for the stop-and-start operation. Two different realizations of combined hybrids have been introduced recently. The first (Toyota THS, Ford Hybrid System, see Fig. 4.4) has a planetary gear set (PGS). The second (Nissan Tino, see Fig. 4.5) combines the chain-driven generator of mild parallel hybrids and a crankshaft-mounted motor, as in full parallel hybrids, coupled at the DC link level.

4.2.2 Power Flow

Because of the different HEV configurations, various operating modes are possible. This section illustrates the power flows among the various components of a hybrid vehicle [45, 116].

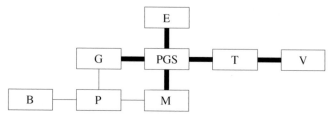

Fig. 4.4. Configuration of a combined hybrid with a planetary gear set. B: battery, E: engine, G: generator, M: motor, P: power converter, PGS: planetary gear set, T: transmission (including clutch and gears), V: axles and vehicle. Bold lines: mechanical link, solid lines: electrical link.

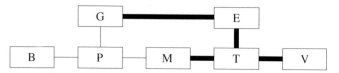

Fig. 4.5. Configuration of a combined hybrid without a planetary gear set. B: battery, E: engine, G: generator, M: motor, P: power converter, T: transmission (including clutch and gears), V: axles and vehicle. Bold lines: mechanical link, solid lines: electrical link.

Series HEVs

In the standard series hybrid configuration, the link between the engine path and the battery path is electrical, i.e., at the power link level, with the battery output voltage feeding the motor and the generator. The motor and generator currents balance the battery terminal current. The power balance at the power link is regulated by the torque distribution controller, which selects the operating mode and the ratio u between the power from/to the battery and the total power at the link. Section 4.6 describes how this ratio is implemented, whereas the control strategies for the energy management are described in detail in Chap. 7.

As illustrated in Fig. 4.6, series hybrid vehicles basically have four modes of operation. In urban driving, when the battery is sufficiently charged, the purely electric, zero emission (ZEV) driving mode ($u = 1$) is usually selected. When the battery charge is too low, the engine is turned on and typically set to its maximum efficiency operating point. The power resulting from the difference between the engine power and the power at the link recharges the battery ($u < 0$) via the generator. Such a combination of battery discharge and charge represents a *duty-cycle operation*, which is typical of series hybrid vehicles. In principle, when the fuel-optimal engine power is below the power at the link, the missing power could be provided by the battery ($0 < u < 1$), though this mode of operation is seldom used in practice. Of course, during braking or deceleration some energy is recuperated in the battery by the motor being used as a generator ($u = 1$).

(a) battery drive, u=1

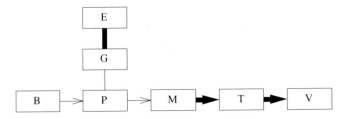

(b) battery recharging, u<0

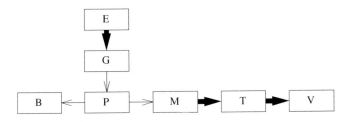

(c) hybrid drive, 0<u<1

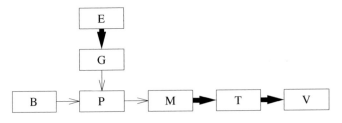

(d) regenerative braking, u=1

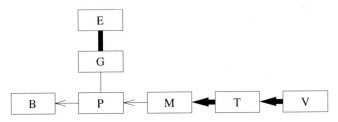

Fig. 4.6. Power flow for the modes of operation of series hybrid vehicles.

Parallel HEVs

In parallel hybrid vehicles the link between the engine path and the electric path is mechanical. The simplest configuration is the one in which the two powertrains drive separate wheel axles. In mild hybrids, the coupling is represented by the belt that typically drives the starter/generator, usually together with auxiliary loads. In full hybrids, usually the motor is mounted on the crankshaft between the engine and the transmission [116]. In many cases, an additional clutch between the engine and the motor is inserted. Generally speaking, the two power flows may be regarded to be combined in a "torque coupler." The power balance at the torque coupler is regulated by the power distribution controller, which selects the operating mode and the ratio u between the power from/to the motor and the total power at the coupler. The modeling of torque couplers is described in Sect. 4.7, whereas the control strategies, again, are presented in Chap. 7.

Depending on the value of u, different operating modes are possible. During startup or acceleration, the engine provides only a fraction of the total power at the coupler, the rest being delivered by the motor ($0 < u < 1$) which realizes the so-called *power assist* concept. During braking or deceleration, the motor recuperates energy in the battery acting like a generator ($u = 1$). At light load, the engine may be required to provide more power than strictly demanded, the extra power charging the battery via the electric machine ($u < 0$). Both the pure engine operation ($u = 0$) and the purely electric operation ($u = 1$) are also possible in principle. Figure 4.7 illustrates these scenarios for a full parallel hybrid configuration.

Combined HEVs

Combined hybrids have the ability to operate as series or as parallel hybrids. Thus the possible modes of operation result from the combination of the modes already discussed in the previous sections. However, the use of a planetary gear set (Fig. 4.4) adds some constraints to the possible energy paths. As will become clear in the following (see Sect. 4.8), the purely ICE operation is often associated with a power flow through the generator and the motor. The other modes, comprising ZEV, regenerative braking, battery recharging, and power assist, are of course possible, as illustrated in Fig. 4.8.

4.2.3 Concepts Realized

In recent years many passenger cars have been demonstrated that use one of the hybrid configurations discussed in this chapter [279, 119, 7]. The models that already have entered mass production or that are considered reasonably ready for the market are of the series-parallel type, preferably with a planetary gear set as a torque coupler, and some parallel hybrids.

(a) power assist, 0<u<1

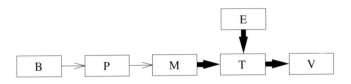

(b) regenerative braking, u=1

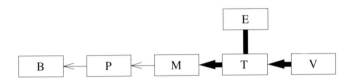

(c) battery recharging, u<0

(d) ZEV, u=1

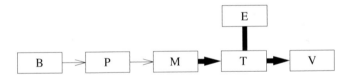

(e) conventional vehicle, u=0

Fig. 4.7. Power flow for the modes of operation of parallel hybrid vehicles.

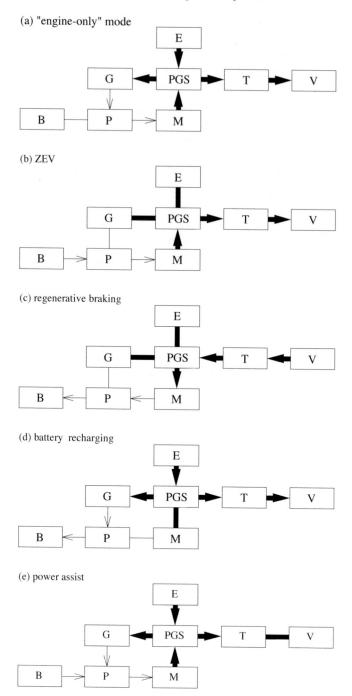

Fig. 4.8. Power flow for the operating modes of combined hybrid vehicles with a planetary gear set.

Combined hybrid concepts (Nissan Tino [175], Toyota THS-I [219], Toyota THS-II [250], Toyota THS-C [251], Ford's Hybrid System [83], Two-mode Hybrid System from Global Hybrid Cooperation [177]) and parallel hybrid concepts (Audi Duo [109], Honda IMA [113], PSA Hybrid HDi [200], DaimlerChrysler's ESX3 [57]) are characterized by very different degrees of hybridization (the ratio between the electric motor power and the engine power) ranging from 15% to 55% or more. Those concepts showing a lower degree of hybridization are often regarded as mild hybrids, even though their architecture is the same as that of full parallel hybrids as depicted in Fig. 4.2. True mild hybrids (Fig. 4.3) have also been demonstrated (GM's Belt Alternator Starter [244], Toyota THS-M [121], PSA Stop & Start [201]). These vehicles have a low degree of hybridization (2–15%), with an electric system that is mainly aimed at implementing a stop-and-start concept.

Usually gasoline engines, permanent-magnet synchronous AC (brushless DC) motor/generators, and nickel–metal hydride batteries are used, though there are a few prototypes equipped with a Diesel engine or an asynchronous AC motor or a lithium-ion battery. Also often used are continuously variable transmissions (CVT). In combined hybrids, the power flow is regulated by a planetary gear set or by some clutching mechanism. Parallel hybrids realize the mechanical coupling with separate axles or directly on the same transmission shaft.

Only few examples of series hybrids have been demonstrated by major car manufacturers, usually as an improvement (range extender) of some older purely electric vehicle.

4.2.4 Modeling of Hybrid Vehicles

The model of a complete HEV can be split into a number of different submodels representing the various components of the system. A good and useful modeling practice consists of making these submodels "autonomous," so that each submodel interfaces only with the submodels of the components which are actually linked by a power flow. This approach yields a modular system description, in which each module has clear input and output variables which may be combined with input and output variables of other submodels to represent a complex configuration. Another advantage of this approach is that it allows various arrangements of the HEV components (i.e., battery, engine, motor) while keeping the same basic submodels. In other words, the same "library" of submodels can be used to represent series, parallel, and combined hybrids.

The difference between quasistatic and dynamic modeling approaches was discussed in Chap. 2. The flow of power factors is illustrated in Figs. 4.9 – 4.11 for series, full parallel, and PGS-based combined hybrid configurations, as well as for both modeling approaches. Each block in these graphs can be extracted to represent a submodel, keeping the causality of the respective input and output power factors. The power flow modeling of mild hybrids

is the same as that of full parallel hybrids, and that of non-PGS combined
hybrids is the same as that of PGS-based ones. Of course, what changes is the
physical realization of the respective models, in particular the nature and the
position of the "torque coupler."

(a) quasistatic approach

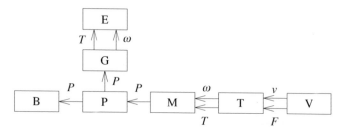

(b) dynamic approach

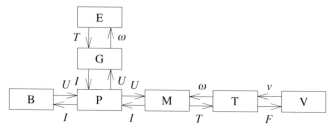

Fig. 4.9. Flow of power factors for a series hybrid configuration, with the quasistatic
approach (a) and the dynamic approach (b). F: force, I: current, P: power, T: torque,
v: speed, U: voltage, ω: rotational speed.

4.3 Electric Motors

Electric machines find a place in conventional vehicles as starters and alterna-
tors. The former boost the engine to reach its idle speed and to start delivering
torque. The latter produce electricity to charge the 12 V battery and to feed
the electric auxiliary loads. In electric and hybrid-electric vehicles, the electric machine is a key component. Usually it is a reversible machine, which can
operate in different ways: (1) convert the electrical power from the battery
into mechanical power to drive the vehicle, (2) convert the mechanical power
from the engine into electrical power to recharge the battery, and (3) recuperate mechanical power available at the drive train to recharge the battery
(regenerative braking). The latter two modes are generator modes. In parallel
hybrid vehicles and in electric vehicles the two functions can be fulfilled in

(a) quasistatic approach

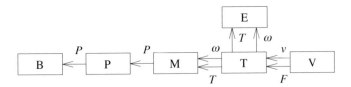

(b) dynamic approach

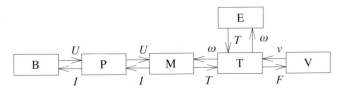

Fig. 4.10. Flow of power factors for a parallel hybrid configuration, with the quasistatic approach (a) and the dynamic approach (b). F: force, I: current, P: power, T: torque, v: speed, U: voltage, ω: rotational speed.

(a) quasistatic approach

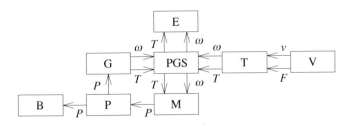

(b) dynamic approach

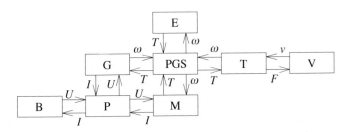

Fig. 4.11. Flow of power factors for a combined hybrid configuration, with the quasistatic approach (a) and the dynamic approach (b). F: force, I: current, P: power, T: torque, v: speed, U: voltage, ω: rotational speed.

principle by a single machine. In series hybrid vehicles and in combined hybrid vehicles two different machines are needed, the generator usually being a second reversible machine, smaller than the traction motor.

Characteristics of a good HEV motor include generally high efficiency, low cost, high specific power, good controllability, fault tolerance, low noise, and uniformity of operation (low torque fluctuations).

Electric motors basically can be organized in two main categories: direct-current (DC) motors and alternating-current (AC) motors (see Fig. 4.12). All types of motors have a stationary part, called the stator, and a rotating part, called the rotor. The latter is connected to the output shaft on which the motor torque is acting. The electricity provided by the DC supply through the motor controller is applied at the motor terminals. Electromechanical energy conversion takes place as a consequence of Faraday's law and of Lorentz' law. The former describes the induction of an electromotive force (emf) in conductors being in relative motion with respect to a magnetic field. The latter describes the force generated on a current-carrying conductor lying in a magnetic field.

In DC motors, the rotor surface hosts a number of conductors (rotor windings) which terminate with a collector. As a consequence of the application of DC voltage to the rotor windings by means of carbon brushes, which are in contact with the collector, a magnetic field is generated whose polarity is continuously changed by contact commutation. At the same time, a stationary magnetic field is generated in the stator using permanent magnets or field windings. The interaction of the two magnetic fields causes a rotation of the rotor.

In AC motors, a rotating magnetic field is generated in the stator by loops of wires (stator windings). Three-phase motors have one or more sets of three windings on their stator. The number of these sets is called the number of poles of the motor. When three-phase AC voltage is applied to the stator, a magnetic field is generated, which changes its orientation according to the sign of the current flowing in the windings. Since this is continuously varying, the orientation of the magnetic field keeps varying, resulting in a rotating magnetic field. The speed of the rotating magnetic field is called the synchronous speed. It equals the pulsation of the three-phase AC voltage divided by the number of poles.

In synchronous AC motors the rotor operates at the same speed as the rotating magnetic field. This synchronization is achieved often using a permanent-magnet rotor. The permanent magnets generate their own magnetic field which interacts with the rotating magnetic field generated by the stator windings. These motors are often referred to as brushless DC motors, as will become clearer in the following.

In asynchronous AC motors, also called induction motors, the rotor hosts a set of conductors with end rings, an arrangement known as "squirrel cage." Electromotive force and thus current is induced in the rotor windings by the interaction of the conductors with the rotating magnetic field generated

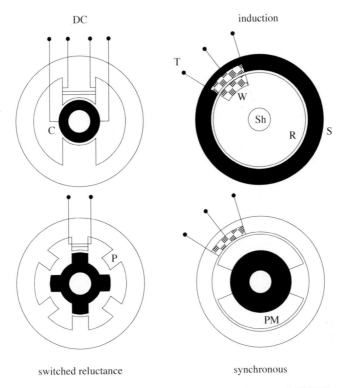

Fig. 4.12. Schematics of four types of motors: DC, induction AC, PM synchronous AC, switched reluctance motors. C: collector, P: pole, PM: permanent magnets, R: rotor, S: stator, Sh: shaft, T: terminal, W: windings.

by the stator. The rotor becomes an electromagnet with alternating poles, attracted by those of the stator rotating magnetic field. In order for torque to be produced, the speed of the rotor must be different from the speed of the rotating magnetic field.

In switched reluctance (SR) motors, both stator and rotor are designed with "notches" or "teeth" referred to as salient poles. Each stator pole carries an excitation coil. Opposite coils are connected to form one "phase," while the rotor has no windings. When DC voltage is supplied to a phase, the rotor rotates in order to minimize the reluctance of the magnetic path. Among various topologies adopted, the most popular is the one with six stator poles (i.e., three-phase) and four rotor poles.

Current motor technologies for HEV applications include separately excited DC, permanent-magnet synchronous AC, induction motors, though switched reluctance motors are being regarded as a very promising opportunity for the near future [270, 157]. A rough comparison among these types shows DC motors to be simpler and less expensive, since they need relatively uncomplicated control electronics to be fed using the DC supply already present on a vehicle. Their main disadvantage is the high maintenance requirement, since brushes must be changed periodically.

AC motors are in general less expensive, but they require more sophisticated control electronics (inverters), which cause the overall cost to be higher than that of DC motors. However, they have higher power density and higher efficiency than DC motors. The majority of vehicle applications therefore use AC motors. Among them, induction motors are generally characterized by a higher specific power than permanent-magnet motors.

For example, an analysis of five induction motors and eleven PM motors of the ADVISOR database [272] yielded an average power density of 0.76 kW/kg for induction motors and 0.66 kW/kg for PM motors. In contrast, PM motors showed a higher peak efficiency of 0.925 compared to 0.905 for induction motors (the efficiency of the respective controllers is included). These facts could explain why induction motors are preferred for high power applications, i.e., electric and series hybrid vehicles, while PM motors are the preferred choice for parallel hybrids, where the fuel economy is a key point. Another difference between induction and PM motors is that the former can bear generally higher rotational speeds. The same motor database cited above shows a maximum speed reached by induction motors ranging from 7500 to 13000 rpm, and PM motors from 4000 to 8500 rpm. Similar conclusions can be drawn for the forthcoming concept of integrated starter/alternator (i.e., "mild" parallel hybrids) [130, 36], although for these applications a lighter construction is often preferred to higher efficiency due to the low degree of utilization of the machine. Thus induction machines are usually adopted.

The SR motor is gaining much interest for its simple and rugged, cost-effective construction and hazard-free operation [206]. Moreover, SR motors have a high efficiency and a wider operating speed range. A certain difficulty in controlling, some noise, and its nonuniformity of operation due to torque ripples (depending on the number of phases) are its main disadvantages.

4.3.1 Quasistatic Modeling of Electric Motors

The causality representation of a motor/generator in quasistatic simulations is sketched in Fig. 4.13. The input variables are the torque $T_2(t)$ and the speed $\omega_2(t)$ required at the shaft. The output variable is the power at the DC link, $P_1(t) = I_1(t) \cdot U_1(t)$. A positive value of $P_1(t)$ is absorbed by the machine operating as a motor, a negative value of $P_1(t)$ is delivered by the machine operating as a generator.

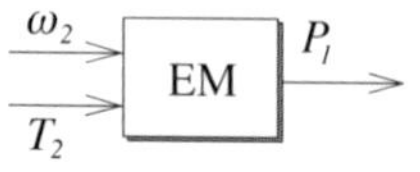

Fig. 4.13. Motor/generators: causality representation for quasistatic modeling.

Motor Efficiency

The relationship between $P_1(t)$ and $P_2(t) = T_2(t) \cdot \omega_2(t)$ can be calculated without a detailed model of the system when a stationary map of the machine efficiency η_m as a function of the input variables $T_2(t)$, $\omega_2(t)$ is available. In such a case, the input power required is evaluated as

$$P_1(t) = \frac{P_2(t)}{\eta_m\left(\omega_2(t), T_2(t)\right)}, \qquad P_2(t) > 0, \tag{4.1}$$

$$P_1(t) = P_2(t) \cdot \eta_m\left(\omega_2(t), -T_2(t)\right), \qquad P_2(t) < 0. \tag{4.2}$$

Two examples of efficiency maps are shown in Fig. 4.14 for a 32 kW permanent-magnet synchronous motor and a 30 kW induction motor, respectively [272]. Also shown is the maximum torque that the motor can deliver under normal conditions.[4] This curve typically is constant up to a certain speed, then it decreases hyperbolically with the speed. This dependency indicates that the quantities limited in a motor are first the current and then the power. At low speed, the current limitation is active, thus the torque is limited. At higher speeds, the power limitation is active, and from the equality $P_2(t) = \omega_2(t) \cdot T_2(t)$, the hyperbolical dependency may be derived. As a consequence, the shape of the maximum torque curve is totally different from that of engines, which typically is quadratic.

The efficiency map $\eta_m(\omega_2, T_2)$ is usually well defined only for the first quadrant (motor mode). Two ways are conceptually possible in order to extend the data available to the second quadrant (generator mode). The first method consists of mirroring the efficiency, assuming that

$$\eta_m\left(\omega_2, -T_2\right) = \frac{1}{\eta_m\left(\omega_2, T_2\right)}. \tag{4.3}$$

The second method consists of mirroring the power losses. These are evaluated from an energy balance in the well-defined motor range,

$$P_l(\omega_2, T_2) = \frac{1 - \eta_m(\omega_2, T_2)}{\eta_m(\omega_2, T_2)} \cdot \omega_2 \cdot T_2, \tag{4.4}$$

and then mirrored to the second quadrant, so that $P_l(\omega_2, -T_2) = P_l(\omega_2, T_2)$. Using the definition (4.2), the efficiency in the generator range is given by

$$\eta_m(\omega_2, -T_2) = 2 - \frac{1}{\eta_m(\omega_2, T_2)}. \tag{4.5}$$

Clearly, the two methods yield different results. In general, the result of a mirroring operation does not coincide with the data that can be obtained

---

[4] Electric motors can often be operated for a short period of time at higher-than-rated power.

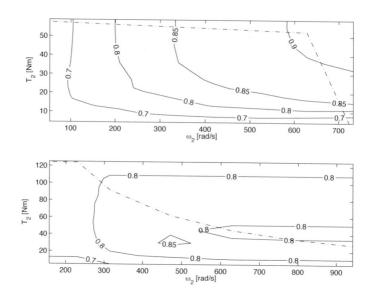

Fig. 4.14. Efficiency maps for a 32 kW PM motor (top) and a 30 kW AC motor (bottom), with curves of maximum torque (dashdot).

by measuring the motor efficiency also in the generator range, and the difference is typically more important for induction motors [10]. For example, the measured efficiency map shown in Fig. 4.15 illustrates a case in which neither the efficiency nor the power losses are mirrored when passing from the motor mode to the generator mode. A possible way to manage this general case consists of deriving a physical expression for the efficiency and the power losses, both in the motor and in the generator ranges, using phenomenological models of the electric machine. With this aim, it is now convenient to distinguish the following derivation according to the type of machine. Switched reluctance motors will be excluded from this analysis since, unlike DC, induction, and synchronous motors, their mathematical characterization is quite difficult to determine. Flux and torque in SR motors are in fact complex, nonlinear functions of motor current and position. Consequently, these motors are conveniently described by means of tabular data [47].

DC Motors

Brush-type DC motors are classified according to the stator excitation. In permanent-magnet DC motors a static magnetic field excitation is generated on the stator using permanent magnets. In separately excited DC motors, the excitation is obtained using field windings which have a separate supply from

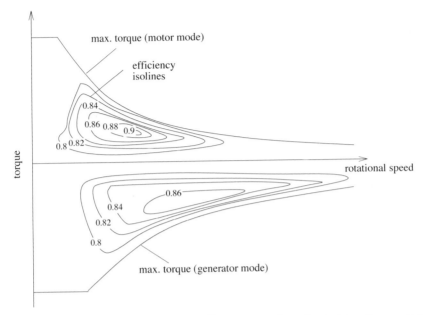

Fig. 4.15. Two-quadrant measured efficiency map for a typical traction motor.

the rotor windings. In both types, a back emf is induced in the rotor, which is also called armature.

The equivalent circuit of a separately excited DC motor is shown in Fig. 4.16 [89]. The Kirchhoff voltage equation for the armature circuit is written as

$$U_a(t) = L_a \cdot \frac{d}{dt} I_a(t) + R_a \cdot I_a(t) + U_i(t) \,, \tag{4.6}$$

where $I_a(t)$ is the armature current, $U_i(t)$ is the induced voltage or back emf, R_a is the armature resistance, L_a is the armature inductance, and $U_a(t)$ is the armature voltage. The armature resistance and inductance can be directly measured and are usually provided by the manufacturer. The armature voltage is the DC voltage applied to the rotor windings. It depends on the input DC voltage $U_1(t)$ and is regulated to fulfill the output requirements, usually with chopper converters (see below).

For the field circuit, the voltage equation is similarly written as

$$U_f(t) = L_f \cdot \frac{d}{dt} I_f(t) + R_f \cdot I_f(t) \,, \tag{4.7}$$

with obvious meanings of the variables. The field voltage $U_f(t)$ is also regulated with a chopper converter as a function of the DC supply.

Newton's second law applied at the motor shaft yields

$$\frac{d}{dt} \omega_2(t) = \frac{T_a(t) - T_2(t)}{\Theta_m} \,, \tag{4.8}$$

where $T_a(t)$ is the armature torque, Θ_m is the moment of inertia of the motor, and $T_2(t)$ is the load torque acting on the motor shaft.

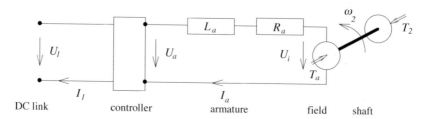

Fig. 4.16. Equivalent circuit of a separately excited DC motor.

The induced voltage (back emf) is proportional to the field current and the rotor speed, $U_i(t) = L_m \cdot I_f(t) \cdot w_2(t)$, with L_m being the field–armature mutual inductance. The armature torque is proportional to the field current and the armature current, $T_a(t) = L_m \cdot I_f(t) \cdot I_a(t)$. In separately excited motors the field voltage is usually kept constant, at least for speeds below the rated or base speed.[5] For constant field voltage and at steady state, as (4.7) indicates, the field current is constant as well. Thus it is common to express back emf and torque as

$$U_i(t) = \kappa_i \cdot w_2(t), \quad T_a(t) = \kappa_a \cdot I_a(t) . \qquad (4.9)$$

The two factors in (4.9), often called speed constant and torque constant, are equal in principle, $\kappa_i = \kappa_a = L_m \cdot I_f$. In practice, often κ_a is multiplied by a coefficient less than unity which accounts for rotational and other losses. Equation (4.9) is valid also during transients for permanent-magnet DC motors, in which the back emf is proportional to the speed and the torque to the armature current, through a constant magnetic flux.

In the quasi-stationary limit, the system is described by

$$T_2(t) = \kappa_a \cdot I_a(t), \quad U_a(t) = \kappa_i \cdot w_2(t) + R_a \cdot I_a(t) . \qquad (4.10)$$

The combination of the two equations (4.10) yields a linear dependency between output torque and speed,

$$T_2(t) = \frac{\kappa_a \cdot U_a(t)}{R_a} - \frac{\kappa_a \cdot \kappa_i}{R_a} \cdot w_2(t) . \qquad (4.11)$$

The characteristic curve $T_2 = f(w_2)$ of a motor is depicted in Fig. 4.17. This curve is determined solely by the parameters κ_i, κ_a, R_a, and by the control input to the motor U_a. The starting torque at stall, i.e., for $w_2 = 0$, is

---

[5] For speeds above the base speed, field weakening control is applied, which implies a variable field voltage.

$\kappa_a{\cdot}U_a/R_a$, the no-load speed is U_a/κ_i. Unlike a conventional vehicle, where the engine must be motored above its idle speed before it can provide full torque, an electric motor provides full torque at low speeds. This characteristic gives the vehicle excellent acceleration from rest.

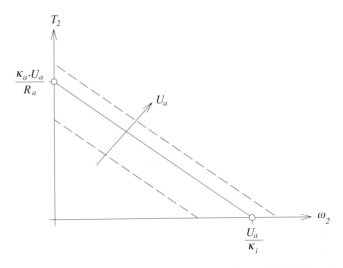

Fig. 4.17. Characteristic curve of a separately excited DC motor.

The dependency between the armature quantities $U_a(t)$, $I_a(t)$ and the input quantities $U_1(t)$, $I_1(t)$ is determined by the motor controller. In DC motors, mostly DC–DC chopper converters are used [158]. A single-quadrant or step-down chopper is depicted in Fig. 4.18. It consists of a fast semiconductor switch and a free-wheeling diode. For low to medium power levels, insulated gate bipolar transistors (IGBT) are common, for higher power levels, gate turn-off thyristors (GTO) are typically used as switches. The switch controls the armature voltage by "chopping" the supply DC voltage into segments. When the switch is on, the armature is directly fed by the supply voltage. When the switch is off, the armature current flows through the free-wheeling diode and the armature voltage is zero. The average value of the voltage is then regulated by the ratio of the time periods during which the switch is on or off ("duty cycle"). Practical chopper converters are more complicated than the simple scheme of Fig. 4.18. For instance, they often include a low-pass filter between the switch network (switch plus diode) and the motor, to smooth the high-frequency switching harmonics.

Simple models of chopper converters are discussed in the section on dynamic motor models. At this point, it is sufficient to take into account the power balance at the two sides of the device. If all the power losses are lumped in the term $P_{l,c}(t)$, the balance is written as

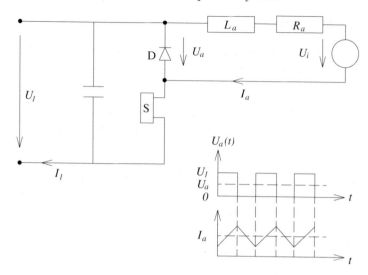

Fig. 4.18. Basic scheme of a DC chopper. D: freewheeling diode, S: semiconductor switch.

$$P_1(t) = I_a(t) \cdot U_a(t) + P_{l,c}(t) \,, \tag{4.12}$$

where $P_1(t)$ is of course the product $U_1(t) \cdot I_1(t)$.

Combining (4.10)–(4.12) yields an expression for the motor efficiency. The input power as a function of $w_2(t)$, $T_2(t)$ is

$$P_1(t) = \frac{T_2(t)}{\kappa_a} \cdot \left(w_2(t) \cdot \kappa_i + \frac{R_a \cdot T_2(t)}{\kappa_a} \right) + P_{l,c}(t) \,, \tag{4.13}$$

and thus the machine efficiency η_m is

$$\eta_m(w_2(t), T_2(t)) = \frac{1}{\dfrac{\kappa_i}{\kappa_a} + \dfrac{R_a \cdot T_2(t)}{\kappa_a^2 \cdot w_2(t)} + \dfrac{P_{l,c}(t)}{w_2(t) \cdot T_2(t)}} \,. \tag{4.14}$$

The power losses $P_l(t) = P_1(t) - P_2(t)$ are calculated as

$$P_l(t) = w_2(t) \cdot T_2(t) \left(\frac{\kappa_i}{\kappa_a} - 1 \right) + \frac{R_a}{\kappa_a^2} \cdot T_2^2(t) + P_{l,c}(t) \,. \tag{4.15}$$

The expression above clearly shows that the overall power losses are due to ohmic resistance, losses in the controller, and other sources, if $\kappa_i \neq \kappa_a$. If $\kappa_i = \kappa_a$ and $P_{l,c}(t) = 0$, then the power losses depend only on the torque squared. They can thus be mirrored from the motor to the generator quadrant, as in (4.4).

Induction AC Motors

In induction AC motors the stator carries p sets of three-phase windings fed by external AC voltage, usually supplied by a DC source through an inverter. The rotor has three-phase, shorted windings ("squirrel cage") and no external connections.

Instead of modeling the single phases, it is convenient to model the three-phase systems using a two-phase reference. In each reference frame, each electrical quantity can be described by its direct (d) and quadrature (q) component. The most convenient reference frame is the synchronous reference frame, which rotates at the same frequency as the stator magnetic field. Generally the transformation into a new reference set simplifies the mathematical manipulation, although the new quantities are not directly measurable. Moreover, the components in the synchronous reference frame can be treated as DC quantities [145, 160, 47, 154, 63, 178, 17, 263, 118].

The Kirchhoff voltage laws for the stator and rotor d-q axes are written as follows [145]. The stator is described by

$$U_q(t) = \sigma \cdot L_s \cdot \frac{d}{dt} I_q(t) + \left(R_s + \frac{L_m^2 \cdot R_r}{L_r^2} \right) \cdot I_q(t) +$$
$$+ \omega(t) \cdot \sigma \cdot L_s \cdot I_d(t) - \frac{L_m \cdot R_r}{L_r^2} \cdot \varphi_q(t) + \tag{4.16}$$
$$+ \frac{L_m}{L_r} \cdot p \cdot \omega_2(t) \cdot \varphi_d(t) ,$$

$$U_d(t) = \sigma \cdot L_s \cdot \frac{d}{dt} I_d(t) + \left(R_s + \frac{L_m^2 \cdot R_r}{L_r^2} \right) \cdot I_d(t) -$$
$$- \omega(t) \cdot \sigma \cdot L_s \cdot I_q(t) - \frac{L_m \cdot R_r}{L_r^2} \cdot \varphi_d(t) - \tag{4.17}$$
$$- \frac{L_m}{L_r} \cdot p \cdot \omega_2(t) \cdot \varphi_q(t) ,$$

The rotor is described by

$$0 = \frac{d}{dt} \varphi_q(t) - \frac{L_m \cdot R_r}{L_r} \cdot I_q(t) + \frac{R_r}{L_r} \cdot \varphi_q(t) +$$
$$+ (\omega(t) - p \cdot \omega_2(t)) \cdot \varphi_d(t) , \tag{4.18}$$

$$0 = \frac{d}{dt} \varphi_d(t) - \frac{L_m \cdot R_r}{L_r} \cdot I_d(t) + \frac{R_r}{L_r} \cdot \varphi_d(t) -$$
$$- (\omega(t) - p \cdot \omega_2(t)) \cdot \varphi_q(t) . \tag{4.19}$$

In these equations, $I_d(t)$, $I_q(t)$ are the d-q axis stator currents, $U_d(t)$, $U_q(t)$ are the d-q axis stator voltages, $\varphi_d(t)$, $\varphi_q(t)$ are the d-q axis stator resolved rotor fluxes, L_r, L_s are the stator resolved rotor and stator inductance, R_r, R_s are the stator resolved rotor and stator resistance, L_m is the magnetizing (mutual) inductance, $\sigma = 1 - L_m^2/(L_s \cdot L_r)$, p is the number of pole pairs, $\omega(t)$ is the frequency of the stator voltage (i.e., the speed of the d-q frame), and $p \cdot \omega_2(t)$ is the frequency of the magnetic field induced in the rotor.

The torque generated at the rotor shaft is found using an energy balance

$$T_m(t) = \frac{3}{2} \cdot p \cdot \frac{L_m}{L_r} \left(\varphi_d(t) \cdot I_q(t) - \varphi_q(t) \cdot I_d(t)\right) . \qquad (4.20)$$

Newton's second law applied at the motor shaft yields

$$\frac{d}{dt}\omega_2(t) = \frac{T_m(t) - T_2(t)}{\Theta_m} , \qquad (4.21)$$

where $T_2(t)$ is the torque of the mechanical load and Θ_m is the moment of inertia of the motor.

Within the quasi-stationary limit, it is possible to solve the set of equations above by expressing $I_q(t)$ and $I_d(t)$ as a function of $U_d(t)$, $U_q(t)$, $\omega(t)$ and $\omega_2(t)$. This procedure requires the inversion of a steady-state matrix [145], leading to

$$T_2(t) = \frac{3}{2} \cdot p \cdot \frac{R_r \cdot L_m^2 \cdot (\omega(t) - p \cdot \omega_2(t)) \cdot}{[R_r \cdot L_s \cdot \omega(t) + R_s \cdot L_r \cdot (\omega(t) - p \cdot \omega_2(t))]^2 +}$$
$$\frac{\cdot (U_d^2(t) + U_q^2(t))}{+[R_r \cdot R_s - \omega(t) \cdot \sigma \cdot L_s \cdot L_r \cdot (\omega(t) - p \cdot \omega_2(t))]^2} . \qquad (4.22)$$

The characteristic curve $T_2 = f(\omega_2)$ of an induction motor is shown in Fig. 4.19. At the synchronous frequency, $\omega_2 = \omega/p$, the motor does not generate any torque. The starting or breakaway torque is obtained by setting $\omega_2 = 0$ in (4.22). As the rotor speed increases, the torque reaches its maximum value, then it decreases to zero when the rotor rotates at the synchronous speed ($\omega = p \cdot \omega_2$). For higher speeds, the machine operates as a generator ($T_2 < 0$). Notice the instability region from rest (or from a slightly higher speed corresponding to a local minimum of torque) to the maximum torque speed, where the slope of the torque curve is positive.

The stator voltage components $U_q(t)$ and $U_d(t)$ in (4.22), as well as the frequency $\omega(t)$, are determined by the electronic frequency converter that feeds the motor. Such power electronic devices, referred to as inverters, convert the DC supply voltage $U_1(t)$ to the variable-frequency three-phase AC voltage required by the motor. Like in chopper drives, the power level determines whether transistor or thyristor switches are to be used. The generation and the waveform of the three stator phase voltages depend on the particular sequence at which the (usually, six) switches are fired [158]. An advanced class of switching schemes is the sinusoidal pulse-width modulation (PWM), whose detailed description is beyond the scope of this book [145].

A balance of power at the two sides of an inverter yields in general

$$P_1(t) = \frac{3}{2} \cdot (U_q(t) \cdot I_q(t) + U_d(t) \cdot I_d(t)) + P_{l,c}(t) , \qquad (4.23)$$

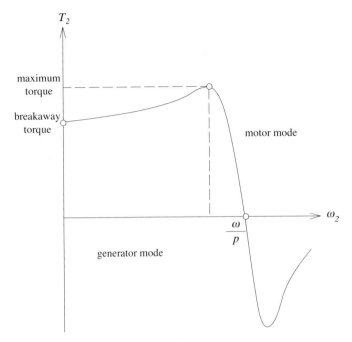

Fig. 4.19. Characteristic curve of an induction AC motor.

where $P_{l,c}(t)$ is the term accounting for the losses in the inverter and the factor $3/2$ reflects the three-phase nature of the machine and its two-phase description (in the d-q reference frame).

A physical expression for the efficiency of an induction motor is now derived from (4.22)–(4.23). The efficiency depends explicitly on the speed and only slightly (via the power losses in the controller) on the torque,

$$\eta_m(\omega_2(t), T_2(t)) = \left(\frac{\omega(t)}{p \cdot \omega_2(t)} + \frac{R_r \cdot R_s}{L_m^2 \cdot (\omega(t) - p \cdot \omega_2(t)) \cdot p \cdot \omega_2(t)} + \right.$$
$$\left. + \frac{R_s}{R_r} \cdot \frac{L_r^2}{L_m^2} \cdot \frac{\omega(t) - p \cdot \omega_2(t)}{p \cdot \omega_2(t)} + \frac{P_{l,c}(t)}{\omega_2(t) \cdot T_2(t)} \right)^{-1} . \qquad (4.24)$$

The corresponding expression of the power losses is

$$P_l(t) = \omega_2(t) \cdot T_2(t) \cdot \left(R_s \cdot \frac{R_r^2 + L_r^2 \cdot (\omega(t) - p \cdot \omega_2(t))^2}{R_r \cdot L_m^2 \cdot p \cdot \omega_2(t) \cdot (\omega(t) - p \cdot \omega_2(t))} + \right.$$
$$\left. \frac{\omega(t) - p \cdot \omega_2(t)}{p \cdot \omega_2(t)} \right) + P_{l,c}(t) . \qquad (4.25)$$

Thus $P_l(t)$ is the sum of three contributions due to ohmic resistance, slip (the difference between $\omega(t)$ and $p \cdot \omega_2(t)$), and controller efficiency, respectively.

The latter equations would imply that, even neglecting $P_{l,c}(t)$, neither the efficiency nor the power losses could be mirrored from the motor to the generator quadrant. The experimental evidence is in agreement with the analysis carried out using the simple models discussed in this section. However, typical experimental efficiency maps for induction machines, like the one in Fig. 4.14a, clearly show that the efficiency depends also on the torque, especially at low load.

Permanent-Magnet Synchronous Motors and Brushless DC Motors

Permanent-magnet synchronous motors (PMSM) are often confused in the literature with the so-called brushless DC motors (BLDC). Basically, these two machines are identical as for the torque generation principle. Both types are synchronous machines that realize the excitation on the rotor with permanent magnets. The external three-phase AC voltage is applied to the armature windings on the stator, which is similar to the stator of an induction machine. The term brushless DC reflects the fact that this motor approximates the behavior of a brush-type DC motor with the power electronics taking the place of the brushes. On the other hand, the term permanent-magnet synchronous motor emphasizes the fact that the machine uses permanent magnets to create the field excitation. The main difference is the waveform of the stator currents, which are rectangular in BLDC motors and sinusoidal in PMSMs. Consequently, the back emf is trapezoidal in BLDC motors and sinusoidal in PMSMs. Historically, the first type is described similarly to DC motors, using a torque constant and a back emf constant. The second type is more often described in terms of synchronous reactance and d-q axis voltages.

For convenience, in the following, both types of motors will be described in the rotor natural d-q reference frame [145, 192, 164, 47, 138, 147], assuming that the back emf is sinusoidal. The equivalent circuit is depicted in Fig. 4.20. Kirchhoff's voltage law for the stator is written as

$$U_q(t) = R_s \cdot I_q(t) + L_s \cdot \frac{d}{dt} I_q(t) + p \cdot w_2(t) \cdot \varphi_m(t) + L_s \cdot p \cdot w_2(t) \cdot I_d(t) , \quad (4.26)$$

$$U_d(t) = R_s \cdot I_d(t) + L_s \cdot \frac{d}{dt} I_d(t) - L_s \cdot p \cdot w_2(t) \cdot I_q(t) , \quad (4.27)$$

where R_s is the stator resistance, L_s is the stator inductance, $\varphi_m(t)$ is the mutual flux linkage, $U_q(t)$, $U_d(t)$ are the d-q axis stator voltage components, and $I_q(t)$, $I_d(t)$ are the d-q axis stator current components.

The torque generated at the rotor shaft is

$$T_m(t) = \frac{3}{2} \cdot p \cdot \varphi_m(t) \cdot I_q(t) , \quad (4.28)$$

where p is the number of pole pairs.

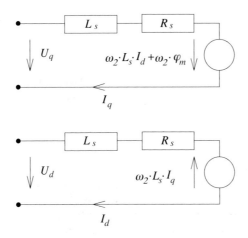

Fig. 4.20. Equivalent circuit of a brushless DC motor.

Newton's second law applied at the motor shaft yields

$$\frac{d}{dt}\omega_2(t) = \frac{T_m(t) - T_2(t)}{\Theta_m} ,\tag{4.29}$$

where $T_2(t)$ is the load torque and Θ_m is the moment of inertia of the motor.

The characteristic curve $T_2 = f(\omega_2)$ can be derived solving (4.26)–(4.29) at the steady-state limit [145]

$$T_2(t) = \frac{3}{2} \cdot p \cdot \varphi_m \cdot \frac{R_s \cdot U_q(t) - p \cdot \omega_2(t) \cdot R_s \cdot \varphi_m - p \cdot \omega_2(t) \cdot L_s \cdot U_d(t)}{R_s^2 + p^2 \cdot \omega_2^2(t) \cdot L_s^2} .\tag{4.30}$$

If the stator inductance can be neglected, the expression above becomes an affine function of the rotor speed, similarly to that expressed by (4.11) for separately excited DC motors. A typical characteristic curve of a brushless DC motor is illustrated in Fig. 4.21.

The expression derived above shows that the torque of a brushless DC motor depends on each of the two d-q axis components of the stator voltage. Since the brushless DC motor does not have an internal commutation like the permanent-magnet DC motor, it needs an inverter to convert the DC voltage $U_1(t)$ supplied by an external source to the operating three-phase AC voltage, whose frequency corresponds to the speed of the rotor [156]. This is usually achieved with a three-phase voltage-source inverter (VSI) in pulse-width modulation (PWM) mode. The two types of inverters that are used for BLDC motors are the 180° and the 120° types [145, 239]. In the former, also called continuous current inverter, each phase is always connected to either the positive or the negative terminal of the DC source. In the latter, also called discontinuous current inverter, each phase is open-circuited for essentially 120° of the cycle. A practical model of both types of inverter is that expressed by (4.23).

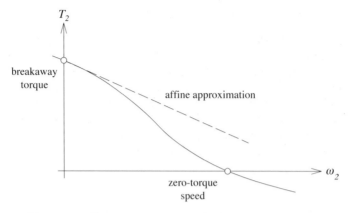

Fig. 4.21. Characteristic curve of a brushless DC motor.

An expression for the efficiency can be derived from (4.23)–(4.30) once the rotor resolved currents are expressed in terms of stator voltage using (4.26)–(4.27). This requires that the values of $U_q(t)$ and $U_d(t)$ as determined by the firing of the inverter switches are known [47]. In the following, a special but common case is considered, i.e., the so-called common operating mode of the three-phase inverter, such that $U_d(t) = 0$. Under this circumstance the efficiency is calculated as

$$\eta_m(\omega_2(t), T_2(t)) = \left(1 + \frac{R_s}{\frac{3}{2} \cdot p^2 \cdot \varphi_m^2} \cdot \frac{T_2(t)}{\omega_2(t)} + \right.$$
$$\left. + \frac{L_s^2}{R_s} \cdot \frac{\omega_2(t) \cdot T_2(t)}{\frac{3}{2} \cdot \varphi_m^2} + \frac{P_{l,c}(t)}{\omega_2(t) \cdot T_2(t)} \right)^{-1} . \qquad (4.31)$$

For a given speed, the power loss term $P_l(t) = P_1(t) - P_2(t)$ is a quadratic function of the torque,

$$P_l(t) = \frac{2 \cdot T_2^2(t)}{3 \cdot \varphi_m^2} \cdot \left(\frac{R_s}{p^2} + \frac{L_s \cdot \omega_2^2(t)}{R_s} \right) + P_{l,c}(t) . \qquad (4.32)$$

As a consequence of (4.32), the power losses can be mirrored from the motor to the generator quadrant (if $P_{l,c}$ is an even function of $T_2(t)$).

Willans Approach

As already discussed in Chap. 3, the use of efficiency maps to evaluate the energy required at the input stage of a power converter presents some disadvantages. At low loads in particular, the efficiency is not well defined, being the ratio of two quantities approaching zero, which makes it hard to measure or estimate the efficiency correctly. Like for other energy converters, an alternative quasistatic description, known as the Willans approach [210], can be used. For an electric motor, this approach takes the form

$$P_2(t) = e \cdot P_1(t) - P_0 , \qquad (4.33)$$

where P_0 represents the power losses occurring after the energy conversion (friction, heat losses, etc.) and e is the "indicated" efficiency, i.e., the maximum efficiency that can be obtained when P_0 is zero. Thus e represents the efficiency of the energy conversion process only (electrical to mechanical energy and vice versa), while η_m also takes into account the "friction losses" P_0.

This model is valid both for the motor and the generator ranges. It implicitly states that neither the efficiency nor the power losses are mirrored from the motor to the generator range. This is easily proved considering that

$$\eta_m(P_2) = \frac{e \cdot P_2}{P_2 + P_0} , \qquad (4.34)$$

$$P_l(P_2) = \left(\frac{1}{e} - 1 \right) \cdot P_2 + \frac{P_0}{e} , \qquad (4.35)$$

and showing that $\eta_m(P_2) \neq \eta_m^{-1}(-P_2)$ and $P_l(P_2) \neq P_l(-P_2)$.

The values of e and P_0 typically depend on the motor speed w_2. Otherwise, the efficiency would be a function of the output power only, and the constant efficiency lines would be hyperbolae in the "efficiency map" of the motor. Despite this difficulty, it is possible to find with good approximation some average values for e and P_0 which are valid over the entire range of speeds for a given motor. This can be done by considering a large number of operating points, i.e., torque and speed T_2, w_2, evaluating P_1 for each of them, then fitting the dependency $P_1 = f(P_2)$ in an affine approximation (approach A). A second approach (approach B) consists of considering a certain number of driving profiles. For each of them the energy use values resulting from the integration of $P_1(t)$ and $P_2(t)$ are evaluated, then the respective average powers. Finally, the dependency between the average values of P_1 and P_2 is fitted.

If the approach A is used to identify e and P_0 for the same motors as in Fig. 4.14, the resulting affine relationship is that shown in Fig. 4.22a–b. The inner efficiency e is 0.78 for the induction motor and 0.91 for the permanent-magnet synchronous motor. For both motors, the friction losses are less than $1\,\mathrm{kW}$. Similar numerical results can be obtained for e with motors of different size. With the approach B, for a permanent-magnet motor the values $e = 0.96$, $P_0 = 1.4\,\mathrm{kW}$ were obtained by considering energy use values over ten driving profiles.

The Willans model can be made scalable by normalizing all the torque and speed quantities involved. Torque is substituted by electric mean effective pressure, in analogy to what is usually done for internal combustion engines. The electric mean effective pressure is defined as the motor mean tangential force divided by the rotor external surface. The mean speed $c_m(t)$, equivalent to the mean piston speed of engines, is defined as the tangential speed at the rotor radius r, $c_m(t) = w_2(t) \cdot r$. The rotor dimensions r and l and the mean effective pressure are related through the energy balance

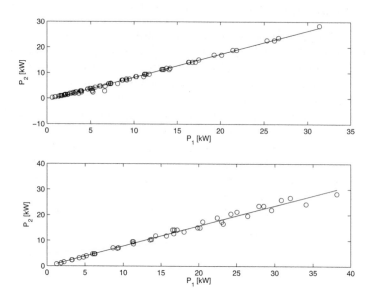

Fig. 4.22. Evaluation of the Willans parameters for a 32 kW permanent-magnet synchronous motor (top) and a 30 kW induction motor (bottom) of the ADVISOR database [272].

$$p_{me}(t) \cdot (2\pi \cdot r \cdot l) \cdot c_m(t) = T_2(t) \cdot \omega_2(t) , \qquad (4.36)$$

where $2\pi \cdot r \cdot l$ is the rotor external surface. Based on this balance,

$$p_{me}(t) = \frac{T_2(t)}{2 \cdot V_r} , \qquad (4.37)$$

where $V_r = \pi \cdot r^2 \cdot l$ is the rotor volume, i.e., the equivalent to the swept volume of engines.

In terms of mean effective pressure the scalable Willans model is written as

$$p_{me}(t) = e \cdot p_{ma}(t) - p_{mr}(t) , \qquad (4.38)$$

where the available mean pressure and the loss term $p_{mr}(t)$ are defined as

$$p_{ma}(t) = \frac{P_1(t)}{2 \cdot V_r \cdot \omega_2(t)}, \quad p_{mr}(t) = \frac{P_0}{2 \cdot V_r \cdot \omega_2(t)} . \qquad (4.39)$$

This description is formally the same as the one for engines, although the numerical values of the variables are quite different. Motors have a much lower peak p_{me} (0.5–1 bar) and a much higher peak c_m (50–100 m/s) than engines (p_{me} of more than 10 bar, mean speed of 10–15 m/s) [65, 274].

4.3.2 Dynamic Modeling of Electric Motors

Already in the previous sections it was shown that models of electric machines are essentially dynamic, although only their quasistatic limit has been described in detail. However, dynamic models are used mainly for specific control and diagnostics purposes and only rarely are embedded in a hybrid-electric vehicle simulator [198, 15].

In dynamic models, the correct physical causality should be used. The physical causality representation of a dynamic model of a motor/generator is sketched in Fig. 4.23. The model input variables are the DC link voltage $U_1(t)$ and the motor shaft rotational speed $\omega_2(t)$. The model output variables are the torque at the motor shaft $T_2(t)$ and the current exchanged at the DC link $I_1(t)$. A positive $I_1(t)$ is absorbed by the machine operating as a motor, a negative $I_1(t)$ is delivered by the machine operating as a generator.

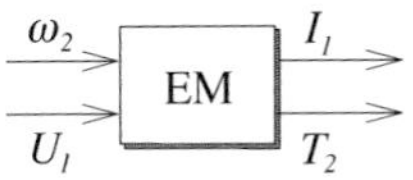

Fig. 4.23. Motor/generators: physical causality for dynamic modeling.

The implementation of a dynamic motor model does not pose any particular problem if the equations introduced in the previous section are correctly coupled with the equations that describe the behavior of the motor controller.

For a DC motor, the armature voltage $U_a(t)$ is determined by the control strategy to be a function of the DC voltage $U_1(t)$ and the desired motor performance (e.g., current control). Since $\omega_2(t)$ and its derivative with respect to time are input variables, (4.6)–(4.8) can be integrated to yield the armature current $I_a(t)$ and the output torque $T_2(t)$. A power balance across the motor controller, as in (4.12), yields $I_1(t)$. The relationship between $U_a(t)$ and $U_1(t)$ depends on the type of chopper that is used and generally can be reconducted to a practically constant gain [191, 92]. If $\alpha(t)$ is the chopper duty cycle, for the single-quadrant or step-down chopper the armature voltage is

$$U_a(t) = \alpha(t) \cdot U_1(t) . \tag{4.40}$$

For the two-quadrant operation, both motoring and regenerating, half-bridge chopper converters are used. The motor operation is the same as in step-down choppers, while in the generator range the average armature voltage is

$$U_a(t) = (1 - \alpha(t)) \cdot U_1(t) . \tag{4.41}$$

To obtain an average value $U_a(t)$ which is higher than the supply voltage, step-up (or boost) choppers are used. These devices use an inductor in parallel with the switch, while the free-wheeling diode is inserted in series with the load. The transfer function of the step-up chopper converter is

$$U_a(t) = \frac{1}{1 - \alpha(t)} \cdot U_1(t) \ . \tag{4.42}$$

For induction motors, the quantities $U_d(t)$, $U_q(t)$, and $\omega(t)$ are determined by the control strategy to be a function of the DC voltage $U_1(t)$ and the desired motor performance (e.g., current control, flux control, speed control). Again, (4.16)–(4.19) can be integrated to yield the currents $I_d(t)$, $I_q(t)$ and the fluxes $\varphi_d(t)$, $\varphi_q(t)$. Equations (4.20)–(4.21) yield the output torque $T_2(t)$. A power balance across the motor controller, (4.23), finally yields $I_1(t)$. The analysis of all possible relationships between the d-q axis voltages and DC voltage is a complex task whose study is beyond the scope of this book [145, 42, 77].

For brushless DC motors the same steps as for induction machines should be followed. As already discussed, in this case the control of $U_d(t)$ and $U_q(t)$ also affects the characteristic curve of the motor. A common control strategy consists of imposing $U_d(t) = 0$ and modulating $U_q(t)$ as a function of $U_1(t)$. Another possible control strategy (maximum-torque control) is to force the system to satisfy

$$\frac{U_d(t)}{U_q(t)} = -\frac{L_s}{R_s} \cdot p \cdot \omega_2(t) \ . \tag{4.43}$$

It is easy to verify that with (4.43) the torque given by (4.30) is maximized for each motor speed.

4.3.3 Causality Representation of Generators

The second electric machine that is used in series and combined hybrids, primarily acting as a generator, must be modeled according to a causality that is the reverse of that of the main traction motor. This becomes clear observing the flow of power factors in Figs. 4.9 – 4.11.

In quasistatic simulations, the input variable is the electric power $P_2(t)$, while the output variables are the rotational speed $\omega_1(t)$ and torque $T_1(t)$. One relationship between these variables is given by the definition of efficiency or by the Willans approach. The missing degree of freedom is typically a control signal. In fact, in series hybrids the rotational speed of the APU (engine plus generator) is determined by the control unit that tries to optimize the operation of the engine. In combined hybrids, the rotational speed of the generator is the variable that regulates the transmission ratio of the planetary gear set. Therefore, in any case a speed command $\hat{\omega}(t)$ is supplied to the generator model, so that $\omega_1(t) = \hat{\omega}(t)$ and from a power balance the torque $T_1(t)$ also can be calculated [258].

In dynamic models, the input variables are the voltage $U_2(t)$ and the torque $T_1(t)$, while the output variables are the speed $\omega_1(t)$ and the current $I_2(t)$. In this case the motor equations introduced above still can be used without any particular problems.

Alternatively, the same causality of Figs. 4.13 – 4.23 can be still used, provided that a model of the supervisor controller is explicitly inserted between the engine and the generator in a series hybrid (Fig. 4.9) or between the power split device and the generator in a combined hybrid (Fig. 4.11). An example of such causality representation is illustrated in the case study 7 and particularly in Fig. 8.28.[6]

4.4 Batteries

Electrochemical batteries are a key component both of electric vehicles (EV) and of hybrid-electric vehicles (HEV). Batteries are devices that transform chemical energy into electrical energy and vice versa. They represent a reversible electrical energy storage system.

Desirable attributes of traction batteries for EV and HEV applications are high specific power, high specific energy, long calendar and cycle life, low initial and replacement costs, high reliability, and high robustness. Among other current technical challenges,[7] a key point is developing accurate techniques to determine the capacity or the state of charge (SoC) of batteries during their operation.

The capacity of a battery, usually expressed in Ah, is the integral of the current that can be delivered under certain conditions. A dimensionless parameter is the state of charge, which describes the amount of charge remaining in the battery, expressed as a percentage of its nominal capacity. Another key design parameter is the specific energy, i.e., the energy that can be stored in the battery per unit mass, typically expressed in Wh/kg. The specific energy affects the mass of batteries that must be carried on board, thus the range of a purely electric vehicle. For HEVs, possibly more important is the specific power, typically expressed in W/kg, which is related to the acceleration and the grade performance levels that the vehicle can achieve.

Batteries are composed of a number of individual cells in which three main components are recognizeable: two electrodes, where half-reactions take place resulting in the circulation of electrons through an external load, and a medium that provides the ion transport mechanism between the positive and negative electrodes. The cathode is the electrode where reduction (gain of electrons) takes place. When discharging, it is the positive electrode, when charging, it becomes the negative electrode. The anode is the electrode where oxidation (loss of electrons) takes place. While discharging, it is the negative electrode, while charging it becomes the positive electrode (see Fig. 4.24).

Batteries used in automotive applications are all rechargeable (secondary batteries). They can be divided in two categories, according to the type of

---

[6] In that case, the causality representation of the electric power link for quasistatic modeling changes with respect to what described in Sect. 4.6. However, the modifications to the model equations are straightforward and not discussed here.

[7] Examples of open issues are: developing new materials to improve the battery performance and tolerance to off-design operation, developing methods and designs to balance the packs electrically and thermally, improving the battery robustness, recycleability, etc.

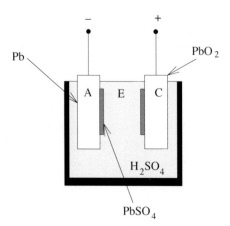

Fig. 4.24. Schematics of a battery cell. A: anode, C: cathode, E: electrolyte. The example illustrates a lead–acid cell. Reduction semi-reaction: $PbO_2 + HSO_4^- + 3H^+ + 2e^- \rightarrow PbSO_4 + 2H_2O$. Oxidation semi-reaction: $Pb + HSO_4^- \rightarrow PbSO_4 + H^+ + 2e^-$.

the electrolyte. Ambient-temperature operating batteries have either aqueous (flooded) or non-aqueous electrolytes. High-temperature operating batteries have molten or solid electrolytes. Presently, more than ten different technologies have been proposed. The most commonly used are: (i) lead–acid, (ii) nickel–cadmium, (iii) nickel–metal hydride, and (iv) lithium-ion. Table 4.1 summarizes the electrochemical aspects of these technologies.

Table 4.1. Electrochemical features of various traction battery technologies (for KOH[a] the solution is in separator sheets).

battery	anode	cathode	electrolyte	cell voltage
lead–acid	Pb	PbO_2	H_2SO_4	2 V
nickel–cadmium	Cd	$Ni(OH)_2$	KOH	1.2 V
nickel–metal hydride	metal hydride	$Ni(OH)_2$	KOH[a]	1.2 V
lithium-ion	carbon	lithium oxide	lithiated solution	3.6 V

Lead–acid batteries have a sponge metallic lead anode, a lead dioxide (PbO_2) cathode, and an aqueous sulfuric acid (H_2SO_4) electrolyte. The cell reactions imply the formation of water. Above a certain voltage, called gassing voltage, this water dissociates into hydrogen and oxygen. This is a heavy limitation in the lead–acid battery's operation, since water has to be continuously replaced. This problem was solved with sealed batteries. Because they cannot be sealed completely, they are also called, more appropriately, valve regulated lead–acid (VRLA) batteries. Hydrogen and oxygen are converted back into water, which no longer needs to be replaced. Lead–acid batteries are widely used within the 12 V network of conventional vehicles, but also in some HEV applications (Audi Duo), mainly due to their low cost, robustness, and reliability. Their main disadvantages are the low cycle life and the low energy density, which limit their use in EVs. In contrast, power densities of up to 600 W/kg are obtained with some advanced high-power lead–acid batteries that are being developed for HEV applications.

Nickel–cadmium batteries have a cadmium (Cd) anode, a nickel oxyhydroxide (NiOOH) cathode, and an aqueous potassium hydroxide (KOH) electrolyte. They are used in many electronic consumer products, since they have a higher specific energy and a longer life cycle than lead–acid batteries. This caused them to be considered for some electric vehicle as well (Citroën, Peugeot). The disadvantages are that nickel–cadmium batteries cost more than the lead–acid batteries and have lower power densities, which limits their use in HEV applications. But their best-known limitation is the so-called "memory effect." This term refers to a temporary loss of cell capacity, which occurs when a cell is recharged without being fully discharged. This can cause the battery life to be shortened.

Nickel–metal hydride (NiMH) batteries work with an anode made of alloys with metals that can store hydrogen atoms. A typical rare earth–nickel alloy used is $LaNi_5$. The cathode is nickel oxyhydroxide, and the electrolyte is potassium hydroxide. Compared to nickel–cadmium batteries, nickel–metal hydride batteries have a higher cost and a lower life cycle (that is still higher than in lead–acid batteries), but they also have a higher energy density and power density. The undesirable memory effect is present only under certain conditions. For these reasons, NiMH batteries have been used successfully in production EVs and recently in HEVs (Toyota, Honda, GM, Ford).

Lithium-ion technology has not yet reached full maturity and is continually improving. Nevertheless, lithium-ion batteries are rapidly penetrating laptop and cell-phone markets because of their high specific energy and specific power. These characteristics make lithium-ion batteries also suitable for HEV applications (Nissan Tino, Dodge ESX3). However, to make them commercially viable for HEVs, further development is needed, including improvements in cycle life and reduction of costs.

Today's cells have a carbon-based anode, usually made of graphite, in which lithium ions are intercalated in the interstitial spaces of the crystal. The anode is generally well optimized and hardly any improvements are expected in terms of design changes. The cathode material, however, shows promise for further enhancements. Most lithium-ion batteries for portable applications are based on a cobalt oxide ($LiCoO_2$) cathode. One of the main advantages of the cobalt-based battery is its high energy density. The main drawback is a relatively low specific power. Another disadvantage is a fast ageing that shortens cycle life, caused by the increase of internal resistance. Battery research is therefore focusing on the cathode material, and several alternatives to cobalt cathodes have been proposed. Lithium manganese oxide cathodes in the form of a *spinel* structure yield high specific power. Multi-metal cathodes based on lithium nickel cobalt oxide, with and without aluminium or manganese doping, are expected to increase the specific energy. Novel systems based on the addition of phosphates in the cathode (e.g., $LiFePO_4$) also promise advanced performance. In general, these alternative technologies pose still many concerns for cycle life, safety, and costs. Another material that has potential to improve the cell performance is the electrolyte, typically made of a lithiated liquid solution, e.g., $LiPF_6$ dissolved in an organic solvent. A promising alternative consists of polymer electrolytes.

Sodium-sulphur batteries have a molten sodium anode, a molten sulfur cathode, and a solid ceramic (beta-alumina) electrolyte. This cell has been studied extensively for EVs because of its inexpensive materials, high cycle life, and high specific energy and power. Specific energy has reached levels of 150 Wh/kg and specific power has attained 200 W/kg. Despite these advantages, the adoption of sodium sulfur batteries is limited by two facts. The cell must operate at high temperatures of around 350°C to keep the sulfur in liquid form. This is achieved through insulation or heating through the cells' own power, which lowers the energy density. Moreover, the electrolyte can develop microfissures, allowing the contact between liquid sodium and sulfur, with potentially very dangerous results.

The sodium–nickel chloride battery, also known as the "zebra" battery, is based on another advanced technology. Similarly to the sodium-sulphur battery, the anode consists of (solid) sodium, while the cathode is made of nickel-chloride. The electrolyte is a solution of $NaAlCl_4$. Specific energy and specific power are comparable with those of sodium-sulphur batteries. Advantages over the latter are that the zebra battery has fewer problems concerning cycle life (1000 cycles expected) and safety.

A comparison of the features of different types of batteries is shown in Table 4.2. The data shown are intended to be average values of present-day technology. More accurate comparative studies can be found in the literature [266, 71, 50, 46, 119, 97].

The specific energy values shown in Table 4.2 refer to advanced *single* battery modules. When arrays of battery modules or battery "packs" are considered, the energy density may be lower. For example, the average energy density of the lead–acid batteries serving in current EVs or HEVs, as calculated from the ADVISOR database [272], is around 30 Wh/kg, for the NiCd 47 Wh/kg, for the NiMH 55 Wh/kg, and for the Li-ion batteries 90 Wh/kg, as of 2005.

Table 4.2. Comparison of battery systems for electrical and hybrid propulsion (various sources).

battery	Wh/kg	W/kg	cycles
lead–acid	40	180	600
nickel–cadmium	50	120	1500
nickel–metal hydride	70	200	1000
lithium-ion	130	430	1200

4.4.1 Quasistatic Modeling of Batteries

The causality representation of a battery in quasistatic simulations is sketched in Fig. 4.25. The input variable is the terminal power $P_2(t)$. The output variable is the battery charge $Q(t)$.

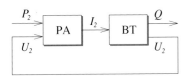

Fig. 4.25. Batteries: causality representation for quasistatic modeling.

The charge variation can be calculated directly from the terminal power P_2 when semi-empirical data are available from the manufacturer [8]. More generally, charge variation is related to terminal current, which in turn is calculated from the terminal power. Besides the trivial equality

$$I_2(t) = \frac{P_2(t)}{U_2(t)} ,\qquad(4.44)$$

a relationship between the terminal voltage U_2 and the terminal current I_2 must be used.

Capacity and C-rate

The battery capacity, usually expressed in Ah rather than in coulomb, is a function of the terminal current and is determined with constant-current discharge/charge tests.

In a typical constant-current discharge test, the battery is initially fully charged and the voltage equals the open-circuit voltage U_{oc}. A constant discharge current I_2 is then applied. After a certain time t_f (called discharge time), the voltage reaches a value (e.g., the 80% of the open-circuit voltage) called *cut voltage* at which the battery is considered as discharged. The discharge time is thus a function of the discharge current. A widely used dependency is that of the Peukert equation

$$t_f = \text{const} \cdot I_2^{-n} \, , \qquad (4.45)$$

where n is the so-called Peukert exponent, which varies between 1 and 1.5 (1.35 for typical lead–acid batteries). The dependency expressed by (4.45) means that the battery capacity depends on the discharge current. If the capacity Q_0^* for a given discharge current I_2^* is known, then the capacity at a different discharge current is given by

$$\frac{Q_0}{Q_0^*} = \left(\frac{I_2}{I_2^*}\right)^{1-n} \, . \qquad (4.46)$$

Other, more sophisticated models for battery discharge have been developed [234, 134], including neural network-based estimations [41]. A modified Peukert equation for low currents is [39]

$$\frac{Q_0}{Q_0^*} = \frac{K_c}{1 + (K_c - 1) \cdot \left(\frac{I_2}{I_2^*}\right)^{n-1}} \, , \qquad (4.47)$$

where K_c is a constant.

Instead of the discharge current, a non-dimensional value called C-rate is often used. This is defined as

$$c(t) = \frac{I_2(t)}{I_0}, \quad I_0 = \frac{Q_0}{1\text{h}} \, , \qquad (4.48)$$

where I_0 is the current that discharges the battery in one hour, which has the same numerical value as the battery capacity Q_0. The C-rate is often written as C/x, where x is the number of hours needed to discharge the battery with a C-rate $c = 1/x$, i.e., a current x times lower than I_0. For example, C/3 corresponds to a current that discharges the battery in three hours, C/5 in five hours.

State of Charge

The state of charge q is the ratio of the electric charge Q that can be delivered by the battery to the nominal battery capacity Q_0,

$$q(t) = \frac{Q(t)}{Q_0} \, . \qquad (4.49)$$

A parameter which is often used in alternative to q is the depth of discharge $1 - q(t)$.

Direct measurement of Q is usually not possible with automotive battery systems. However, the variation of the battery charge can be approximately related to the discharge current I_2 by charge balance,

$$\dot{Q}(t) = -I_2(t) \, . \tag{4.50}$$

In case of charge, the evaluation of the state of charge must take into account the fact that a fraction of the current I_2 is not transformed into charge. This fraction is due to irreversible, parasitic reactions taking place in the battery. Often [65, 149, 129] such an effect is modeled by a charging or coulombic efficiency η_c,

$$\dot{Q}(t) = -\eta_c \cdot I_2(t) \, . \tag{4.51}$$

The combination of (4.50)–(4.51) yields a method to determine the state of charge by measuring the terminal current. This method, known as "Coulomb counting" has the advantage of being easy and reliable as long as the current measurement is accurate. Practically, however, this method requires frequent recalibration points, to compensate the effects neglected by (4.50)–(4.51) [193]. Modern system for SoC determination attempt in some cases to estimate some of these effects, namely, the charge "efficiency" during discharge, which is due to reaction kinetics and diffusion processes, battery self-discharge, and capacity loss as the battery ages [195].

More advanced methods of SoC determination include adaptive methods based on physical models of batteries [194, 193, 195]. If the state of charge is a state of the model, it can be estimated by comparing the measurements available (terminal voltage, current), with the model outputs, using well-known techniques such as Kalman filtering.

Equivalent Circuit

A basic physical model of a battery can be derived by considering an equivalent circuit of the system such as the one shown in Fig. 4.26. In this circuit, the battery is represented by an ideal open-circuit voltage source in series with an internal resistance. Kirchhoff's voltage law for the equivalent circuit yields the equation

$$U_{oc}(t) - R_i(t) \cdot I_2(t) = U_2(t) \, . \tag{4.52}$$

The steady-state battery equivalent circuit has been applied mainly for various lead–acid batteries [65, 75, 238, 129, 149], but also for nickel–cadmium, nickel–metal hydride and lithium-ion batteries [129].

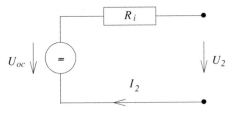

Fig. 4.26. Equivalent circuit of a battery.

The open-circuit voltage U_{oc} represents the equilibrium potential of the battery. Since it is a function of the battery charge, a possible parameterization is the affine relationship [65] written as

$$U_{oc}(t) = \kappa_2 \cdot q(t) + \kappa_1 . \tag{4.53}$$

The coefficients κ_2 and κ_1 depend only on the battery construction and the number of cells, but not on operative variables, thus they can be considered as constant with time. As can be easily proven with physical considerations, they must be the same both for charging and discharging. More complex parameterizations may be derived from the nonlinear Nernst equation [149, 277]. Alternatively, U_{oc} can be tabulated as a function of the state of charge [129, 238, 75].

The linear dependency of (4.53) is equivalent to a constant voltage source in series with a capacitor with constant capacitance. In fact, in a capacitor the voltage is proportional to the charge stored. A bulk capacitance C_b is thus an alternative way to represent the battery potential when it varies linearly with the state of charge [65].

As for the internal resistance R_i, it takes into account several phenomena. In principle, it is the combination of three contributions. The first is the ohmic resistance R_o, i.e., the series of the ohmic resistance in the electrolyte, in the electrodes, and in the interconnections and battery terminals. The second contribution is the charge-transfer resistance R_{ct}, associated with the "charge-transfer" (i.e., involving electrons) reactions taking place at the electrodes. The third contribution is the diffusion or concentration resistance R_d, associated with the diffusion of ions in the electrolyte due to concentration gradients. Thus, in principle, the internal resistance of a battery is calculated as

$$R_i = R_d + R_{ct} + R_o . \tag{4.54}$$

The internal resistance can be evaluated as a function of the state of charge (and possibly temperature) only [129, 238, 149], conveniently distinguishing between charge and discharge. The fact that the resistance does not depend on the battery current is a serious limitation of these models, since the processes described by (4.54) are in fact highly nonlinear. Nonlinear models are derived in literature by fitting experimental data with semi-physical equations such as the Tafel equation [65, 75].

Instead of modeling the various electrochemical processes of a battery, often experimental data from a constant-current discharge test are used to derive a black-box model. Simple fitting techniques or more sophisticated neural networks are used to derive these input/output representations of the battery behavior. A typical voltage profile as a function of time for a constant discharge current test is shown in Fig. 4.27. Since for a constant current the state of charge varies linearly with time, often the voltage profile is given as a function of q instead of time, with the C-rate or the current as a parameter. The voltage variation in Fig. 4.27 shows an initial voltage drop, occurring when the current is applied, which can be considered as instantaneous in a first approximation. Subsequently, the voltage varies linearly with the state of charge. The initial voltage drop clearly increases with the discharge current.

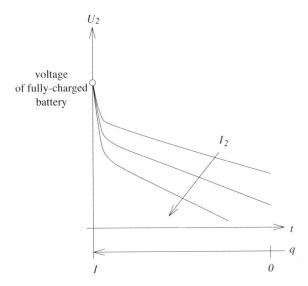

Fig. 4.27. Voltage profile of a battery for constant discharge current.

The behavior shown by the battery voltage during the discharge tests is in agreement with the steady-state model of (4.52), with U_{oc} expressed by (4.53), and with a similar affine relationship for the internal resistance,

$$R_i(t) = \kappa_4 \cdot q(t) + \kappa_3 \ . \tag{4.55}$$

In fact, combining (4.52), (4.53) and (4.55), the following equation is obtained for the battery voltage

$$U_2(t) = (\kappa_1 - \kappa_3 \cdot I_2(t)) + (\kappa_2 - \kappa_4 \cdot I_2(t)) \cdot q(t) \ . \tag{4.56}$$

This equation clearly expresses U_2 as the result of a fully-charged battery voltage $\kappa_1 + \kappa_2$, a voltage drop occurring when the current is applied at $q = 1$, $(\kappa_3 + \kappa_4) \cdot I_2$, and finally a voltage drop which increases as the state of charge decreases, $(\kappa_2 - \kappa_4 \cdot I_2) \cdot (1 - q)$. This description is in agreement with the trend observed in Fig. 4.27. The parameters κ_3 and κ_4 may vary from charging to discharging, as in some models for NiMH batteries described in literature [274, 65].

Now, substituting (4.44) into (4.56), a quadratic equation is obtained for the battery voltage

$$U_2^2(t) - (\kappa_2 \cdot q(t) + \kappa_1) \cdot U_2(t) + P_2(t) \cdot (\kappa_4 \cdot q(t) + \kappa_3) = 0 \ . \tag{4.57}$$

The solution of this equation in terms of terminal voltage is

$$U_2(t) = \frac{\kappa_1 + \kappa_2 \cdot q(t)}{2} \pm \sqrt{\frac{(\kappa_1 + \kappa_2 \cdot q(t))^2}{4} - P_2(t) \cdot (\kappa_4 \cdot q(t) + \kappa_3)} \ . \quad (4.58)$$

Alternatively, the substitution of (4.44) into (4.52) yields the equivalent quadratic equation

$$U_2^2(t) - U_{oc}(t) \cdot U_2(t) + P_2(t) \cdot R_i(t) = 0 \ . \quad (4.59)$$

whose solution in terms of terminal voltage is calculated as

$$U_2(t) = \frac{U_{oc}(t)}{2} \pm \sqrt{\frac{U_{oc}^2(t)}{4} - P_2(t) \cdot R_i(t)} \ . \quad (4.60)$$

The terminal variables are subjected to power and current limitations which, for the discharge case, are $P_2 > 0$ and $U_2 < U_{oc}$. The power as a function of the voltage is calculated as

$$P_2(t) = \frac{-U_2^2(t) + U_2(t) \cdot U_{oc}(t)}{R_i(t)} \ . \quad (4.61)$$

The power is zero both for $U_2 = 0$ and for $U_2 = U_{oc}$ (see Fig. 4.28). For values of voltage between these two limits, the power is positive and thus the curve $P_2(U_2)$ has a maximum in that range. The condition for maximum power is obtained by setting to zero the derivative of P_2 with respect to U_2,

$$\frac{dP_2}{dU_2} = \frac{U_{oc} - 2 \cdot U_2}{R_i} = 0 \ . \quad (4.62)$$

This equation leads to a maximum power available from the battery

$$P_{2,max}(t) = \frac{U_{oc}^2(t)}{4 \cdot R_i(t)} \ , \quad (4.63)$$

to which the following values of voltage and current correspond

$$U_{2,P}(t) = \frac{U_{oc}(t)}{2}, \quad I_{2,P}(t) = \frac{U_{oc}(t)}{2 \cdot R_i(t)} \ . \quad (4.64)$$

In practice, the battery voltage is limited to a relatively narrow band around U_{oc}, $U_2 \in (U_{2,min}, U_{2,max})$. Typical values of $U_{2,min}$ are higher than $U_{2,P}$. Therefore, (4.63) is in practice substituted by

$$P_{2,max}(t) = \frac{U_{oc}(t) \cdot U_{2,min} - U_{2,min}^2}{R_i(t)} \ . \quad (4.65)$$

The corresponding limit for the discharge current is

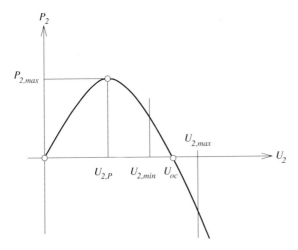

Fig. 4.28. Dependency between battery power and voltage.

$$I_{2,max}(t) = \frac{U_{oc}(t) - U_{2,min}}{R_i(t)} . \qquad (4.66)$$

In case of charge, the terminal power would increase (in absolute value) indefinitely with the terminal voltage $U_2 > U_{oc}$. In this case, the power is limited solely by the maximum battery voltage allowed $U_{2,max}$,

$$P_{2,min}(t) = -\frac{U_{2,max}^2 - U_{oc}(t) \cdot U_{2,max}}{R_i(t)} . \qquad (4.67)$$

The corresponding limit for the charge current (negative) is

$$I_{2,min}(t) = -\frac{U_{2,max} - U_{oc}(t)}{R_i(t)} . \qquad (4.68)$$

Battery Efficiency

The efficiency of a battery is not an obvious concept, since the battery is not an energy converter, but rather an energy storage system. Nevertheless, at least two definitions can be formulated for practical purposes. The global efficiency is defined on the basis of a full charge/discharge cycle as the ratio of total energy delivered to the energy that is necessary to recharge the device. Such a definition is dependent on the features of the charge/discharge cycle, i.e., whether the battery is charged/discharged at constant current (Peukert test) or at constant power (Ragone test). Assuming that the battery has a coulombic efficiency of 1, that it can be represented by the steady-state equivalent circuit model of (4.52), with a constant open-circuit voltage source and an internal resistance that varies with the sign of the current (charge or discharge), the global efficiency can be evaluated in the "Peukert" case. At constant-current discharge, the battery is depleted in a time $t_f = Q_0/I_2$. The discharge energy is therefore

$$E_d = \int_0^{t_f} P_2(t)\, dt = t_f \cdot (U_{oc} - R_i \cdot I_2) \cdot I_2 \,. \tag{4.69}$$

Charging the battery with a current of the same intensity, $I_2 = -|I_2|$, requires an energy that is evaluated as

$$|E_c| = \int_0^{t_f} |P_2|(t)\, dt = t_f \cdot (U_{oc} + R_i \cdot |I_2|) \cdot |I_2| \,. \tag{4.70}$$

The ratio of E_d to E_c is by definition the global efficiency which is a function of I_2,

$$\eta_b = \frac{E_d}{E_c} = \frac{U_{oc} - R_i \cdot |I_2|}{U_{oc} + R_i \cdot |I_2|} \,. \tag{4.71}$$

The "Ragone" charge/discharge cycle with constant power P_2 can be treated in the same way. The current I_2 to be inserted in (4.71) is a function of P_2 and may be calculated from (4.44), (4.60).

If the assumption of constant U_{oc} and R_i is removed, then the calculation of η_b made by integrating the power is no longer valid, since now P_2 varies with time alongside with q. If the affine parameterization of (4.53)–(4.55) is used to express U_2 as a function of the state of charge q, it can be proven that (4.71) still holds, but with the values of U_{oc} and R_i calculated at the middle point $q = 0.5$, i.e., for a battery half full. However, for nonlinear parameterizations or for the "Ragone" case of constant power, (4.71) cannot be used in this way either. An example of numerical integration of E_d, E_c is shown in Fig. 4.29 for two batteries, a 25 Ah lead–acid battery and a 45 Ah NiMH battery, whose data are found in the ADVISOR database [272]. The global efficiency for the first battery is $\eta_b = 0.89$, for the second $\eta_b = 0.92$.

In alternative to global efficiency, which always depends on the cyclic pattern used, a *local efficiency* can be also defined, as a power ratio instead of an energy ratio. This yields

$$\eta_b(t) = \frac{P_{2,d}(t)}{|P_{2,c}|(t)} = \frac{U_{oc}(t) - R_i(t) \cdot |I_2|(t)}{U_{oc}(t) + R_i(t) \cdot |I_2|(t)} \,, \tag{4.72}$$

which is formally the same relationship as that of (4.71). The difference is that in (4.72) the open-circuit voltage and internal resistance may depend on the state of charge and on the current I_2 as well. In other terms, the local efficiency is an instantaneous function of both current and charge, $\eta_b(I_2, q)$. This concept can be easily represented in terms of "efficiency maps" similar to the maps used to describe other energy converters. Figure 4.30 shows the local efficiency maps of the same two batteries of Fig. 4.29. The figures clearly show that the local efficiency is strongly dependent on the C-rate and only very weakly on the state of charge.

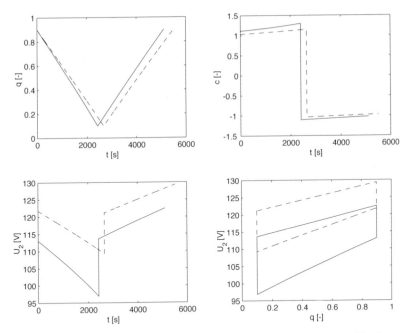

Fig. 4.29. Simulated state of charge q, C-rate c, terminal voltage U_2 for a discharge/charge cycle at constant power. Solid lines: 25 Ah lead–acid battery ($\kappa_1 = 11.7$, $\kappa_2 = 1.18$, $\kappa_3^{(d)} = 0.0369$, $\kappa_4^{(d)} = -0.0314$, $\kappa_3^{(c)} = 0.0272$, $\kappa_4^{(c)} = 0.0047$, $|P_2| = 3.14\,\text{kW}$). Dashed lines: 45 Ah NiMH battery ($\kappa_1 = 12.7$, $\kappa_2 = 1.40$, $\kappa_3 = 0.0140$, $\kappa_4 = -0.0051$, $|P_2| = 5.65\,\text{kW}$).

The observed limitations of steady-state models are that the voltage response to load changes is too prompt and that the internal resistance does not change as a function of the current [129]. For a 1 Ah lead–acid battery, it was observed that the internal resistance substantially varies (by a factor of eight) with the discharge rate varying from 1 A to 100 A. Therefore, for more accurate calculations, dynamic models are required.

4.4.2 Dynamic Modeling of Batteries

The physical causality representation of a dynamic model of a battery is sketched in Fig. 4.31. A battery is a passive electric source. Thus, the model input variable is the terminal current $I_2(t)$. A positive $I_2(t)$ discharges the battery, a negative $I_2(t)$ charges it. The model output variables are the terminal voltage $U_2(t)$ and the battery charge $Q(t)$.

Dynamic models describe the transient behavior of a battery, including the rate of change of the battery terminal voltage. Dynamic equivalent-circuit models implicitly take into account the capacitive and inductive effects that are neglected in the steady-state circuits. With a different approach, elec-

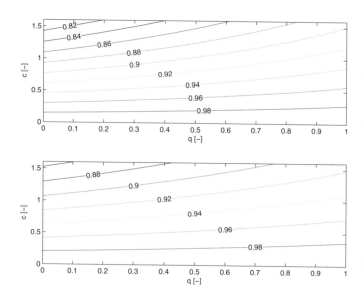

Fig. 4.30. Local efficiency map for a 25 Ah lead–acid battery (top) and a 45 Ah NiMH battery (bottom).

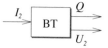

Fig. 4.31. Batteries: physical causality for dynamic modeling.

trochemical models explicitly represent the variations in concentration of the relevant chemical species. Here only lumped-parameter electrochemical models are described. Both dynamic equivalent circuits and lumped-parameter electrochemical models are suitable for the model-based determination of the state of charge (SoC observers).

Dynamic Equivalent Circuit

The simplest of the dynamic models of practical use is the Randles or Thevenin model [149, 217, 38]. This circuit represents a modification of the "steady-state battery" of (4.52), i.e., a voltage source in series with an internal resistance. In this equivalent circuit, depicted in Fig. 4.32, the ohmic voltage drop is distinguished from the non-ohmic overpotential or polarization.[8] The latter is

---

[8] The terms "overpotential" and "polarization" are both traditionally used in electrochemistry to denote a voltage drop caused by the passage of current. Here the former is used for batteries, the latter for fuel cells.

the sum of *charge-transfer (or surface) overpotential* and *diffusion overpotential*. It drives two parallel branches, in which the capacitive current and the charge-transfer current flow, respectively. The capacitive current flows across a *double-layer capacitor* C_{dl}, which describes the capacitive effects of the charge accumulation/separation that occurs at the interface between electrodes and electrolyte. The charge-transfer current, caused by chemical reactions, crosses the diffusion resistance and the charge-transfer resistance. The dynamic equations for this circuit are derived from Kirchhoff's voltage and current laws,

$$U_2(t) = U_{oc} - R_o \cdot I_2(t) - U_o(t)$$

(4.73)

$$R_o \cdot C_{dl} \cdot \frac{d}{dt} U_o(t) = U_{oc} - U_2(t) - U_o(t) \cdot \left(1 + \frac{R_o}{R_d + R_{ct}}\right) \;,$$

where U_o is the non-ohmic overpotential. Clearly, these equations coincide at steady state with (4.52), i.e., with $R_i = R_o + R_{ct} + R_d$.

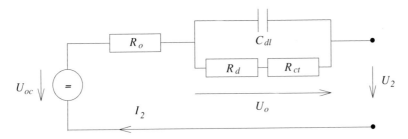

Fig. 4.32. Randles model for batteries.

Many authors adopted refinements to the Randles model to take into account additional phenomena or to improve the dynamic description of the basic processes. Among these refinements, the diffusion layer is described by the so-called *Warburg impedance* Z_d instead of a purely resistive element R_d [32, 21, 162, 243, 22]. In some cases the diffusion impedance is moved for simplicity from the charge-transfer branch to the main circuit branch crossed by the whole battery current [32, 21, 222, 162]. Vice versa, in some other cases all the resistances are placed in the charge-transfer branch [43]. Often a parallel current branch is included to simulate battery *self-discharge* or parasitic current, due to impurity and side reactions other than the cell reaction, mainly taking place during charging, thus leading to the coulombic efficiency of (4.51) [32, 43, 39, 38, 217, 9]. A bulk capacity is used sometimes to represent an open-circuit voltage which varies with the state of charge, as observed while discussing (4.53) [43, 21, 222, 38, 217, 243, 9, 129]. More complex circuit schemes may have multiple RC parallel elements in series [32, 224, 21, 243], or additional capacity paths in parallel with some of the resistances [222, 9]. In many cases, some of the resistances introduced are actually functions of

the current itself, representing the nonlinear nature of the processes involved
[32, 43, 21, 222, 162]. In some other cases [91] it is the battery potential which
depends on the current.

Another approach consists of representing the battery transient behavior
by means of "black-box" dynamic circuits, where the resistive and capacitive
elements have no physical meaning but are identified from transient battery
response. The resulting topology is usually different from that of the Randles
circuit consisting of various capacitor parallel branches. Example topologies
include two capacitors and three resistors [129], two capacitors and two resistors along with a voltage source [199], or two capacitors and four resistors
[65]. Johnson's model is included in the ADVISOR simulator [272]. Its state
equations are (see Fig. 4.33):

$$\frac{d}{dt}U'(t) = \frac{U''(t) - U'(t) - R'' \cdot I_2(t)}{C' \cdot (R' + R'')}$$

$$\frac{d}{dt}U''(t) = \frac{U'(t) - U''(t) - R' \cdot I_2(t)}{C'' \cdot (R' + R'')} \tag{4.74}$$

$$U_2(t) = \frac{R''}{R' + R''} \cdot U'(t) + \frac{R'}{R' + R''} \cdot U''(t) - R''' \cdot I_2(t) - \frac{R' \cdot R''}{R' + R''} \cdot I_2(t)$$

All the resistances are tabulated functions of the state of charge of the battery, which is calculated as a weighted sum of the charge on both capacitors,
$Q' = C' \cdot U'$, $Q'' = C'' \cdot U''$.

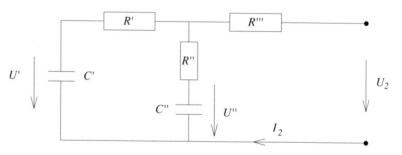

Fig. 4.33. Johnson's capacitive battery model.

Dynamic models have been widely validated for lead–acid batteries [32,
43, 224, 39, 199, 162, 38, 217, 65], but also for nickel–metal hydride batteries
[224, 21, 129], lithium-ion batteries [91, 21, 129, 243], as well as for other,
non-automotive types [22, 222].

Lumped Parameter Electrochemical Models

Batteries can be also modeled from a purely electrochemical point of view,
representing the complex physical processes that take place in the various

constituting regions (electrodes, electrolyte, gas reservoir, etc.). This type of modeling is inherently multi-species and multi-phase, since in a battery the solid, liquid, and gaseous phases may coexist. Several detailed electrochemical models can be found in the literature, see [186, 95] for NiMH batteries and [70, 94] for lithium batteries. Due to their complexity, these methods are not suitable for vehicle system-level simulation, nor for the online determination of the SoC. However, simpler lumped-parameter counterparts of electrochemical models can be derived under certain assumptions. The advantages of this lumped parameter approach are that the resulting models derive from first-principle equations, they are simple to solve, and they can be easily integrated in system simulators or SoC observers. In contrast, the drawbacks are that model assumptions may not be valid under some situations, and the effect of several cell parameters cannot be evaluated [277].

Considering a discharging NiMH battery, the terminal electric quantities and the variations of the SoC mainly depend on the concentrations of five chemical species, namely nickel hydroxide $Ni(OH)_2$, metal hydride MH, oxygen O_2, OH^- ions and nickel oxyhydroxide NiOOH. The reactions involving these species can be summarized as

$$NiOOH + H_2O + e^- \rightarrow Ni(OH)_2 + OH^- , \tag{4.75}$$

$$\frac{1}{2}O_2 + H_2O + 2e^- \rightarrow 2OH^- , \tag{4.76}$$

$$MH + OH^- \rightarrow H_2O + e^- + M , \tag{4.77}$$

$$2OH^- \rightarrow \frac{1}{2}O_2 + H_2O + 2e^- . \tag{4.78}$$

Reaction (4.75) along with the side reaction (4.76) take place at the nickel cathode. Reaction (4.77) along with the side reaction (4.78) take place at the MH anode.

The kinetics of reactions (4.75)–(4.78) are described by the Butler-Volmer equations, which involve the concentrations of the five species considered. For the z-th reaction, $z = 1, \ldots, 4$, the rate of reaction, i.e., the charge-transfer current density J_z is given by [64]

$$J_z = J_{z,0} \left[\prod_i \left(\frac{c_i}{c_{i,ref}} \right)^{\kappa_i} \cdot e^{\alpha_{a,z} \cdot K \cdot \eta_z} - \prod_j \left(\frac{c_j}{c_{j,ref}} \right)^{\kappa_j} \cdot e^{-\alpha_{c,z} \cdot K \cdot \eta_z} \right] , \tag{4.79}$$

where $J_{z,0}$ is the *exchange current density* at reference reactant concentrations (positive in the direction of oxidation), $\alpha_{a,z}$ and $\alpha_{c,z}$ are the anodic and cathodic *transfer coefficients*, η_z is the surface overpotential that drives the charge-transfer reaction, $K = F/(R \cdot \vartheta_b)$, F is Faraday's constant, ϑ_b is the cell temperature, and R is the gas constant. The subscripts i and j denote reducers and oxidizers of the reaction considered, respectively. The c's are the species concentrations, the κ's are their respective numbers of moles participating in the reaction (stoichiometric coefficients), and subscript ref denotes reference concentrations.

Frequently used simplifying assumptions consist of neglecting the variations of the OH^- concentration in the electrolyte and relating the concentration of $NiOOH$ to the $Ni(OH)_2$ concentration. Moreover, due to large electrochemical driving force, reaction (4.78) is considered as limited only by the oxygen concentration. Under these assumptions, (4.79) is specified as [277, 95, 18]

$$J_1(t) = J_{1,0} \cdot \left\{ \left(\frac{c_n(t)}{c_{n,ref}} \right) \cdot \left(\frac{c_e}{c_{e,ref}} \right) \cdot e^{0.5 \cdot K \cdot \eta_1(t)} - \left(\frac{c_{n,max} - c_n(t)}{c_{n,max} - c_{n,ref}} \right) \cdot e^{-0.5 \cdot K \cdot \eta_1(t)} \right\}, \tag{4.80}$$

$$J_2(t) = J_{2,0} \cdot \left\{ \left(\frac{c_e}{c_{e,ref}} \right)^2 \cdot e^{K \cdot \eta_2(t)} - \left(\frac{p_o(t)}{p_{o,ref}} \right)^{1/2} \cdot e^{-K \cdot \eta_2(t)} \right\}, \tag{4.81}$$

$$J_3(t) = J_{3,0} \cdot \left\{ \left(\frac{c_m(t)}{c_{m,ref}} \right)^\mu \cdot \left(\frac{c_e}{c_{e,ref}} \right) \cdot e^{0.5 \cdot K \cdot \eta_3(t)} - e^{-0.5 \cdot K \cdot \eta_3(t)} \right\}, \tag{4.82}$$

$$J_4(t) = -J_{4,0} \cdot \left(\frac{p_o(t)}{p_{o,ref}} \right), \tag{4.83}$$

where $c_n(t)$ is nickel hydroxide concentration, c_e is the constant concentration of KOH electrolyte representing the concentration of OH^- ions, $c_m(t)$ is the concentration of hydrogen in metal hydride material and μ its stoichiometric coefficient, and $p_o(t)$ is oxygen partial pressure.

The surface overpotentials are

$$\eta_1(t) = \Delta\Phi_{pos}(t) - \phi_1(t), \quad \eta_2(t) = \Delta\Phi_{pos}(t) - \phi_2(t), \tag{4.84}$$

$$\eta_3(t) = \Delta\Phi_{neg}(t) - \phi_3(t), \tag{4.85}$$

where $\Delta\Phi_{pos}(t)$ and $\Delta\Phi_{neg}(t)$ are the potential differences at the solid–liquid interface on the positive and negative electrodes, respectively, while $\phi_1(t), \ldots, \phi_3(t)$ are the equilibrium potentials at the reference state of reactions (4.75)–(4.77). The latter terms are conveniently parameterized as a function of the species concentrations and the temperature. The typical hysteresis behavior of NiMH batteries can be simulated by distinguishing $\phi_1(t)$ during charging and discharging phases and relaxing the switching with an exponential term [277].

The charge balance at the electrodes imposes the two constraints

$$I_2(t) = S_{pos} \cdot (J_1(t) + J_2(t)), \tag{4.86}$$

$$I_2(t) = -S_{neg} \cdot (J_3(t) + J_4(t)), \tag{4.87}$$

where S_{pos}, S_{neg} are the equivalent surfaces of the positive and negative electrode, respectively. The combination of (4.80)–(4.87) yields a solution for the four current densities $J_1(t), \ldots, J_4(t)$, $\Delta\Phi_{pos}(t)$, and $\Delta\Phi_{neg}(t)$.

The terminal voltage results from the contribution of equilibrium potential, surface overpotential, ohmic losses, and concentration overpotential. The lumped-parameter approach does not consider the latter term, while the first two terms are combined in the quantities $\Delta\Phi_{pos}$ and $\Delta\Phi_{neg}$ through (4.84)–(4.85). Consequently, the terminal voltage can still be calculated with an equation of the type (4.52), however, $U_{oc}(t)$ is replaced by $\Delta\Phi_{pos}(t) - \Delta\Phi_{neg}(t)$ and R_i is represented only by the ohmic resistance R_o.

Once the current densities are known, the mass balance of the nickel active material under the lumped parameter assumption yields

$$\frac{d}{dt}c_n(t) = -\frac{J_1(t)}{l_{y,pos} \cdot F} \ , \tag{4.88}$$

where $l_{y,pos}$ is the effective thickness of the nickel active material. The mass balance of metal hydride material reads

$$\frac{d}{dt}c_m(t) = -\frac{J_3(t)}{l_{y,neg} \cdot F} \ , \tag{4.89}$$

where $l_{y,neg}$ is the effective thickness of the metal hydride material. The mass balance of oxygen reads

$$\frac{d}{dt}p_o(t) = -\frac{R \cdot \vartheta_b}{V_{gas}} \cdot \frac{S_{pos} \cdot J_2(t) + S_{neg} \cdot J_4(t)}{F} \ , \tag{4.90}$$

where V_{gas} is the gas volume of the cell.

Finally, this model allows the direct calculation of the SoC

$$q(t) = 1 - \frac{c_n(t)}{c_{n,max}} \ . \tag{4.91}$$

Battery Thermal Models

Detailed dynamic simulations require a battery submodel that evaluates how the battery temperature ϑ_b varies during vehicle operation. Temperature variations in general affect many aspects of a battery's operation, including efficiency, cycle life, and capacity. From the point of view of energy use, temperature variations are important to evaluate the thermal power to be removed via the cooling system. In battery modeling, temperature is often used to correct ambient-temperature data for capacity, open-circuit voltage, and internal resistance.

Practical thermal models are of the lumped-capacitance type [188]. The battery pack is treated as a single reservoir with thermal capacitance $C_{t,b}$. Inflow thermal power is heat generation in the battery core mainly due to resistive heating. In the simplest battery equivalent circuit, this term is approximated by $R_i \cdot I_2^2$. Another source of heat generation are the parasitic reactions modeled with coulombic efficiency η_c, see (4.51). Outflow thermal power is due to thermal conduction through the battery case and convective heat transfer to the cooling air.

The thermal balance for the battery is written as

$$\frac{d}{dt}\vartheta_b(t) = \frac{q_{in}(t) - q_{out}(t)}{C_{t,b}} \ ,$$
(4.92)

with

$$q_{in}(t) = R_i \cdot I_2^2(t), \quad q_{out}(t) = \frac{\vartheta_b(t) - \vartheta_{air}(t)}{R_{th}} \ .$$
(4.93)

The effective air temperature is conveniently taken as the average between air flow inlet and outlet:

$$\vartheta_{air}(t) = \vartheta_a + \frac{1}{2} \cdot \frac{q_{out}(t)}{\overset{*}{m}_a \cdot c_{p,a}} \ .$$
(4.94)

where $\overset{*}{m}_a$ is the mass flow rate of cooling air, and $c_{p,a}$ and ϑ_a its specific heat and inlet temperature. The equivalent thermal resistance R_{th} is the sum of two contributions, a conductive term s/kA and a convective term $1/hA$, where A is the effective battery surface, and s, h and k are the case thickness, the convective heat transfer coefficient, and the thermal conductivity. The most difficult parameter to estimate is h, for which various empirical correlations are available, depending on the type of air flow (natural convection, forced convection), cooling air flow rate and temperature, geometrical features, etc. [196].

4.5 Supercapacitors

Supercapacitors (also termed electrochemical capacitors or ultracapacitors) store energy in the electric field of an electrochemical double layer. While their specific power is much higher than in batteries, their specific energy is substantially lower. As principal energy storage systems, these devices are being developed for power assist during acceleration and hill climbing, as well as for the recovery of braking energy [213, 144, 106]. Another possible application is in "mild" hybrids together with an integrated starter/alternator, as a low-voltage (42V) energy buffer that is also capable of high power recuperation [130, 235]. Supercapacitors are also potentially useful as secondary energy storage systems in HEVs, providing load-leveling power to electrochemical batteries which may be downsized [166]. Another advantage in this case would be the additional degree of freedom they add to the vehicle energy management, which allows for an optimization of the operating conditions of the main energy storage system.

A supercapacitor differs from conventional capacitors both in the materials of which it is made and in the physical processes involved. In a supercapacitor, the dielectric is an ion-conducting electrolyte interposed between conducting electrodes. The energy is stored by the charge separation taking place in the layers that separate the electrolyte and the electrodes. Since the voltage that can be applied is limited to a few volts by the physical characteristics of the electrolyte, the storage capacity is increased by raising the capacitance, i.e., increasing the surface and decreasing the thickness of the electrolyte. The surface is increased by using electrodes made of a porous material. Electrode materials with the required very high specific area are active carbon ($10^3 \, \mathrm{m^2/g}$) and some metallic oxides (ruthenium, iridium). The porous carbon electrodes are connected to metallic plates that collect charge. The electrodes are separated by an insulating, ion-conducting membrane, referred to as the separator. The separator also has the function of storing and immobilizing the liquid electrolyte. The electrolyte may be an aqueous acid solution or an organic liquid filling the porous electrodes.

Compared to electrochemical batteries, supercapacitors show a very high specific power of 500–2500 W/kg, but a very low specific energy of 0.2–5 Wh/kg, according to the material of the electrodes (carbon, metallic oxides) and of the electrolyte (aqueous, polymer) [76]. In automotive applications, most of the attention has been focused on carbon-based cells with polymer electrolyte, which seem to offer the best performance at the lowest cost [50]. The future use of this technology seems indeed to be dependent on cost issues in comparison with high-power battery systems [97].

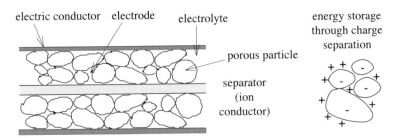

Fig. 4.34. Schematic of a supercapacitor.

4.5.1 Quasistatic Modeling of Supercapacitors

The causality representation of a supercapacitor in quasistatic simulations is sketched in Fig. 4.35. The input variable is the terminal power $P_2(t)$. The output variable is the charge $Q(t)$.

Similarly to batteries, the state of charge is evaluated from the terminal current I_2 and the nominal capacity Q_0. The former may be calculated from the terminal power P_2 using the trivial equality

$$I_2(t) = \frac{P_2(t)}{U_2(t)} \tag{4.95}$$

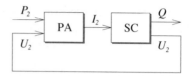

Fig. 4.35. Supercapacitors: causality representation for quasistatic modeling.

and a relationship between current and voltage U_2.

Equivalent Circuit

A basic physical model of a supercapacitor can be derived from a description of the system in terms of equivalent circuit. The simplest equivalent circuit consists of a capacitor and a resistor in series [50, 48, 129]. More complex equivalent circuits describe the distributed nature of the resistance and of the charge stored in a porous electrode [167, 187]. The basic equivalent circuit is depicted in Fig. 4.36. Kirchhoff's voltage law yields

$$R_{sc} \cdot I_2(t) - \frac{Q_{sc}(t)}{C_{sc}} + U_2(t) = 0, \quad I_2(t) = -\frac{d}{dt} Q_{sc}(t) . \tag{4.96}$$

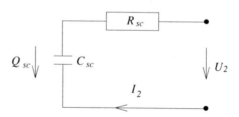

Fig. 4.36. Equivalent circuit of a supercapacitor.

Substitution of (4.95) into (4.96) yields a quadratic equation for the supercapacitor voltage,

$$U_2^2(t) - \frac{Q_{sc}(t)}{C_{sc}} \cdot U_2(t) + P_2(t) \cdot R_{sc} = 0 . \tag{4.97}$$

To let Q_{sc} vanish, both terms of (4.97) are differentiated and the second of (4.96) is used. This leads to the following differential equation for U_2

$$\left(1 - \frac{R_{sc} \cdot P_2(t)}{U_2^2(t)}\right) \cdot \frac{d}{dt} U_2^2(t) = -\frac{2 \cdot P_2(t)}{C_{sc}} . \tag{4.98}$$

Based on (4.95) the current I_2 and consequently the charge Q_{sc} may be calculated.

An alternative approach is quite similar to the one discussed for electrochemical batteries. After having substituted the open-circuit voltage U_{oc} with the ratio Q_{sc}/C_{sc}, (4.60) can be used. The resulting equation for the voltage is

$$U_2(t) = \frac{Q_{sc}(t)}{2 \cdot C_{sc}} + \sqrt{\frac{Q_{sc}^2(t)}{4 \cdot C_{sc}^2} - P_2(t) \cdot R_{sc}} \ . \tag{4.99}$$

Using numerical integration methods, this equation may be evaluated at any time using the value of Q_{sc} at the previous time step. The maximum power available may be found with the approach used to derive the same quantity for batteries, (4.62)–(4.66).

Supercapacitor Efficiency

The definition of the efficiency of supercapacitors is similar to that of batteries, both elements being energy storage systems rather than energy converters. On the basis of a full charge/discharge cycle, the global (or "round-trip") efficiency is defined as the ratio of total energy delivered to the energy that is necessary to charge the device [68]. Such a definition is dependent on the features of the charge/discharge cycle, i.e., whether the battery is charged/discharged at constant current (Peukert test) or at constant power (Ragone test) [48, 49]. If the supercapacitor is represented by the equivalent circuit model of (4.96) with constant capacitance and internal resistance, the global efficiency can be evaluated in both "Peukert" and "Ragone" cases.

At constant-current discharge, the supercapacitor is depleted in a time $t_f = Q_0/I_2$. The charge varies linearly with time, $Q_{sc}(t) = Q_0 - I_2 \cdot t$. The terminal voltage thus varies according to (4.96). The discharge energy is therefore

$$E_d = \int_0^{t_f} U_2(t) \cdot I_2 \, dt = I_2 \cdot \left(\frac{Q_0^2}{2 \cdot C_{sc} \cdot I_2} - R_{sc} \cdot Q_0 \right) . \tag{4.100}$$

Charging the supercapacitor with a current of the same magnitude, i.e., $I_2 = -|I_2|$, the charge varies as $Q_{sc} = |I_2|t$. The charge energy is evaluated as

$$|E_c| = \int_0^{t_f} U_2(t) \cdot |I_2| \, dt = |I_2| \cdot \left(\frac{Q_0^2}{2 \cdot C_{sc} \cdot |I_2|} + R_{sc} \cdot Q_0 \right) . \tag{4.101}$$

By definition the ratio of E_d to E_c is the global efficiency, which is a function of I_2:

$$\eta_{sc} = \frac{E_d}{E_c} = \frac{Q_0 - 2 \cdot R_{sc} \cdot C_{sc} \cdot |I_2|}{Q_0 + 2 \cdot R_{sc} \cdot C_{sc} \cdot |I_2|} . \tag{4.102}$$

At constant-power discharge (Ragone test), the current varies with time, thus (4.100)–(4.101) cannot be used. Instead, (4.98) may be solved for constant power, yielding an implicit dependency of the terminal voltage on time:

$$t = \frac{C_{sc}}{2 \cdot P_2} \cdot \left(R_{sc} \cdot P_2 \cdot \ln\left(\frac{U_2}{U_0}\right)^2 + U_0^2 - U_2^2(t) \right). \tag{4.103}$$

The initial voltage U_0 follows from (4.99) with $Q_{sc} = Q_0$. From (4.99), the discharge ends when $Q_{sc} = 2 \cdot C_{sc} \cdot \sqrt{P_2 \cdot R_{sc}}$ which, based on (4.96), corresponds to a terminal voltage $U_f = \sqrt{P_2 \cdot R_{sc}}$. Note that, in contrast with the Peukert discharge, the supercapacitor voltage is not zero at the time t_f. From (4.103), the final time is calculated as

$$t_f = \frac{C_{sc}}{2 \cdot P_2} \cdot \left(R_{sc} \cdot P_2 \cdot \ln\left(\frac{R_{sc} \cdot P_2}{U_0^2}\right) + U_0^2 - R_{sc} \cdot P_2 \right). \tag{4.104}$$

The discharge energy is given by

$$E_d = t_f \cdot P_2 = \frac{C_{sc}}{2} \cdot \left(-R_{sc} \cdot P_2 \cdot \ln\left(\frac{U_0^2}{R_{sc} \cdot P_2}\right) + U_0^2 - R_{sc} \cdot P_2 \right). \tag{4.105}$$

For a charge with a constant power of the same intensity, $P_2 = -|P_2|$, the initial voltage is $\sqrt{R_{sc} \cdot |P_2|}$, while the final voltage equals the value of U_0 calculated previously. The charge energy is evaluated as

$$|E_c| = t_f \cdot |P_2| = \frac{C_{sc}}{2} \cdot \left(R_{sc} \cdot |P_2| \cdot \ln\left(\frac{U_0^2}{R_{sc} \cdot |P_2|}\right) + U_0^2 - R_{sc} \cdot |P_2| \right). \tag{4.106}$$

The efficiency is therefore calculated from

$$\eta_{sc} = \frac{E_d}{E_c} = \frac{U_0^2 - R_{sc} \cdot |P_2| - R_{sc} \cdot |P_2| \cdot \ln\left(\dfrac{U_0^2}{R_{sc} \cdot |P_2|}\right)}{U_0^2 - R_{sc} \cdot |P_2| + R_{sc} \cdot |P_2| \ln\left(\dfrac{U_0^2}{R_{sc} \cdot |P_2|}\right)}. \tag{4.107}$$

A third possibility is to charge and discharge the supercapacitor with maximum power. Equation (4.99) states that discharge power is limited by the state of charge, $P_2(t) < Q_{sc}^2(t)/4/C_{sc}^2/R_{sc}$. This limit is thus varying in time, in contrast to what happens during charge, when the power is limited only by the constant maximum current. The efficiency cannot be evaluated analytically in this case, thus a numerical integration is needed. Examples of supercapacitor discharges at constant current, at constant power, and at maximum power are illustrated in Fig. 4.37.

The local definition of supercapacitor efficiency is based on a power ratio rather than an energy ratio. If the discharge and charge powers are expressed in terms of charge and current, then the local efficiency is evaluated as

$$\eta_{sc}(I_2) = \frac{P_{2,d}}{|P_{2,c}|} = \frac{Q_{sc} - R_{sc} \cdot C_{sc} \cdot |I_2|}{Q_{sc} + R_{sc} \cdot C_{sc} \cdot |I_2|}. \tag{4.108}$$

Note that, if in (4.108) an average charge $Q_{sc} = Q_0/2$ is used, the result is the efficiency of (4.102). The advantage of the local definition is that it

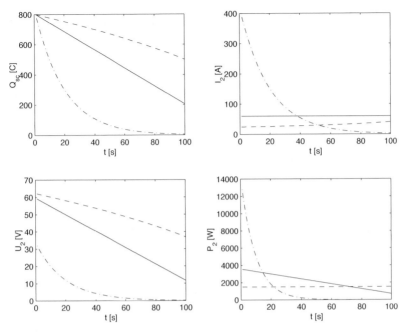

Fig. 4.37. Calculated discharge tests for a supercapacitor ($C_{sc} = 12.5\,\mathrm{F}$, $R_{sc} = 0.08\,\Omega$, $Q_0 = 800\,\mathrm{C}$). Solid lines: discharge with constant current $I_2 = 60\,\mathrm{A}$. Dashed lines: discharge with constant power $P_2 = 1500\,\mathrm{W}$. Dashdot lines: discharge at maximum power.

does not require any assumption on the type of charge/discharge, while global expressions like (4.102) or (4.107) are strictly valid for Peukert and Ragone cycles, respectively. Similarly to batteries and energy converters, the local supercapacitor efficiency can be easily represented in terms of "efficiency maps" as well.

4.5.2 Dynamic Modeling of Supercapacitors

The physical causality representation of a supercapacitor is the same as that of a battery and is sketched in Fig. 4.38. The model input variable is the terminal current $I_2(t)$. A positive $I_2(t)$ discharges the supercapacitor, a negative $I_2(t)$ charges it. The model output variables are the terminal voltage $U_2(t)$ and the charge $Q(t)$.

The basic equivalent circuit model, (4.96) can be used to calculate the terminal voltage as a function of the terminal current. Integrating the current yields the supercapacitor charge and thus the non-dimensional state of charge.

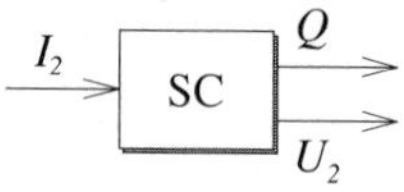

$$I_2 \rightarrow \boxed{\text{SC}} \begin{array}{c} Q \rightarrow \\ \rightarrow \\ U_2 \end{array}$$

Fig. 4.38. Supercapacitors: physical causality for dynamic modeling.

4.6 Electric Power Links

In a conventional vehicle architecture, all the electric loads are supplied by a 14 V DC link connected to a 12 V battery and to an engine-driven alternator. The most common electric loads are: ignition system, lighting system, and electric starter motor. Additional electric loads include power steering motors, anti-lock motors, fans and pumps, air conditioning, active suspension actuators, catalyst heaters, throttle actuators, etc. In common passenger cars today the electric power demand is around 1 kW. Luxury cars may have a maximum electrical load of 2 kW. However, current predictions suggest that the electrical load in automobiles will increase in the next years up to 6–10 kW [133].

In this context of "more electric" cars, the conventional architecture might be complemented with a higher voltage DC link, typically at 42 V. In this way it will be possible to introduce new electric loads, e.g. the air conditioning compressor or the coolant pump, that in the conventional architecture are driven by the engine. The payback usually is a higher flexibility in the operation as well as the possibility of gaining higher efficiency and reliability. The advent of a 42 V system would also facilitate the possibility of introducing the integrated starter/alternator (ISA) concept, i.e., a "mild" hybrid configuration for effective stop-and-start operation and some degree of regenerative braking [235].

To manage the presence of electric sources at different voltage levels, various solutions have been proposed. They are roughly classifiable in: (1) multi-level systems, in which a single high-voltage battery (36 V) is used to feed directly the 42 V loads and, through a 14 V link, the other low-voltage loads, and (2) dual-voltage systems, in which two batteries are used, a 12 V battery for the 14 V link and a 36 V battery for the 42 V link [133]. Of course electric traction in EVs and HEVs requires much higher power and voltage levels, typically hundreds of volts. However, in this case also the two concepts mentioned above are valid.

Multi-input power converters are used when multiple electric sources are present in a vehicle. Examples are series hybrids with a generator and a battery, or purely electric hybrids with a battery and a supercapacitor, or hybrid fuel-cell vehicles with a battery or a supercapacitor. In all these cases, each power source is connected to the DC link by means of a bi-directional DC/DC converter [67].

4.6.1 Quasistatic Modeling of Electric Power Links

The causality representation of a power link (with m power sources and a single load) in quasistatic simulations is sketched in Fig. 4.39 for $m = 2$ (usually, a battery and a generator). The input variable is the power $P_{m+1}(t)$ at the load port. The output variables are the power $P_j(t)$ at each source port, $j = 1, \ldots, m$.

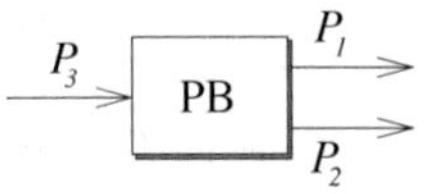

Fig. 4.39. Electric power links: causality representation for quasistatic modeling.

In the general case (in most practical cases $m = 1, 2$), one equation is given by the power balance across the link,

$$P_{m+1}(t) = \sum_{j=1}^{m} P_j(t) - P_l(t) , \tag{4.109}$$

where $P_l(t)$ is the power to the electric loads, which generally is a quantity that varies in time according to the vehicle operation (from a control point of view, it may be regarded as a disturbance). There are still $m - 1$ degrees of freedom available. These can be represented by $m - 1$ control variables u_j, which are conveniently defined as the power-split ratios

$$u_j(t) = \frac{P_j(t)}{P_{m+1}(t) + P_l(t)} . \tag{4.110}$$

The combination of (4.109) and (4.110) allows the evaluation of the power at the source ports,

$$P_j(t) = u_j(t) \cdot (P_{m+1}(t) + P_l(t)) , \quad j = 1, \ldots, m - 1 , \tag{4.111}$$

$$P_m(t) = \left(1 - \sum_{j=1}^{m-1} u_j(t) \right) \cdot (P_{m+1}(t) + P_l(t)) . \tag{4.112}$$

Of course, in the common case of $m = 1$ (EVs and parallel HEVs), this model simply implies the equality of the power delivered by the battery and the power absorbed by the motor and the electric loads.

4.6.2 Dynamic Modeling of Electric Power Links

The physical causality representation of a power link is sketched in Fig. 4.40. The model input variables are the voltage at the first main port (typically the

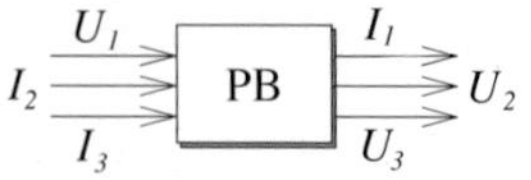

$$U_1 \qquad I_1$$
$$I_2 \overset{}{\Longrightarrow} \boxed{\text{PB}} \Longrightarrow U_2$$
$$I_3 \qquad U_3$$

Fig. 4.40. Electric power links: physical causality for dynamic modeling.

battery), $U_1(t)$, and the current at the other ports, $I_j(t)$, $j = 2, \ldots, m+1$. The model output variables are the current at the main source port, $I_1(t)$, and the voltage at the other ports, $U_j(t)$, $j = 2, \ldots, m+1$.

Power links have capacitive dynamics (see Fig. 4.41). However, the link capacitors usually have such a negligible capacitance ($\approx$ mF) that models aimed at the evaluation of vehicle energy consumption do not consider any dynamic effects. Consequently, m equations are given by the equalization of voltage

$$U_j(t) = R_j \cdot U_1(t), \quad j = 2, \ldots, m+1 , \qquad (4.113)$$

where the R_j's are DC/DC conversion ratios. Another equation is given by the balance of currents flowing across the link

$$I_1(t) = R_{m+1} \cdot I_{m+1}(t) - \sum_{j=2}^{m+1} R_j \cdot I_j(t) + \frac{P_l(t)}{U_1(t)} . \qquad (4.114)$$

Since these are $m+1$ equations in the $m+1$ model variables, the variables u_j introduced in the previous section are outputs of the model rather than inputs.

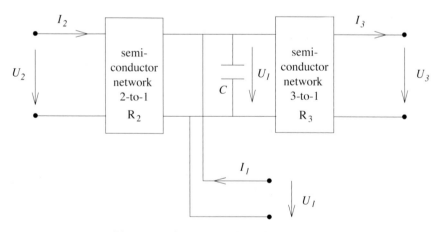

Fig. 4.41. Schematic of a DC power link.

4.7 Torque Couplers

In mechanical-hybrid and parallel hybrid-electric vehicles, the mechanical power outputs from different power sources are combined using various devices that can be generically called "torque couplers." These include three-sprocket gears driven by belts or chains or direct coupling on the same shaft. The case of different prime movers that power different wheel axles is straightforward and is not considered here.

The entry level to hybridization is the belt-driven architecture typically implemented by starter/generator "mild" hybrids. In this configuration, the traditional alternator is replaced by an electric machine capable of not only generating electricity but also acting as a motor to crank the engine, assisting it in powering, and driving auxiliary loads when the engine is off. Such an implementation requires a relatively low amount of modification to an ICE-based vehicle [116]. The engine and transmission do not require any modifications, nor does the space between them. Modifications are required to mount a larger electric machine: larger pulleys are needed on all belt-driven accessories, and an inverter.

Another concept for torque coupling is implemented in full parallel hybrids. It consists of mounting the rotor of the electric machine on the same engine shaft, between the engine and the final drive. The stator is mounted on the outer transmission housing or on a separate intervening housing [116]. This mounting allows the electric machine to be used as a damper to cancel oscillations at the drive train. This concept is realized in the ETH-Hybrid III [65, 274], a hybrid-electric-inertial vehicle with three power sources, the third being a flywheel accumulator, see case study in Sect. 8.3.

4.7.1 Quasistatic Modeling of Torque Couplers

The causality representation of a generic torque coupler (with m power sources and a single load) in quasistatic simulations is sketched in Fig. 4.42 for $m = 3$. The input variables are the torque $T_{m+1}(t)$ and the speed $\omega_{m+1}(t)$ required at the load shaft. The output variables are the torque $T_j(t)$ and the speed $\omega_j(t)$ at each power shaft, $j = 1, \ldots, m$.

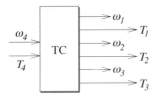

Fig. 4.42. Torque couplers: causality representation for quasistatic modeling.

In the general case (in all practical cases $m = 2, 3$), m equations are given by the equalization of rotational speed

$$\omega_j(t) = R_j \cdot \omega_{m+1}(t), \quad j = 1, \ldots, m, \tag{4.115}$$

where the R_j's are the transmission ratios of the power sources to the output gear. Usually, the main power sources have a unitary transmission ratio to the output. In the case of direct coupling on the same crankshaft, all the transmission ratios are unitary. Another equation is given by the power balance across the device

$$T_{m+1}(t) = \sum_{j=1}^{m} R_j \cdot T_j(t) - T_l(t), \tag{4.116}$$

where $T_l(t)$ is a loss term that, when not neglected, is usually modeled as a constant [272]. There are still $m - 1$ degrees of freedom available. These can be represented by $m - 1$ control variables u_j which are conveniently defined as the torque-split ratios

$$u_j(t) = \frac{R_j \cdot T_j(t)}{T_{m+1}(t) + T_l(t)}. \tag{4.117}$$

The combination of (4.116) and (4.117) permits the estimation of the torque at the power shafts,

$$T_j(t) = \frac{u_j(t)}{R_j} \cdot (T_{m+1}(t) + T_l(t)), \quad j = 1, \ldots, m - 1. \tag{4.118}$$

$$T_m(t) = \frac{1 - \sum_{j=1}^{m} u_j(t)}{R_j} \cdot (T_{m+1}(t) + T_l(t)). \tag{4.119}$$

4.7.2 Dynamic Modeling of Torque Couplers

The physical causality representation of a torque coupler is sketched in Fig. 4.43. The model input variables are the torque at the power shafts, $T_j(t)$, $j = 1, \ldots, m$ and the rotational speed at the output shaft, $\omega_{m+1}(t)$. The model output variables are the rotational speed at the input shafts, $\omega_j(t)$, $j = 1, \ldots, m$ and the torque at the output shaft, $T_{m+1}(t)$.

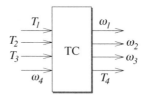

Fig. 4.43. Torque couplers: physical causality for dynamic modeling.

Usually in modeling torque couplers, no dynamic effects are considered. The rotational inertia is conveniently attributed to the power and the load shafts, respectively. Moreover, shaft elasticity is usually neglected. Consequently, the same quasistatic equations, (4.115)–(4.116) can be used. In particular, they constitute $m+1$ equations in the $m+1$ model variables. Therefore, the variables u_j introduced in the previous section are outputs of the model rather than inputs.

4.8 Power Split Devices

Power split devices (PSDs) are widely used in automatic transmissions. A power split device is also found in many hybrid vehicles to combine mechanical power from various (usually, two) power sources to various (usually, two) mechanical loads. In the typical configuration of a combined hybrid vehicle, a PSD connects an engine, a motor, a generator, and the drive train.

The core of most PSDs is a *planetary gear set*, often referred to also as *epicyclic gearing*. A basic planetary gear has three main rotating parts (Fig. 4.44). The inner part is the *sun*, the outer part is the *ring*. The intermediate part carries rotating elements (planets) and is the *carrier*. More complex configurations are possible but they are not considered here. Each of the three parts can be connected to the input shaft, the output shaft, or can be held stationary. Choosing which piece plays which role determines the gear ratio for the gear set.

Common automatic transmissions use a compound planetary gear set in combination with an hydraulic torque converter. In contrast, several PSDs for HEVs use one or two planetary gear sets in combination with two electric machines. The latter case is analyzed in detail in the next section, however, the modeling of these devices can be applied in both cases.

4.8.1 Quasistatic Modeling of Power Split Devices

The causality representation of a power split device in quasistatic simulations is sketched in Fig. 4.45. The input variables are the torque and speed at the load shafts, $\omega_f(t)$, $\omega_g(t)$, $T_f(t)$, $T_g(t)$, where the subscripts f and g stand for "final driveline" and "generator", respectively. The output variables are the torque and speed at the power source shafts, $\omega_e(t)$, $\omega_m(t)$, $T_e(t)$, $T_m(t)$, where the subscripts e and m stand for "engine" and "motor", respectively. Often in combined hybrids, the roles of the electric machines as generator and motor are interchangeable.

Simple Power Split Devices

The basic equation to consider in analyzing the quasistatic behavior of a planetary gear set is the relationship between the speed of the three main

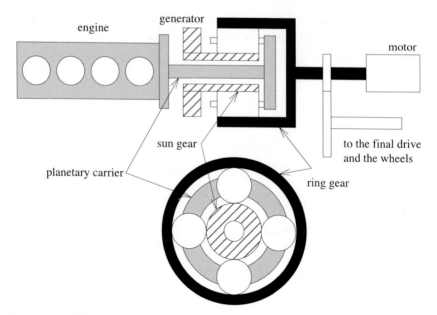

Fig. 4.44. Schematics of a planetary gear set arrangement (as in Toyota Prius [250]).

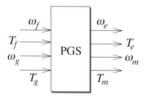

Fig. 4.45. Power split devices: causality representation for quasistatic modeling.

parts, which can be derived according to the Willis formula. The planetary gear set can be considered as an ordinary gear set in a rotating frame that is attached to the carrier. Thus, the ratio of the relative speeds of the ring and of the sun can be written as

$$\frac{\omega_r(t) - \omega_c(t)}{\omega_s(t) - \omega_c(t)} = -z, \quad z = \frac{n_s}{n_r}, \tag{4.120}$$

where n is the number of teeth and the subscripts c, s, and r refer to the carrier, the sun, and the ring, respectively. A typical value of the epicyclic gear ratio z is 0.385 [219].

The connection of the machines to the various ports may differ from one system to another. In the Toyota Hybrid System (THS-II) [250], the sun gear is connected to the generator, the planetary carrier to the engine, and the ring gear to the motor shaft. On the output shaft, the power transmitted by the ring gear and the power from the motor are combined. In other hybrid-electric vehicles, the connections can be different [161, 265].[9]

Assuming $\omega_e(t) = \omega_c(t)$, $\omega_g(t) = \omega_s(t)$, $\omega_f(t) = \omega_r(t)$, (4.120) is specified as [219]

$$\omega_m(t) = \omega_f(t) , \qquad (4.121)$$

$$\omega_e(t) = \frac{z \cdot \omega_g(t) + \omega_f(t)}{1 + z} . \qquad (4.122)$$

The balance of power applied to the four ports,

$$T_f(t) \cdot \omega_f(t) + T_g(t) \cdot \omega_g(t) = T_e(t) \cdot \omega_e(t) + T_m(t) \cdot \omega_m(t) , \qquad (4.123)$$

combined with the torque balance $T_f(t) + T_g(t) = T_e(t) + T_m(t)$ and with (4.121)–(4.122) yields

$$T_e(t) = \frac{1 + z}{z} \cdot T_g(t) , \qquad (4.124)$$

$$T_m(t) = T_f(t) - \frac{T_g(t)}{z} , \qquad (4.125)$$

which are the remaining two equations.

In contrast to torque couplers, planetary gear sets do not have any available degrees of freedom to control. In other words, planetary gear sets should be regarded as passive elements which are inherently self-controlled. Notice also that if the generator shaft is not connected, i.e., $T_g(t) = 0$, it follows from (4.124) that also $T_e(t) = 0$.

Equation (4.122) may be regarded as expressing the relationship between engine speed and output axle speed, the generator speed being a parameter. In other words, the planetary gear set represents a sort of continuously variable transmission, where the transmission ratio is regulated by the generator controller. Four typical modes of operation are illustrated in Fig. 4.46, which clearly shows the linear dependency between ω_e and ω_f [250].

A simple PSD may also have an output stage gearing to combine the torque from the motor to the torque from the ring gear of the planetary gear set. An example of such case is found in the Ford Hybrid System (FHS) [83], for which (4.121)–(4.125) are modified to

$$\omega_m(t) = R_m \cdot \omega_f(t) , \qquad (4.126)$$

$$\omega_e(t) = \frac{z \cdot \omega_g(t) + R_r \cdot \omega_f(t)}{1 + z} , \qquad (4.127)$$

$$T_e(t) = \frac{1 + z}{z} \cdot T_g(t) , \qquad (4.128)$$

$$T_m(t) = \frac{1}{R_m} \cdot T_f(t) - \frac{R_r}{R_m} \cdot \frac{T_g(t)}{z} . \qquad (4.129)$$

---

[9] Usually, in combined hybrids, the two electric machines are simply referred to as motor–generator 1 and 2, without distinguishing their prevalent functions.

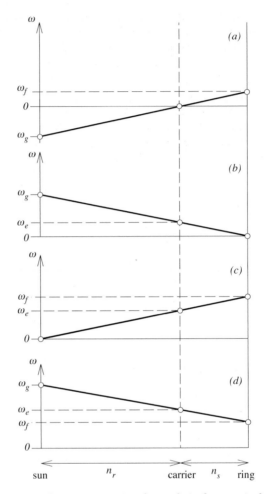

Fig. 4.46. Relationship between rotational speeds in four typical operation modes: (a) purely electric (at start-up, $\omega_e = 0$), (b) engine/generator operation (engine start, $\omega_m = \omega_f = 0$), (c) electric assist (normal drive, $\omega_g = 0$), (d) acceleration.

where R_r and R_m are the transmission ratios of the motor and ring speeds, respectively, to the final shaft. These ratios are easily calculated as a function of the number of teeth of the three torque-coupling gears.

Compound Power Split Devices

Compound power split devices are increasingly used, especially in large hybrid sport utility vehicle applications, since they have the advantage over simple PSDs of a reduced power of the electric machines.

An example of a compound PSD is found in GM's Two-mode Hybrid System (previously known as AHS-II) shown in Fig. 4.47a [177]. The gears of two planetary gearings are mutually connected and linked to the engine, to the two electric machines, and to the final driveline. Choosing the locking status of a pair of controlled clutches E1 and E2 permits the realization of two different operating modes.

In the first operating mode, used at low loads and relatively low vehicle speeds, E1 is disengaged while E2 is engaged, thus the mechanical connections can be schematized as in Fig. 4.47b. Applying (4.120) and (4.123) to both planetary gear sets, one obtains for the speeds

$$\omega_e(t) = -z_1 \cdot \omega_g(t) + \omega_f(t) \cdot (1 + z_1) \tag{4.130}$$

and

$$\omega_m(t) = \frac{1 + z_2}{z_2} \cdot \omega_f(t) , \tag{4.131}$$

where z_1 and z_2 are the epicyclic ratios of the first and of the second planetary gear set, respectively. The corresponding relationships between the torques are

$$T_e(t) = -\frac{1}{z_1} \cdot T_g(t) \tag{4.132}$$

and

$$T_m(t) = \frac{z_2}{1 + z_2} \cdot \left(T_f(t) + \frac{1 + z_1}{z_1} \cdot T_g(t) \right) . \tag{4.133}$$

The second operating mode is used for heavy loads and high vehicle speeds. The clutch E1 is engaged, while E2 is disengaged, see Fig. 4.47c. The relationships between the speeds are

$$\omega_e(t) = -z_1 \cdot \omega_g(t) + \omega_f(t) \cdot (1 + z_1) \tag{4.134}$$

and

$$\omega_m(t) = \frac{1 + z_2}{z_2} \cdot \omega_f(t) - \frac{1}{z_2} \cdot \omega_g(t) . \tag{4.135}$$

The corresponding relationships between the torques are

$$T_e(t) \cdot \left(\frac{1}{z_2} - z_1 \right) = \frac{1 + z_2}{z_2} \cdot T_g(t) + \frac{T_f(t)}{z_2} \tag{4.136}$$

and

$$T_m(t) \cdot \left(z_1 - \frac{1}{z_2} \right) = z_1 \cdot T_f(t) + (1 + z_1) \cdot T_g(t) . \tag{4.137}$$

Understanding the role of mode-switching of AHS-II requires a dynamic analysis that is introduced in the next section. A similar operating principle is used in other compound PSDs, such as the Timken eVT [1] and the Renault e-IVT [262], whose quasistatic modeling is obtained by using the methods illustrated above.

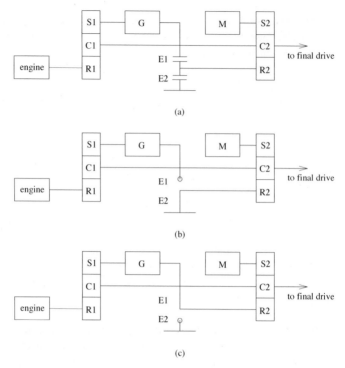

Fig. 4.47. Schematic of a Two-Mode Hybrid System (a) and of its two modes of operations (b) and (c). The figure show the connections of the engine, the "speeder" machine or generator (G), the "torquer" machine or motor (M), via two planetary gear sets, for which S: sun gear, C: planetary carrier, R: ring gear. Two control clutches are labelled E1 and E2.

4.8.2 Dynamic Modeling of Power Split Devices

The physical causality representation of a power split device is sketched in Fig. 4.48. The model input variables are the torque at the power source shafts, $T_e(t)$ and $T_m(t)$, and the rotational speeds at the load shafts, $\omega_f(t)$ and $\omega_g(t)$. The model output variables are the rotational speed at the power source shafts, $\omega_e(t)$ and $\omega_m(t)$, and the torques at the load shafts, $T_f(t)$ and $T_g(t)$.

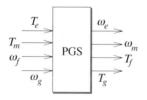

Fig. 4.48. Planetary gear sets: physical causality for dynamic modeling.

Dynamic models of PSDs include the inertia effects of the gears. Equations (4.121)–(4.122) are still valid and already in the required causality representation. Conversely, (4.124)–(4.125) are substituted by the more general equations

$$T_f(t) = T_m(t) + \frac{1}{1+z} \cdot T_e(t) - \dot\omega_g(t) \cdot \frac{z \cdot \Theta_c}{(1+z)^2} - \dot\omega_f(t) \cdot \left(\frac{\Theta_c}{(1+z)^2} + \Theta_r \right)$$
(4.138)

and

$$T_g(t) = \frac{z}{1+z} \cdot T_e(t) - \dot\omega_g(t) \cdot \left(\frac{z^2 \cdot \Theta_c}{(1+z)^2} + \Theta_s \right) - \dot\omega_f(t) \cdot \frac{z \cdot \Theta_c}{(1+z)^2} , \quad (4.139)$$

where Θ_c, Θ_s, and Θ_r are the moment of inertia of the carrier, the sun, and the gear ring, respectively. Despite the generality of (4.138)–(4.139), in planetary gear sets the dynamic effects are normally not considered, as in the case of torque couplers. The rotational inertia of the gears is indeed negligible when compared with those of the machines and shafts connected.

However, the physical causality representation is still useful to describe the role and the operation of a PSD within a generic "variator" architecture, i.e., a system to split an input power between a main mechanical path and a parallel path that may not be mechanical. Indeed several variator technologies exist, including hydraulic, toroidal [87], and belt/chain CVT [264]. A hybrid-electric powertrain as a whole can be seen as an electric variator [169, 262, 54].

Since PSDs are full four-port converters, their system equations can be expressed in the general form

$$\begin{pmatrix} \omega_e(t) \\ \omega_m(t) \end{pmatrix} = \begin{bmatrix} A & B \\ C & D \end{bmatrix} \cdot \begin{pmatrix} \omega_f(t) \\ \omega_g(t) \end{pmatrix} = \mathbf{M} \cdot \begin{pmatrix} \omega_f(t) \\ \omega_g(t) \end{pmatrix}$$
(4.140)

and

$$\begin{pmatrix} T_f(t) \\ T_g(t) \end{pmatrix} = \mathbf{M}^T \cdot \begin{pmatrix} T_e(t) \\ T_m(t) \end{pmatrix} = \begin{bmatrix} A & C \\ B & D \end{bmatrix} \cdot \begin{pmatrix} T_e(t) \\ T_m(t) \end{pmatrix} ,$$
(4.141)

where the parameters A, B, C, D depend on the system configuration. For example, the simple planetary arrangement of THS-II has $A = 1/(1+z)$, $B = z/(1+z)$, $C = 1$, and $D = 0$. Compound PSDs usually are designed such that $A > 0$, $B < 0$, $C > 0$. The fourth term D can be zero as in the first mode of AHS-II, negative as in its second mode, or positive.

The PSD operation is described by a few parameters, among which the speed transmission ratio $K(t) = \omega_f(t)/\omega_e(t)$ and the variator speed ratio $K_v(t) = \omega_m(t)/\omega_g(t)$. The relationship between these two parameters is calculated from (4.140)–(4.141) as

$$K_v(t) = D \cdot \frac{1 - \dfrac{K(t)}{K_1}}{1 - \dfrac{K(t)}{K_2}}$$
(4.142)

where $K_1 = (A - B \cdot C/D)^{-1}$ and $K_2 = 1/A$.

The relationship between $K(t)$ and $K_v(t)$ is shown in Fig. 4.49a. If the system allows K_v varying in a wide range, from negative to positive values, enabling a generator or motor speed equal to zero, then at a given input speed (engine speed), the output speed (vehicle speed) can vary continuously from reverse to forward speed passing through the zero speed. In other words, K may vary "infinitely"[10] without using a torque converter or any equivalent device.

The variator ratio K_v not only controls the speed transmission ratio, but also the split of power between the mechanical path and the electrical path. In variator theory, the power split ratio $r(t) = P_g(t)/P_e(t)$ is conveniently used to describe this effect. In the ideal case of unitary efficiency of the electrical path and no power supply from the battery, $P_m(t) = P_g(t)$.[11] After some manipulation of (4.140)–(4.141), the power split ratio thus can be expressed as a function of $K(t)$ as

$$r(t) = \frac{\left(1 - \dfrac{K(t)}{K_2}\right) \cdot \left(1 - \dfrac{K(t)}{K_1}\right)}{K(t) \cdot \left(\dfrac{1}{K_2} - \dfrac{1}{K_1}\right)}. \tag{4.143}$$

The dependency between $r(t)$ and $K(t)$ is shown in Fig. 4.49b for $K_1 < K_2$. The figure clearly shows that K_1 and K_2 are particular values of the speed transmission ratio called nodes, such that $r(t) = 0$, a condition that corresponds to a purely mechanical transmission. Between the nodes, the electric power exhibits a finite maximum. Outside of this range, however, the electric power would have nonphysically high levels [262]. Therefore, variators are designed to operate properly for values of K between nodes.

The power split ratio of simple PSDs is calculated as a special case of (4.143) with $D = 0$. This condition applies also to represent the first operating mode of a compound PSD such as the AHS-II, see (4.130)–(4.133). In both cases,

$$r(t)^{(\text{simple})} = 1 - \frac{K(t)}{K_2}, \tag{4.144}$$

that is to say, simple PSDs exhibit only one node K_2, while $K_1 \to 0$ as $D \to 0$. For $K = 0$ the system works as a series hybrid, with $r = 1$. In Fig. 4.49b the function $r(t) = r(K(t))$ of a simple PSD is compared with those of a compound PSD characterized by the same K_2. In the nominal operating range of K, i.e., between K_1 and K_2, the power split ratio of the simple PSD is always greater than that of the compound PSD. In other terms, the use of a compound

---

[10] Whence the name Infinitely Variable Transmission (IVT) often given to these systems.

[11] This assumption is obviously not realistic and is used here only for illustrating the general operation of PSDs; however, it has to be handled carefully since, for instance, it leads to the conclusion that for $K \to 0$, $T_m \to \infty$.

PSD permits the reduction of the electric power for the same speed span. Consequently, the sizing of the electric machines can be reduced.

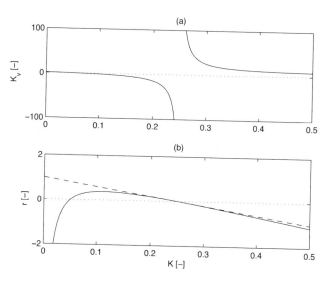

Fig. 4.49. Variator ratio (top) and power split ratio (bottom) as a function of the speed transmission ratio. Dashed curve is representative of a simple PSD with $D = 0$. Solid curves are representative of a compound PSD with $K_1 < K_2$. Numerical values: $K_1 = 0.05$, $K_2 = 0.25$.

The analysis above also explains how dual-mode PSDs allow increasing the range of K without increasing the power split ratio. Some systems, like for instance the e-IVT system, have two operating modes with two nodes each. The two modes share the same node K_2, while the other node K_1 switches from $K_1^{(1)} < K_2$ to $K_1^{(2)} > K_2$. Consequently, the curve $r(t) = r(K(t))$ results from the superimposition of the two curves labeled A and B in Fig. 4.50a, one for each mode, with the switching occurring at K_2. In this way, large values of K are obtained without any excessive increase of r. The operation of a system such as the AHS-II is described as in Fig. 4.50a with $K_1^{(1)} \to 0$, thus with curve A given by (4.144).

The nodes of the PSD also play the role of vehicle speed thresholds at which the directions of energy flows between the linked machines are inverted and the system operation mode is changed [1, 262]. This behavior is observed in Fig. 4.50b, which shows ω_g and ω_m as a function of K for a dual-mode compound PSD. For $K < K_1^{(1)}$, the motor and the generator rotate in the same negative direction and they operate in their reverse power direction, with the "generator" working as a motor and the "motor" as a generator. As the vehicle speed increases, $K_1^{(1)} < K < K_2$, the machines rotate in opposite

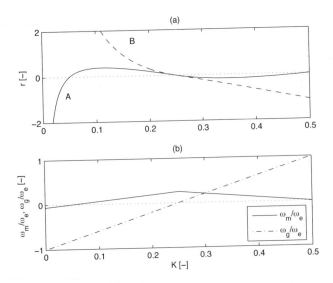

Fig. 4.50. Power split ratio (top), generator and motor speed (bottom) as a function of the speed transmission ratio. In (a) the solid curve is the resulting superimposition of curve A for the first mode and curve B for the second mode, with a switching occurring at K_2. Numerical values: $K_1^{(1)} = 0.05$, $K_2 = 0.25$, $K_1^{(2)} = 0.5$.

directions and in their positive power direction. Thus, the "generator" works as a generator and the "motor" as a motor. For $K_2 < K < K_1^{(2)}$ both machines rotate in the same positive direction; their power directions are inverted again. Finally, for $K > K_1^{(2)}$, the two machines counter-rotate again, while keeping their positive power direction.

5

Non-electric Hybrid Propulsion Systems

The introductory section of this chapter describes several devices that may all be classified as short-term storage systems. The use of short-term storage systems in powertrain applications and the possible powertrain configurations are discussed first.

The next sections analyze the modeling approaches of hybrid-inertial, hybrid-pneumatic, and hybrid-hydraulic powertrains. Specific components such as flywheel accumulators, continuously-variable transmissions, pneumatic and hydraulic accumulators, pumps/motors are all described in accordance with the quasistatic and the dynamic modeling approach.

5.1 Short-Term Storage Systems

Aside from energy carriers encountered in Chap. 3 (fossil fuels) and in Chap. 4 (electrochemical batteries), other methods are suitable for storing energy on-board. Due to their lower specific energy, these systems are referred to as short-term storage systems (3S). Their very limited energy density does not allow the use of short-term storage systems as the sole energy storage devices. Instead, these devices may be used in hybrid vehicles, in combination with a main prime mover, with two main goals. On the one hand, they are aimed at recuperating the energy made available by the vehicle's deceleration and to make it utilizable for subsequent traction phases (regenerative braking). On the other hand, they allow for the implementation of cyclic operations (duty-cycle operation, DCO), in which the main prime mover operates in a high-efficiency full-load point or is turned off, including engine start/stop strategies. In the off phase, the short-term storage system provides the energy for traction, while in the engine-on phase it is recharged. This operation is made convenient by the circumstance that short-term storage systems exhibit a higher specific power than most long-term energy carriers.[1] Of course, the

---

[1] Typically, the energy available for recuperation, $E_v = \frac{1}{2}m_v v^2$ (order of magnitude, 10–100 Wh), is commonly stored at a power of 10–50 kW.

overall benefit obtained with a 3S-based hybridization is partially overcome
by the additional mass installed on-board, which therefore has to be carefully
limited at a reasonable fraction (e.g., 10%) of the vehicle mass, see Chap. 2.

In general, several principles are conceivable for 3S-based power trains, for
instance:

1. electrochemical, generator/motor and battery;
2. electrostatic, generator/motor and supercapacitor;
3. electromagnetic, generator/motor and superconductor coil;
4. inertial, CVT and flywheel;
5. potential, CVT and torsion spring;
6. pneumatic, pneumatic pump/motor and accumulator; and
7. hydraulic, hydraulic pump/motor and accumulator.

Technically, only the solutions (1), (2), (4) and (7) are currently employed,
though also (3) has been proposed [237]. The solution (6) has been analyzed
using simulations [111] and is the topic of ongoing research. The first two
electric solutions are used in hybrid-electric vehicles and have been discussed
already in Chap. 4, while solutions (4), (6), and (7) will be treated in more
detail below. Of course, many other classes of hybrid powertrains have been
proposed in the literature.

Figure 5.1 shows the typical ranges of specific energy and specific power
of the most common short-term storage systems (Ragone plot).

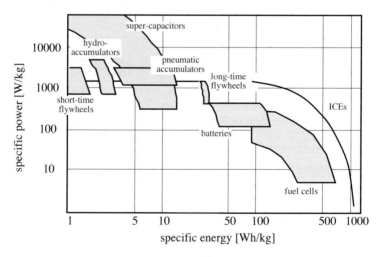

Fig. 5.1. Specific power versus specific energy for various short-term energy storage
systems.

Usually the vehicular concepts with short-term storage systems are series
hybrids, according to the classification of Chap. 4. In engine–supercapacitor

systems the linking energy is electrical, while in engine–accumulator systems it is hydraulic energy or enthalpy. The propulsion system in these cases (see Fig. 5.2) implies three prime movers and two main energy conversion steps. Parallel concepts are also possible and actually there are some applications, especially of hydraulic hybrids, mainly aimed at stop-and-go operation (mild hybrids). In engine–flywheel systems the linking energy is mechanical, so there is no difference in principle between the parallel and the series configuration. However, these concepts also can be conveniently regarded as series hybrids.

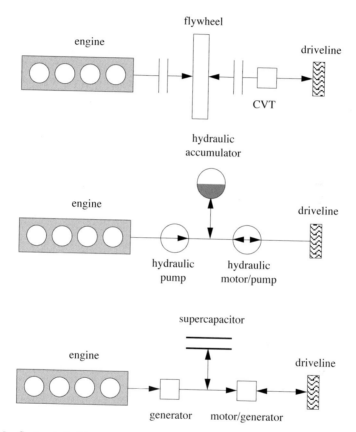

Fig. 5.2. Series hybrid concepts with short-term energy storage systems. For a description of a pneumatic hybrid, see Fig. 5.24

The design of models of hybrid-hydraulic and hybrid-inertial systems is based on the basic ideas introduced in the previous chapters. In particular, the concepts of modularity and of quasistatic versus dynamic models are still valid. Figures 5.3–5.4 show the flow of power factors in quasistatic and dynamic simulations of a hybrid-inertial and a hybrid-hydraulic vehicle. Notice that the power flow of Fig. 5.3 describes also configurations in which the flywheel is mounted on the engine transmission shaft, as shown in Fig. 5.2.

(a) quasistatic approach

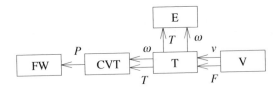

(b) dynamic approach

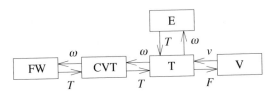

Fig. 5.3. Flow of power factors for a hybrid-inertial propulsion system, with the quasistatic approach (a) and the dynamic approach (b). F: force, P: mechanical power, T: torque, v: speed, ω: rotational speed. Nomenclature of blocks as in Chap. 4, plus CVT: continuously-variable transmission, FW: flywheel.

5.2 Flywheels

Low-speed flywheels have been used in various forms for centuries, and they have a long history of use in automotive applications. Early passenger cars featured a hand crank connected to a flywheel to start the engine, whereas all of today's internal combustion engines use flywheels to store energy and deliver a smooth flow of power despite the reciprocating nature of the combustion torque.

In hybrid-inertial concepts a flywheel is combined with a downsized engine, thus playing the same role as chemical batteries in HEVs [124, 264]. The flywheel rotational speed must vary independently of the vehicle speed, in order to allow for duty-cycle operation. Therefore a continuously variable transmission (CVT) system with a very wide range is necessary between the flywheel shaft and the drive train. The case study in Sect. 8.3 describes the optimization of the duty-cycle parameters. Section 5.3 treats CVT systems in more detail.

(a) quasistatic approach

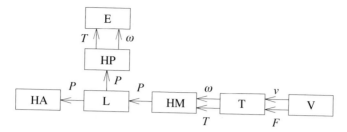

(b) dynamic approach

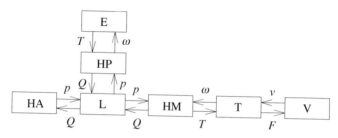

Fig. 5.4. Flow of power factors for a hybrid-hydraulic propulsion system, with the quasistatic approach (a) and the dynamic approach (b). F: force, p: pressure, P: mechanical power, Q: flow rate, T: torque, v: speed, ω: rotational speed. Nomenclature of blocks as in Chap. 4, plus HA: hydraulic accumulator, HM: hydraulic motor, HP: hydraulic pump, L: hydraulic line.

Mixed inertial/electrical hybrid concepts employ a high-speed flywheel[2] to load-level electrochemical batteries [274, 166] or as the only energy storage system [254]. An electric motor/generator is mounted on the rotor shaft both to spin the rotor (charging) and to convert the rotor kinetic energy to electrical energy (discharging), while the traction power is provided by a second electric machine. Regenerative braking is possible through the conversion of mechanical energy to kinetic energy through electric energy. Since typically, the built-in motor/generator is an AC machine, it needs a proper converter to interface with the traction motor when that is a DC machine, as it is usual in battery–flywheel applications.

With respect to electrochemical batteries, the potential of flywheels is comparable (but usually lower) in terms of specific energy and higher (up to 10 times or more) in terms of specific power. The advantages of flywheels are that they contain no acids or other potentially hazardous materials, that they are not affected by extreme temperatures, and that they usually exhibit a longer life. So far, flywheel batteries have only been used in some bus applications [171]. For flywheels to be successful in passenger cars, they would need to

---

[2] Often referred to as "electromechanical batteries" in these applications.

provide a specific energy higher than the levels currently available. In addition, there are some concerns regarding the complexity and the safety of a device that spins mass at very high speeds. The latter aspect strongly limits the flywheel mass that can be installed on board, thus limiting maximum power and energy capacity.

Flywheels store kinetic energy within a rapidly spinning wheel-like rotor or disk. In order to achieve a sufficient amount of specific energy, modern flywheel rotors must be constructed from materials of high specific strength, leading to the selection of composite materials employing graphite fibers rather than metals. This also increases rotor speed, which ranges from typical engine speeds up to very high speeds of about 3000 rev/s. The reduction of the aerodynamic losses associated with such high rotor speeds requires spinning the rotor in a vacuum chamber. This in turn leads to additional design requirements for the rotor bearing. These bearings must have low losses and must be stiff to adequately constrain the rotor and stabilize the shaft. These requirements frequently lead to choosing magnetic bearings with losses on the order of 2% per hour. In order to reduce the gyroscopic forces transmitted to the magnetic bearings during pitching and rolling motions of the vehicle, a gimbal mount is often adopted [108]. The alternative of using counter-rotating rotors, which do not transmit any gyroscopic forces to the outside, is burdened by the fact that internally they transmit very large forces that are not easily supported by the magnetic bearings.

The specific energy of a flywheel is calculated as the ratio of the energy stored E_f to the flywheel mass m_f. The kinetic energy stored is $E_f = \frac{1}{2} \cdot \Theta_f \cdot \omega^2$, with Θ_f being the moment of inertia of the flywheel and ω its rotational speed. Using the notation of Fig. 5.5 the moment of inertia is calculated as [74]

$$\Theta_f = \rho \cdot b \cdot \int r^2 \cdot 2 \cdot \pi \cdot r \, dr = 2 \cdot \pi \cdot \rho \cdot b \cdot \left.\frac{r^4}{4}\right|_{q \cdot d/2}^{d/2} = \frac{\pi}{2} \cdot \rho \cdot b \cdot \frac{d^4}{16} \cdot (1 - q^4) \,, \quad (5.1)$$

where ρ is the material's density and q the ratio between the inner and the outer flywheel ring. The flywheel mass is given by

$$m_f = \pi \cdot \rho \cdot b \cdot \frac{d^2}{4} \cdot (1 - q^2) \,. \quad (5.2)$$

Consequently, the energy-to-mass ratio is evaluated as

$$\frac{E_f}{m_f} = \frac{d^2}{16} \cdot (1 + q^2) \cdot \omega^2 = \frac{u^2}{4} \cdot (1 + q^2) \,, \quad (5.3)$$

where $u = d \cdot \omega / 2$ is the flywheel speed at the outer radius. Equation (5.3) may be written in a more compact way, as

$$\frac{E_f}{m_f} = k_f \cdot u^2 \,, \quad (5.4)$$

where the coefficient k_f typically ranges from 0.5 to 5, according to the type of construction.

The specific energy is limited by several factors, such as the maximum allowable stress and the maximum allowable rotor speed.[3] Small units that are technically feasible today reach 30 Wh/kg, including housing, electronics, etc. Much higher values of up to 140 Wh/kg are predicted by some authors for advanced rotor materials [257].

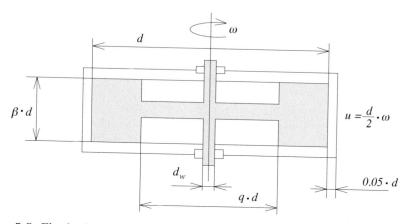

Fig. 5.5. Flywheel accumulator for duty-cycle operation and regenerative braking.

5.2.1 Quasistatic Modeling of Flywheel Accumulators

The causality representation of a flywheel accumulator in quasistatic simulations is sketched in Fig. 5.6. The input variable is the power $P_2(t)$ required at the output shaft. A positive value of $P_2(t)$ discharges the flywheel, a negative value of $P_2(t)$ charges it. The output variable is the flywheel speed $w_2(t)$, sometimes regarded as the "state of charge" of the flywheel.

$$\xrightarrow{\;P_2\;}\boxed{\text{FW}}\xrightarrow{\;w_2\;}$$

Fig. 5.6. Flywheel accumulators: causality representation for quasistatic modeling.

Models of flywheel accumulators may be derived on the basis of Newton's second law for a rotational system. The resulting equation is

$$\Theta_f \cdot w_2(t) \cdot \frac{d}{dt} w_2(t) = -P_2(t) - P_l(t) \,, \tag{5.5}$$

---

[3] This is usually limited by the first critical speed of the flywheel [257].

from which the flywheel speed can be calculated. The term P_l describes the power losses.

For flywheel accumulators, two main loss contributions are usually considered, namely air resistance and bearing losses, $P_l(t) = P_{l,a}(t) + P_{l,b}(t)$. Both terms are functions of $\omega_2(t)$, i.e., of the peripheral velocity $u(t)$. The general expression for the air resistance force is proportional to the air density ρ_a, to $u^2(t)$, and to d^2, through a coefficient that is a function of the Reynolds number and of the geometric ratio $\beta = b/d$. For Reynolds numbers above $3 \cdot 10^5$, an expression for the power losses due to air resistance [74] is

$$P_{l,a}(t) = 0.04 \cdot \rho_a^{0.8} \cdot \eta_a^{0.2} \cdot u^{2.8}(t) \cdot d^{1.8} \cdot (\beta + 0.33) , \tag{5.6}$$

where η_a is the dynamic viscosity of air.

For the bearing losses, a general expression frequently used [74] is

$$P_{l,b}(t) = \mu \cdot k \cdot \frac{d_w}{d} \cdot m_f \cdot g \cdot u(t) , \tag{5.7}$$

where the physical quantities involved are a friction coefficient μ, a corrective force factor k that models unbalance and gyroscopic forces, etc., and the ratio of the shaft diameter d_w to the wheel diameter d.

A first estimation of the power losses may be obtained using the values for the physical parameters listed in Table 5.1. Figure 5.7 shows the variation of the bearing losses and the air resistance losses as a function of the flywheel rotational speed for an optimized flywheel construction [74]. The figure clearly shows the dependency of the power losses on the speed as given by (5.6). With the same data it is possible to obtain a value for the time range of the flywheel, i.e., the time that the flywheel speed remains above a certain threshold without any external torque. Typical values are about 10 min.

The main design task in designing flywheels consists of assigning values to the flywheel dimensions b, d in order to obtain the desired kinetic energy at a given rotational speed (thus, a given moment of inertia), while minimizing weight and power losses. A typical value for the stored energy may be estimated considering the kinetic energy of a vehicle with a mass of 910 kg and a speed of 80 km/h. The energy to be recuperated in the flywheel during a braking until stop is 247 kJ, having assumed a first tentative value of the flywheel mass equal to 10% of the vehicle mass. The moment of inertia is related to dimensions and mass by (5.1)–(5.2). Figure 5.8 shows that the dimension b of the flywheel increases as m_f^2, while the diameter d decreases as $1/\sqrt{m_f}$. Consequently, the power losses and the peripheral speed u decrease as well (the rotational speed is 100 rev/s). A possible compromise between flywheel weight and power losses can be obtained with a flywheel mass of about 50 kg.

5.2.2 Dynamic Modeling of Flywheel Accumulators

The physical causality representation of a flywheel accumulator is sketched in Fig. 5.9. The model input variable is the rotational speed at the output

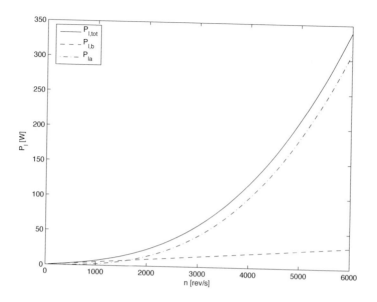

Fig. 5.7. Power losses as a function of the rotational speed. Flywheel data: $d = 0.36$ m, $b = 0.108$ m, $\beta = 0.3$, $q = 0.6$, $m_f = 56.33$ kg.

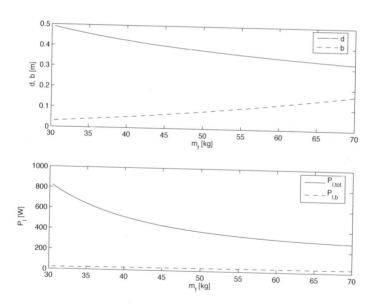

Fig. 5.8. Dimensions and losses of a flywheel as a function of its mass, for a given energy $E_f = 247$ kJ, speed $n = 100$ rev/s, and thus a moment of inertia $J_f = 1.24$ kg m^2.

Table 5.1. Numerical values for the flywheel parameters.

μ	$1.5 \cdot 10^{-3}$
k	4
d_w/d	0.08
ρ_a	$1.3\,\mathrm{kg/m^3}$
η_a	$1.72 \cdot 10^{-5}\,\mathrm{Pa\,s}$
ρ	$8000\,\mathrm{kg/m^3}$

or downstream shaft, $\omega_2(t)$. The model output variable is the torque at the output shaft, $T_2(t)$.

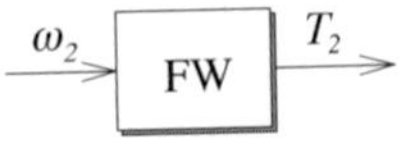

Fig. 5.9. Flywheel accumulators: physical causality for dynamic modeling.

The equations developed in the previous section are suitable also for dynamic modeling. The only dynamic term is related to the derivative of the flywheel speed. It can be easily estimated since $\omega_2(t)$ is an input variable.

5.3 Continuously Variable Transmissions

The use of continuously variable transmissions (CVT) is a promising technology to increase the energy efficiency of conventional and hybrid vehicles [81]. Due to higher transmission losses in the hydraulic part of the CVT system[4] as well as to slip in the CVT power transmitter device, the efficiency of CVTs is inherently lower than that of conventional fixed-ratio gear boxes. Nevertheless, since CVTs allow for the optimization of the power flows, the overall energy consumption of the vehicle can be smaller.

In conventional vehicles, the use of a CVT permits the selection of that combination of engine speed and torque which delivers the output power required with the best efficiency possible. In contrast, multi-step transmissions allow only for a limited number of combinations, usually five or six according to the number of gear ratios (see Chap. 3). A CVT therefore increases the efficiency of the engine and thus the efficiency of the overall propulsion system. An additional advantage of CVTs is the fact that the gear shifting mechanism exhibits a smooth and comfortable behavior that cannot be achieved – or only with additional clutches – by conventional transmissions.

---

[4] For this reason CVTs actuated by electromechanical devices [260] are being studied, with the goal of reducing the power losses by 25%.

Continuously variable transmissions are key components in hybrid-inertial vehicles, where the rotational speed of the flywheel accumulator must be decoupled from that of the drive train. In hybrid-electric vehicles, though not strictly necessary, the use of CVTs is gaining a notable amount of attention as an additional control device [232, 55].

The core of a modern CVT (see Fig. 5.10) consists of a transmitter element and two V-shaped pulleys. The element that transmits power between the pulleys can be a metal belt [81] or a chain [190]. The pulley linked to the prime mover is referred to as primary pulley, while the one connected to the drive train is the secondary pulley. Each pulley consists of a fixed and an axially slidable sheave. Each of the two moving pulley halves is connected to a hydraulic actuation system, consisting of a hydraulic cylinder and piston. In the simplest hydraulic configuration the secondary pressure is the hydraulic supply pressure. The primary pressure is determined by one or more hydraulic valves, actuated by a solenoidal (electromagnetic) valve, which connect the two circuits with the pump return line [264].

In the i²-CVT used in the ETH-III propulsion system [65], the range of the CVT alone (typically 1:5) is substantially increased by combining the CVT core (a chain converter) with a two-ratio gear box. In the "slow" gear arrangement, the overall transmission ratio is given by the product of the CVT ratio and the gear ratio. In the "fast" gear arrangement, the power flow through the CVT is inverted, thus the overall transmission ratio is proportional to the reciprocal of the CVT ratio. This allows for reaching transmission ratios higher than 1:20.

5.3.1 Quasistatic Modeling of CVTs

The causality representation of CVTs in quasistatic simulations is sketched in Fig. 5.11. The input variables are the torque $T_2(t)$ and the speed $\omega_2(t)$ at the downstream shaft. The output variables are the torque $T_1(t)$ and the speed $\omega_1(t)$ required at the input shaft.

A simple quasistatic model of a CVT can be derived neglecting slip at the transmitter but considering torque losses [101, 232, 155]. The definition of the transmission (speed) ratio ν implies that

$$\omega_1(t) = \nu(t) \cdot \omega_2(t) .$$
(5.8)

Newton's second law applied to the two pulleys yields

$$\Theta_1 \cdot \frac{d}{dt}\omega_1(t) = T_1(t) - T_{t1}(t) ,$$
(5.9)

$$\Theta_2 \cdot \frac{d}{dt}\omega_2(t) = T_{t2}(t) - T_2(t) ,$$
(5.10)

where $T_{t1}(t)$ and $T_{t2}(t)$ represent the torque transmitted by the chain or the metal belt to the pulleys. The torque losses may be taken into consideration by applying them to the primary pulley, as

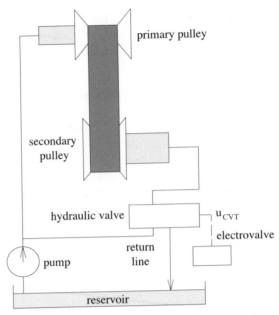

Fig. 5.10. Scheme of a CVT system.

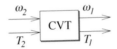

Fig. 5.11. CVTs: causality representation for quasistatic modeling.

$$T_{t2}(t) = \nu(t) \cdot (T_{t1}(t) - T_l(t)) \ . \tag{5.11}$$

Combining (5.8)-(5.11) and recalling that ν is a variable quantity, the final expression for the input torque obtained is

$$T_1(t) = T_l(t) + \frac{T_2(t)}{\nu(t)} + \frac{\Theta_{CVT}}{\nu(t)} \cdot \frac{d}{dt}\omega_2(t) + \Theta_1 \cdot \frac{d}{dt}\nu(t) \cdot \omega_2(t) \ , \tag{5.12}$$

where the secondary reduced inertia is defined as $\Theta_{CVT} = \Theta_2 + \Theta_1 \cdot \nu^2(t)$.

The variation of the gear ratio $\nu(t)$ is determined by the CVT controller, $\dot{\nu}(t) = u_{CVT}(t)$, thus it is an input variable for the model.

The evaluation of the torque losses requires the evaluation of the losses in the hydraulic part of the CVT, which are dominated by the pump losses, and the friction losses at the mechanical contacts between various CVT components. Analytical expressions for $T_l(t)$ are derived by fitting experimental data, but due to the complexity of the processes involved, at least a second-order dependency of $T_l(t)$ on $\nu(t)$, $\omega_1(t)$ and on the input power $P_1(t)$ is required [265, 65].

An alternative approach is based on the definition of the CVT efficiency $\eta_{CVT}(\omega_2, T_2, \nu)$, according to which the torque required at the input shaft is evaluated as

$$T_1(t) = \frac{T_2(t)}{\nu(t) \cdot \eta_{CVT}(\omega_2(t), T_2(t), \nu(t))}, \quad T_2(t) > 0, \tag{5.13}$$

$$T_1(t) = \frac{T_2(t)}{\nu(t)} \cdot \eta_{CVT}(\omega_2(t), T_2(t), \nu(t)), \quad T_2(t) < 0. \tag{5.14}$$

The typical dependency of η_{CVT} on output speed and torque as well as on the transmission ratio is depicted in Fig. 5.12. The efficiency of the CVT increases with torque at constant speed and ν, exhibiting a maximum that is more pronounced at higher speeds. Higher transmission ratios also favorably affect the efficiency. Values around 90% can be reached for high-load, low-speed conditions. Values lower than 70% are typical during low-load operation [265, 190].

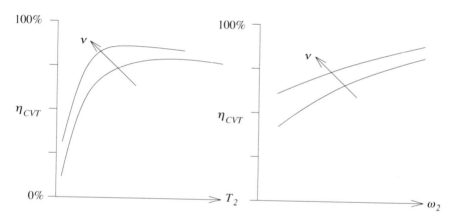

Fig. 5.12. Qualitative dependency of CVT overall efficiency on torque, speed, and transmission ratio.

As an example, Fig. 5.13 shows the result of a detailed model of a push-belt CVT [189] of the type discussed in the next section.

The efficiency of CVTs can be approximated using Eqs. (3.13) and (3.14) as well. However, in this case the factor e_{gb} is smaller than in conventional gear boxes and depends on the speed and gear ratio of the CVT. At a gear ratio $\nu = 1$, typical values are $e_{gb} = 0.96$ and $P_{0,gb} = 0.02 \cdot P_{max}$ at the rated speed and $e_{gb} = 0.94$ and $P_{0,gb} = 0.04 \cdot P_{max}$ at 50% of the rated speed. At the maximum or minimum gear ratio, the idling losses increase, i.e., $P_{0,gb} = 0.04 \cdot P_{max}$ at the rated speed and $P_{0,gb} = 0.06 \cdot P_{max}$ at 50% of the rated speed.

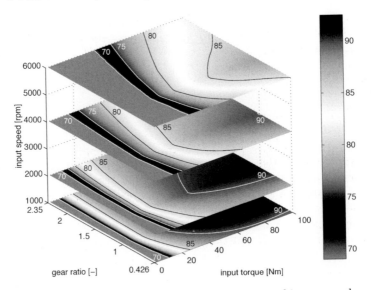

Fig. 5.13. Efficiency of a push-belt CVT as a function of input speed ω_e, input torque T_1, and gear ratio ν [189].

5.3.2 Dynamic Modeling of CVTs

The physical causality representation of a CVT is sketched in Fig. 5.14. The model input variables are the rotational speed at the output or downstream shaft, $\omega_2(t)$, and the torque at the input or upstream shaft, $T_1(t)$. The model output variables are the torque at the output shaft, $T_2(t)$, and the speed at the input shaft, $\omega_1(t)$.

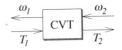

Fig. 5.14. CVTs: physical causality for dynamic modeling.

Dynamic CVT models calculate the rate of change of $\nu(t)$, and consequently the output variables, as a result of a fundamental hydraulic and mechanical modeling of the system. The hydraulic submodel calculates the forces acting on the primary and secondary pulley halves, $F_1(t)$ and $F_2(t)$, which basically depend on pressure in the corresponding circuits and pulley speed (via a centrifugal term). The pressure is modeled according to continuity equations in the various branches of the hydraulic circuit, whose geometrical characteristics depend on the status of the hydraulic valves and ultimately on the electric control signal $u_{CVT}(t)$ [265].

The mechanical submodel calculates the speed transmission ratio and its time derivative as a function of $F_1(t)$ and $F_2(t)$, usually neglecting slip but considering the variable pulley inertia and, in some cases, the lumped belt mass. A model of the latter type [233] results in the following equation for the rate of change of the transmission ratio [265],

$$\frac{d}{dt}\nu(t) = k(\nu(t)) \cdot (F_1(t) - \kappa(\nu(t), T_1(t), F_2(t)) \cdot F_2(t)) \,, \tag{5.15}$$

where $F_1(t)$ and $F_2(t)$ are the forces acting on the primary and the secondary pulley, respectively, while k and the so-called pulley thrust ratio κ are functions that have to be determined experimentally. Other models, mainly proposed for metal-belt CVTs, although based on a different derivation, show a final dependency similar to that of (5.15) [117, 96].

Once $\dot{\nu}(t)$ is calculated or assumed, the output model variables are evaluated using the equations derived in the previous section. In particular (5.8) can be used to evaluate $\omega_1(t)$ and (5.12) to evaluate $T_2(t)$, since both $\omega_2(t)$ and its derivative are known as input variables.

5.4 Hydraulic Accumulators

Besides supercapacitors, flywheels, and pneumatic systems, a fourth type of short-term storage system is represented by hydraulic accumulators. Like the other concepts mentioned, hydraulic accumulators are characterized by a higher power density and a lower energy density than electrochemical batteries. Due to the high power flow that can be recuperated during deceleration and then made available, the application of hydraulic hybrid concepts has been of interest so far for use in heavy vehicles, i.e., sport utility vehicles, urban delivery vehicles, trucks, etc. As the energy storage system, the hydraulic accumulator has the ability to accept both high frequencies and high rates of charging/discharging, both of which are not possible for electrochemical batteries. However, the relatively low energy density of the hydraulic accumulator requires a carefully designed control strategy if the fuel economy potential is to be realized to its fullest.

In this context, both series and parallel hybrid architectures have been proposed. The parallel architecture (EPA's Hydraulic Launch Assist [132], Ford's Hydraulic Power Assist [82]) includes an engine (often a Diesel engine) and a hydraulic reversible machine (pump/motor). Besides regenerative braking, typically the hydraulic motor is used alone at low loads (stop-and-go) and for the first few seconds during acceleration. The hydraulic motor also provides additional torque during hill climbing and heavy acceleration, thus covering many of the possibilities already illustrated for hybrid-electric vehicles. Expected benefits are in terms of fuel consumption in stop-and-go operation (20-30%), exhaust emissions, and acceleration time. The series hybrid configuration (EPA [132]) requires two hydraulic pumps/motors. It is often referred to as "full hybrid," since larger degrees of hybridization are used, with larger improvements (70% and more) in reducing fuel consumption than with parallel "mild" concepts.

All types of hybrid-hydraulic propulsion systems include a high-pressure accumulator and a low-pressure reservoir. The accumulator contains the hydraulic fluid and a gas such as nitrogen (N_2) or methane (CH_4), separated by a membrane (see Fig. 5.15). When the hydraulic fluid flows in, the gas is compressed. During the discharge phase, the fluid flows out through the motor and then into the reservoir. The state of charge is defined as the ratio of instantaneous gas volume in the accumulator, V_g, to its maximum capacity. Typical data of an accumulator designed for a truck are listed in Table 5.2. The reservoir can be regarded as an accumulator working at a much lower pressure, e.g., 8.5–12.5 bar.

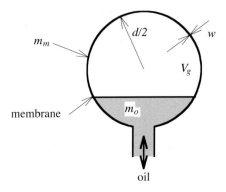

Fig. 5.15. Schematic of a hydraulic accumulator.

Table 5.2. Typical data of hydraulic accumulators.

fluid capacity	50 l
maximum gas volume	100 l
minimum gas volume	50 l
pre-charge pressure (at 320 K)	125 bar
maximum pressure	360 bar

5.4.1 Quasistatic Modeling of Hydraulic Accumulators

The causality diagram of a simple hydraulic accumulator used for quasistatic simulations is sketched in Fig. 5.16. The input variable is the mechanical power $P_2(t) = p_2(t) \cdot Q_2(t)$ required at the shaft. Here $p_2(t)$ is the pressure in and $Q_2(t)$ is the volumetric flow from the accumulator. The output variable is the state of charge of the accumulator, i.e., the volume occupied by the gas, $V_g(t)$.

Fig. 5.16. Hydraulic accumulators: causality representation for quasistatic modeling.

Thermodynamic Model

A basic physical model of a hydraulic accumulator can be derived from the mass and energy conservation laws and the ideal gas state equation for the charge gas [197, 79]. The resulting set of nonlinear equations includes an energy balance

$$m_g \cdot c_v \cdot \frac{d}{dt}\vartheta_g(t) = -p_g(t) \cdot \frac{d}{dt}V_g(t) - h \cdot A_w \cdot (\vartheta_g(t) - \vartheta_w) , \qquad (5.16)$$

a mass balance

$$\frac{d}{dt}V_g(t) = Q_2(t) , \qquad (5.17)$$

and the ideal gas law

$$p_g(t) = \frac{m_g \cdot R_g \cdot \vartheta_g(t)}{V_g(t)} , \qquad (5.18)$$

where A_w is the effective accumulator wall area for heat convection, h is the effective heat transfer coefficient, m_g is the gas mass, c_v is the constant-volume specific heat of the gas, $\vartheta_g(t)$ is the gas temperature, $p_g(t)$ the gas pressure, $V_g(t)$ the gas volume, ϑ_w the wall temperature, and R_g is the gas constant. More detailed models [197] may describe the charge gas with a nonlinear state equation (Benedict–Webb–Rubin equation of state) and may take into account the heat exchange between the charge gas and the elastomeric foam that is usually inserted on the gas side of the accumulator to reduce the thermal loss to the accumulator walls.

In a first approximation the gas pressure $p_g(t)$ equals the fluid pressure at the accumulator inlet, $p_2(t)$. Frictional losses caused by flow entrance effects, viscous shear, and piston-seal friction or bladder hysteresis can hardly be accounted for in an analytical fashion. Usually the pressure loss term is evaluated as a fraction of $p_2(t)$ (e.g., 2%) [197].

A relationship between fluid flow rate, charge volume, and power can be derived by solving (5.16)–(5.18) at steady state. Assuming that $p_g(t) = p_2(t)$, i.e., no pressure losses at the accumulator inlet, the resulting equation for fluid pressure is written as

$$p_2(t) = \frac{h \cdot A_w \cdot \vartheta_w \cdot m_g \cdot R_g}{V_g(t) \cdot h \cdot A_w + m_g \cdot R_g \cdot Q_2(t)} \ . \tag{5.19}$$

Combining (5.19) with the definition of output power, the fluid flow rate is obtained,

$$Q_2(t) = \frac{V_g(t)}{m_g} \cdot \frac{h \cdot A_w \cdot P_2(t)}{R_g \cdot \vartheta_w \cdot h \cdot A_w - R_g \cdot P_2(t)} \ . \tag{5.20}$$

The state of charge is thus given by integrating (5.17). With only one state variable, this quasi-stationary model has a complexity that is equivalent to that of most battery models. Thus it can easily be embedded, for instance, in a dynamic programming algorithm [278] and also generally in quasistatic algorithms.

Accumulator Efficiency

The efficiency of a hydraulic accumulator may be defined as the ratio of the total energy delivered during a complete discharge to the energy that is necessary to charge up the device. This definition is conceptually identical to the "global efficiency" introduced in Chap. 4 for electrochemical batteries. The energy spent to charge the accumulator depends on the type of process assumed for the accumulator charge/discharge cycle. Similarly to electrochemical batteries, constant flow rate or constant power processes may be assumed. For hydraulic accumulators, however, the usual definition of efficiency is based on a reference cycle consisting of isentropic compression and expansion from the maximum to the minimum volume and vice versa.

The thermodynamic transformations followed by the gas in the reference cycle are represented in the temperature/entropy (ϑ–s) diagram of Fig. 5.17. The transformation AB is an isentropic compression from an initial state A, which is characterized by a maximum volume and a temperature that equals that of the surrounding ambient ($\vartheta_A = \vartheta_w$). The transformation BC is an isochoric cooling of the gas, which loses thermal energy[5] to the ambient until $\vartheta_C = \vartheta_A$. The transformation CD is an isentropic expansion that ends when $p_D = p_A$. The transformation DE is a further expansion that ends when $V_E = V_A$.

By definition, the isentropic transformation yields the following equations

$$\frac{p_B}{p_A} = \left(\frac{V_A}{V_B}\right)^{\gamma} = r^{-\gamma}, \quad \frac{\vartheta_B}{\vartheta_A} = \left(\frac{p_B}{p_A}\right)^{\frac{\gamma-1}{\gamma}} = r^{1-\gamma} , \tag{5.21}$$

---

[5] This corresponds to a worst-case scenario in which the device is left to cool for a long time.

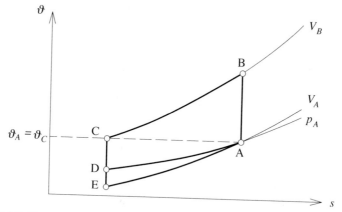

Fig. 5.17. Temperature–entropy (ϑ–s) diagram of the accumulation process.

where r is the expansion ratio and γ is the ratio of the specific heats. The compression work is calculated along the path AB. For a closed, adiabatic system, the work exchanged equals the variations of the internal energy. Thus the work W_{AB} is calculated as

$$W_{AB} = m_g \cdot c_{v,g} \cdot \vartheta_A \cdot \left(\frac{\vartheta_B}{\vartheta_A} - 1 \right) = m_g \cdot c_{v,g} \cdot \vartheta_A \cdot \left(r^{1-\gamma} - 1 \right) . \quad (5.22)$$

The transformation BC is isochoric, i.e., the volume V_C is equal to V_B. Moreover, since $\vartheta_C = \vartheta_A$, the pressure can be calculated as

$$p_C = \frac{m_g \cdot R_g \cdot \vartheta_C}{V_C} = \frac{m_g \cdot R_g \cdot \vartheta_A}{V_B} = \frac{p_A \cdot V_A}{V_B} = \frac{p_A}{r} . \quad (5.23)$$

For the reference cycle ABCDA, the expansion ends at the state D, which is characterized by a pressure $p_D = p_A$, and a temperature ϑ_D evaluated as

$$\frac{\vartheta_D}{\vartheta_C} = \left(\frac{p_D}{p_C} \right)^{\frac{\gamma-1}{\gamma}} = \left(\frac{p_A}{p_A} r \right)^{\frac{\gamma-1}{\gamma}} = r^{\frac{\gamma-1}{\gamma}} . \quad (5.24)$$

The expansion work is thus calculated as

$$W_{CD} = m_g \cdot c_{v,g} \cdot \vartheta_C \cdot \left(\frac{\vartheta_D}{\vartheta_C} - 1 \right) = m_g \cdot c_{v,g} \cdot \vartheta_A \cdot \left(r^{\frac{\gamma-1}{\gamma}} - 1 \right) . \quad (5.25)$$

For the reference cycle ABCEA, the expansion ends at the state E, which is characterized by a volume $V_E = V_A$, and a temperature evaluated as

$$\frac{\vartheta_E}{\vartheta_C} = \left(\frac{V_E}{V_C} \right)^{1-\gamma} = r^{\gamma-1} . \quad (5.26)$$

The expansion work is thus

$$W_{CE} = m_g \cdot c_{v,g} \cdot \vartheta_C \cdot \left(\frac{\vartheta_E}{\vartheta_C} - 1 \right) = m_g \cdot c_{v,g} \cdot \vartheta_A \cdot \left(r^{\gamma-1} - 1 \right) . \qquad (5.27)$$

The accumulator efficiency is the ratio between the discharge energy and the charge energy. For the isochoric reference cycle ABCEA, the efficiency is evaluated as

$$\eta_{ha,V} = \frac{-W_{CE}}{W_{AB}} = \frac{1 - r^{\gamma-1}}{r^{1-\gamma} - 1} . \qquad (5.28)$$

For the "isobaric" reference cycle ABCDA, the efficiency is evaluated as

$$\eta_{ha,P} = \frac{-W_{CD}}{W_{AB}} = \frac{1 - r^{\frac{\gamma-1}{\gamma}}}{r^{1-\gamma} - 1} . \qquad (5.29)$$

The variations of W_{AB}, W_{CD}, W_{DE} and η_{ha} with the expansion ratio r are shown in Fig. 5.18. The plots clearly show that the efficiency of the isobaric cycle is always lower than the efficiency of the isochoric cycle which reaches the 100% value for $r = 1$. In both cases, the efficiency is a monotonically increasing function of r.

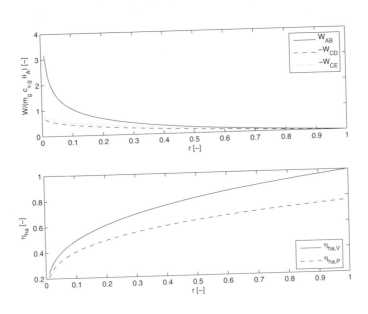

Fig. 5.18. Compression work W_{AB}, expansion works W_{CD}, W_{CE} (top), and efficiency η_{ha} (bottom) as a function of the expansion ratio r.

Specific Energy of Accumulators

The results of the analysis developed in the previous section allow for the evaluation of the specific energy of a hydraulic accumulator. The compression work given by (5.22) must be divided by the accumulator mass. This is given by the sum of three terms, *viz.* the mass of the charge gas, the mass of the hydraulic fluid, and the mass of the housing. The latter term may be evaluated as the mass of a spherical shell designed to bear a pressure p_B, the maximum gas pressure. The thickness of the housing (see Fig. 5.15) may be evaluated using the formula:

$$w = \frac{p_B \cdot d}{4 \cdot \sigma} , \tag{5.30}$$

where d is the diameter of the accumulator shell and σ is the maximum tensile stress. The housing mass is thus

$$m_m = \rho_m \cdot \pi \cdot d^2 \cdot w = \frac{3}{2} \cdot V_A \cdot p_B \cdot \frac{\rho_m}{\sigma} , \tag{5.31}$$

since V_A, the maximum volume occupied by the gas, by definition is the accumulator volume. The fluid mass to be considered in the calculation is the mass that occupies the maximum volume left by the gas, i.e., the volume $V_A - V_B$. The mass of the charge gas is the constant m_g already introduced in the previous section. The orders of magnitude of such masses are rather different. Evaluating the gas mass as $m_g = V_A \cdot p_A/(R_g \cdot \vartheta_A)$, the ratios of fluid to gas mass and of housing to gas mass are

$$\frac{m_o}{m_g} = \rho_o \cdot \frac{R_g \cdot \vartheta_A}{p_A} \cdot (1 - r), \qquad \frac{m_m}{m_g} = \frac{3}{2} \cdot R_g \cdot \vartheta_A \cdot r^{-\gamma} \cdot \frac{\rho_m}{\sigma} . \tag{5.32}$$

Using typical data for the materials, such as the ones listed in Table 5.3, clearly shows that m_g can be neglected when compared with m_o and m_m. For higher pressures, even m_o can be neglected when compared with m_m.

Table 5.3. Typical parameter values for hydraulic accumulator models (methane as compressed gas).

ρ_m/σ	10^{-5} kg/J
ρ_o	900 kg/m^3
ϑ_A	300 K
R_g	520 J/(kg K)
p_B	400–800 bar

In the latter case, the expression for the specific energy is written as

$$\frac{E_{ha}}{m_{ha}} \approx \frac{W_{AB}}{m_m} = \alpha \cdot r^\gamma \cdot \left(r^{1-\gamma} - 1 \right) , \tag{5.33}$$

with $\alpha = 2 \cdot c_{v,g} \cdot \sigma/(3 \cdot \rho_m \cdot R_g)$. The variation of the specific energy as a function of the compression ratio r is shown in Fig. 5.19.

The specific energy of (5.33) can be maximized with respect to r by setting the relevant derivative to zero. This condition yields the optimal compression ratio

$$r_{opt} = \left(\frac{1}{\gamma}\right)^{\frac{1}{\gamma-1}} . \qquad (5.34)$$

With $\gamma = 1.31$ (methane), the optimal compression ratio is $r_{opt} = 0.42$, and the corresponding efficiency is $\eta_{ha,P} = 0.60$.

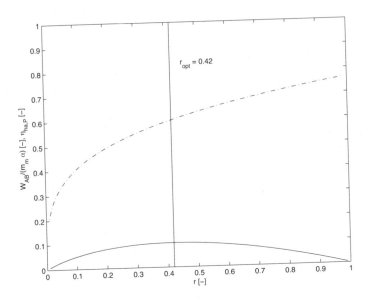

Fig. 5.19. Specific energy of a hydraulic accumulator as a function of the compression ratio with methane as a hydraulic fluid.

5.4.2 Dynamic Modeling of Hydraulic Accumulators

The physical causality representation of a hydraulic accumulator is sketched in Fig. 5.20. The model input variable is the flow rate of the hydraulic fluid $Q_2(t)$. A positive $Q_2(t)$ discharges the accumulator, a negative $Q_2(t)$ charges it. The model output variables are the hydraulic fluid pressure $p_2(t)$ and the state of charge $V_g(t)$.

The state of charge is calculated by directly integrating (5.17), while in a first approximation the fluid pressure equals the gas pressure $p_g(t)$, which may be integrated using (5.16) and (5.18).

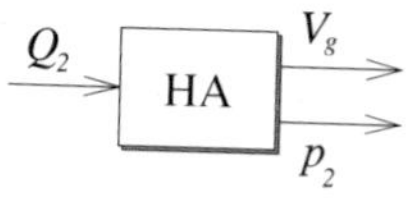

Fig. 5.20. Hydraulic accumulators: physical causality for dynamic modeling.

5.5 Hydraulic Pumps/Motors

In hybrid-hydraulic propulsion systems, hydraulic motors convert pressure energy of a fluid into mechanical energy available at the motor shaft. Conversely, the machine can act as a pump, converting mechanical energy back into hydraulic energy.

Hydraulic motors are rotary or reciprocating volumetric machines with a fixed or a variable displacement volume. Different applications may require different types of machine, according to the maximum pressure required (100–600 bar), the displacement volume, and the rotational speed. Rotary machines handle the fluid in chambers whose volume cyclically changes for the design of the walls (vane machines) or the action of teeth (gear machines). However, most types have a fixed displacement volume. Reciprocating machines are characterized by the cyclic variation of the chamber volume for the action of a piston. Variable displacement operation is usually possible by changing the geometry of the driving mechanism. Reciprocating machines are further classified into radial piston and axial piston machines.

In the former type, the cylinder axes are perpendicular to the shaft. In pump operation, as the cylinder block rotates, the pistons are pressed against the rotor and are forced in and out of the cylinders, thereby receiving fluid and pushing it out into the system. The motor operation is the reverse.

In axial piston machines, the cylinder axes are parallel to the shaft (see Fig. 5.21). In the swash-plate type, the reciprocating motion is created by a plate mounted on the shaft at a fixed or variable angle. One end of each piston rod is held in contact with the plate as the cylinder block and piston assembly rotates with the drive shaft. This causes the pistons to reciprocate within the cylinders. In bent-axis units, it is the pistons that are bent with respect to the shaft and plate axes.

In automotive applications, the faster axial piston machines of the bent-axis or swash-plate types, with typical speeds of 300–3000 rpm, are preferred to slower radial piston machines, which are used instead for very high pressure applications. The displacement per revolution of the machine is adjusted to control the power delivered or absorbed, thus letting the machine operate as a pump or as a motor. Both in bent-axis units and in swash-plate units, the adjusted quantity is the swivel angle, i.e., the angle between the pistons' axes and the connecting plate axis. Typical values of this angle are 20–25°.

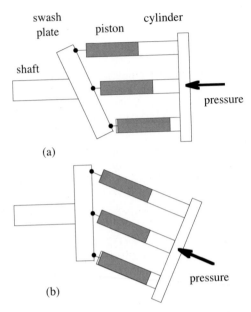

Fig. 5.21. Schematics of piston pumps/motors: swash-plate (a), bent-axis (b).

5.5.1 Quasistatic Modeling of Hydraulic Pumps/Motors

The causality representation of a hydraulic pump/motor in quasistatic simulations is sketched in Fig. 5.22. The input variables are the torque $T_2(t)$ and speed $\omega_2(t)$ required at the shaft. The ouput variable is the hydraulic power $P_1(t) = p_1(t) \cdot Q_1(t)$.

Fig. 5.22. Hydraulic pumps/motors: causality representation for quasistatic modeling.

The relationship between $P_2(t) = T_2(t) \cdot \omega_2(t)$ and $P_1(t)$ may be described by defining a pump/motor efficiency $\eta_{hm}(\omega_2, T_2)$ such that

$$P_1(t) = \frac{P_2(t)}{\eta_{hm}(\omega_2(t), T_2(t))}, \quad P_2(t) > 0 , \tag{5.35}$$

$$P_1(t) = P_2(t) \cdot \eta_{hm}(\omega_2(t), -|T_2|(t)), \quad P_2(t) < 0 . \tag{5.36}$$

Of course, as for other energy converters, an alternative quasi-stationary description for hydraulic pumps/motors is the Willans approach

$$P_1(t) = \frac{P_2(t) + P_0}{e}, \tag{5.37}$$

where e is the efficiency of the energy conversion process and P_0 represents the friction losses of the pump/motor.

A basic physical model of variable-displacement piston pumps/motors can be derived by following the classic Wilson's approach [197]. The volumetric flow rate of the hydraulic fluid through a pump is given by

$$Q_1(t) = -\eta_v(t) \cdot x(t) \cdot D \cdot w_2(t), \quad T_2(t) < 0, \tag{5.38}$$

where D is the maximum displacement and x is the fraction of D available. For a bent-axis unit, $x(t)$ is related to the swivel angle, i.e., the control input of the machine. The negative sign in (5.38) is necessary because usually for pumps $x(t)$ is considered to be a positive quantity. The volumetric efficiency $\eta_v(t)$ accounts for various phenomena, including leakage (turbulent and laminar) and fluid compressibility [197]. For motors, (5.38) is modified as

$$Q_1(t) = -\frac{x(t) \cdot D \cdot w_2(t)}{\eta_v(t)}, \quad T_2(t) > 0, \tag{5.39}$$

with $x(t)$ that is now a negative quantity.

The fluid pressure in a pump ($p_1(t)$ is defined as the differential between fluid pressure in the machine and in the reservoir) is calculated from a basic energy balance

$$p_1(t) = -\frac{\eta_t(t) \cdot T_2(t)}{x(t) \cdot D}, \quad T_2(t) < 0. \tag{5.40}$$

The mechanical efficiency $\eta_t(t)$ accounts for viscous, frictional, and hydrodynamic torque losses. The corresponding equation for the motor mode is

$$p_1(t) = -\frac{T_2(t)}{\eta_t(t) \cdot x(t) \cdot D}, \quad T_2(t) > 0. \tag{5.41}$$

A semi-physical approach to evaluate the volumetric and the torque efficiency is now illustrated [197]. The volumetric efficiency in the pump mode is evaluated as

$$\eta_v(t) = 1 - \frac{C_s}{x(t) \cdot S(t)} - \frac{p_1(t)}{\beta} - \frac{C_{st}}{x(t) \cdot \sigma(t)}, \quad x(t) > 0, \tag{5.42}$$

where C_s and C_{st} are the laminar and the turbulent leakage coefficients, respectively, and β is the bulk modulus of elasticity (1660 MPa for most hydraulic fluids). Cavitation losses, small in most modern pumps, are neglected in this model. The variables $S(t)$ and $\sigma(t)$ are defined by

$$S(t) = \frac{\mu_o \cdot w_2(t)}{p_1(t)}, \tag{5.43}$$

$$\sigma(t) = \frac{\omega_2(t) \cdot D^{1/3} \cdot \rho_o^{1/2}}{(2 \cdot p_1(t))^{1/2}} , \tag{5.44}$$

where μ_o is the fluid viscosity and ρ_o its density. The volumetric efficiency in the motor mode is calculated from simple considerations as

$$\eta_v(T_2, \omega_2) = \frac{1}{2 - \eta_v(-|T_2|, \omega_2)} . \tag{5.45}$$

By accounting for the viscous torque, the frictional torque, and the hydrodynamic loss with the three coefficients C_v, C_f and C_h, respectively, it can be shown that for a motor the mechanical efficiency is

$$\eta_t(t) = 1 - \frac{C_s \cdot S(t)}{x(t)} - \frac{C_f}{x(t)} - C_h \cdot x^2(t) \cdot \sigma^2(t), \quad x(t) < 0 . \tag{5.46}$$

In the pump mode, the following relationship can be easily proven,

$$\eta_t(-|T_2|, \omega_2) = \frac{1}{2 - \eta_t(T_2, \omega_2)} . \tag{5.47}$$

The overall machine efficiency is the product of the volumetric efficiency and the torque efficiency

$$\eta_{hm}(t) = \eta_v(t) \cdot \eta_t(t) . \tag{5.48}$$

Thus (5.45)–(5.47) can be used to estimate the motor efficiency from pump data and vice versa, at least for normal operation. At very low loads, the difference between the loss coefficients of the same unit operating as a motor and a pump may be substantial [197].

5.5.2 Dynamic Modeling of Hydraulic Pumps/Motors

The physical causality representation of a hydraulic pump/motor is sketched in Fig. 5.23. The model input variables are the rotational speed $\omega_2(t)$ and the fluid pressure $p_1(t)$. The model output variables are the shaft torque $T_2(t)$ and the fluid flow rate $Q_1(t)$.

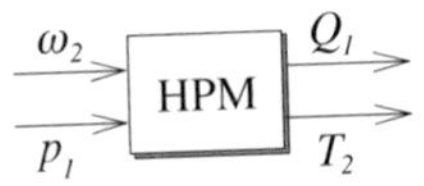

Fig. 5.23. Hydraulic pumps/motors: physical causality for dynamic modeling.

The quasistatic model derived in the previous section can easily be used also for dynamic simulations. The output torque is calculated from (5.40)–(5.41) as a function of the fluid pressure, while (5.38)–(5.39) serve to calculate the fluid flow rate. The behavior of the machine as a pump ($T_2(t) < 0$) or as a motor ($T_2(t) > 0$) is determined by the sign of the fractional capacity $x(t)$, which is the control input to the machine.

5.6 Pneumatic Hybrid Engine Systems

The combination of a conventional IC engine with a pneumatic short-term storage system is an interesting approach to achieve a lower fuel consumption. The key idea is to use the IC engine as pneumatic pump and pneumatic motor [221, 111]. This allows the recuperation of some of the energy otherwise lost when braking and the elimination of the most inefficient engine operating points, i.e., idling and very low loads. Moreover, such a pneumatic hybrid system ideally complements a down-sized and supercharged engine [103, 259].

In fact, turbocharged engines usually have a reduced drivability due to the relatively slow acceleration of the compressor-turbine during load steps. This leads to the choice of small turbines, which minimize the delays but which have a rather low efficiency. In pneumatic hybrid systems, the fresh air available in the pressure tank can be used to provide the air necessary for supercharging the engine in heavy transients and, therefore, the turbines can be designed for optimal fuel economy. Since the air is provided to the cylinder by a fully variable charge valve, the torque can be raised from idling to full-load from one engine cycle to the next, i.e., in the shortest time possible. Moreover, the combustion with the larger amount of fresh air and fuel accelerates the turbocharger much faster, resulting in a fast pressure rise in the intake manifold. Therefore, the additional air from the pressure tank is needed only for a very short time such that relatively small air tanks can be used.

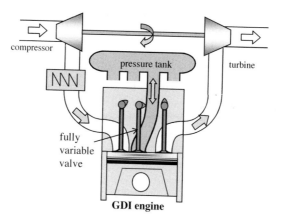

Fig. 5.24. Pneumatic hybrid engine system.

The hardware configuration necessary for a pneumatic hybrid operation includes an additional valve in the cylinder head, which is connected to a pressure tank. A fully variable actuation of this valve is mandatory. To achieve maximum efficiency in the pneumatic modes, fully variable intake and exhaust valves are necessary as well. An illustration of the hardware configuration which corresponds to the discussed system is shown in Fig. 5.24.

The combination of mainly mechanical components makes this engine system a very cost-efficient alternative to hybrid-electric propulsion systems. The additional weight can be kept very small since only a low-pressure air tank is used. Of course the fuel economy potential offered by the additional degrees of freedom can only be realized by a well-designed supervisory control system. In Chap. 8.8 a case study is presented in which this problem is analyzed.

5.6.1 Modeling of Operation Modes

Basically, two independent groups of operation modes of the engine system have to be investigated:

- combustion engine mode, and
- several pneumatic modes.

All the modes can be best explained by plotting the thermodynamic cycle in a p–V diagram [137]. Since for any polytropic expansion $p \cdot v^\gamma$ is constant, a double logarithmic p–V diagram is used in Figs. 5.25 – 5.30. In these figures, all polytropic processes are represented by a straight line. The dashed lines represent operating cycles with a higher torque than those depicted with solid lines.

Conventional ICE Mode

The cyclic operation in conventional ICE mode (throttled operation) is represented in Fig. 5.25 as a succession of eight ideal transformations: (i) a polytropic compression from bottom dead center (BDC) to top dead center (TDC), (ii) an isochoric combustion at TDC, followed by (iii) an isobaric part of the combustion, (iv) a polytropic expansion from end of combustion to BDC, (v) an isochoric expansion at BDC from cylinder pressure to exhaust pressure, (vi) an isobaric exhaust stroke from BDC to TDC, (vii) an isochoric expansion at TDC from exhaust pressure to intake manifold pressure, and (viii) an isobaric intake stroke from TDC to BDC.

The manifold pressure varies from levels below ambient to higher than ambient if the turbocharger supplies enough fresh air. For a comparison with a typical efficiency map of the conventional combustion mode the reader is referred to Fig. 3.2.

Pneumatic Supercharged Mode

The pneumatic supercharged mode is used for operating points at which the turbocharger does not yet provide enough air. As schematically shown in Fig. 5.26, during compression the charge valve is opened (CVO) and closed again (CVC), at the latest when the tank pressure is reached. The tank pressure and the timing of this short opening determines the amount of additional

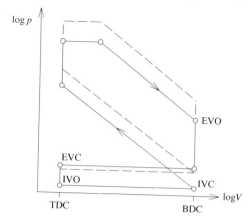

Fig. 5.25. Double logarithmic p–V diagram of the conventional engine mode. IVO: intake valve opening, IVC: intake valve closing, EVO: exhaust valve opening, EVC: exhaust valve closing. Dashed line: cycle with higher torque.

air in the cylinder. The amount of fuel in the cylinder has to be metered to match the amount of air coming both from the intake manifold and from the tank.

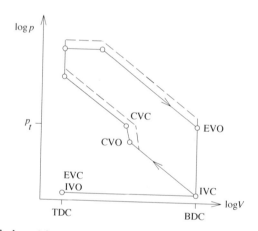

Fig. 5.26. Double logarithmic p–V diagram of the pneumatic supercharged mode. Nomenclature as in Fig. 5.25, plus CVO: charge valve opening, CVC: charge valve closing.

Pneumatic Undercharged Mode

For a demanded engine torque lower than that achieved at naturally aspirated full load conditions, the excess air can be used to fill the tank. As shown schematically in Fig. 5.27, during the compression stroke the charge-valve is opened at a crank angle where the cylinder pressure is equal to the tank pressure. The duration of the opening determines the amount of air remaining in the cylinder. For this mode, a direct fuel injection system is crucial because for safety reasons no fuel is allowed to enter the air tank. Accordingly, fuel injection can start only after CVC.

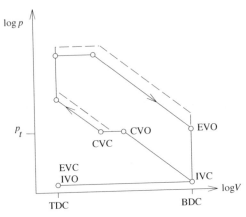

Fig. 5.27. Double logarithmic p–V diagram of the pneumatic undercharged mode. Nomenclature as in Fig. 5.26.

Pneumatic Pump Mode

The pneumatic pump mode is used for recuperating energy during braking. A two-stroke operation, illustrated in Fig. 5.28, yields the highest braking torque. The pressure tank is provided with enthalpy which can be used later in the pneumatic motor mode. The braking torque can be varied by changing the timing of the charge valve and of the intake valve simultaneously. For instance, the dashed-line cycle of Fig. 5.28 provides a less negative torque at the crankshaft than the solid-line cycle, thus a smaller braking torque results. Starting with ambient pressure at BDC, a polytropic compression brings the cylinder pressure up to the pressure of the tank. Then, the charge valve opens and an enthalpy flow into the tank starts. The timing of the charge valve closing (CVC) determines the braking torque. The highest braking torque is reached if CVC is at TDC. Leaving the charge valve open for a longer period reduces the braking torque because the tank pressure acts on the piston downstroke again. When the cylinder pressure again reaches manifold pressure, the intake valve remains open until it closes at BDC.

The thermodynamic performance of the pneumatic pump operation is described by the coefficient of performance (COP). The COP is defined as the

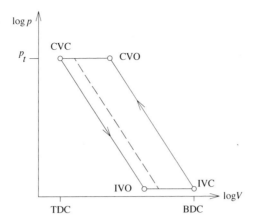

Fig. 5.28. Double logarithmic p–V diagram of the pneumatic pump mode. Nomenclature as in Fig. 5.26.

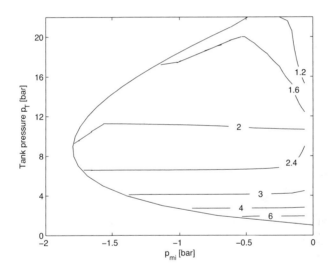

Fig. 5.29. Iso-COP lines for the pneumatic pump mode.

variation of the internal energy in the air tank divided by the engine work output per cycle. The latter is normalized as *mean indicated pressure*, in analogy to the approach of Sect. 3.1.2. Thus the COP is calculated as

$$\text{COP} = \frac{\Delta U_t}{p_{mi} \cdot V_d} \, . \tag{5.49}$$

The variation of the internal energy in the air tank is, by definition,

$$\Delta U_t = c_v \cdot \Delta(\vartheta_t \cdot m_t) \, , \tag{5.50}$$

where ϑ_t is the temperature of the air accumulated in the tank and m_t its mass. Using the ideal gas law $p_t \cdot V_t = m_t \cdot R \cdot \vartheta_t$, the value of ΔU_t can be related to the variations of pressure

$$\Delta U_t = V_t \frac{\Delta p_t}{\gamma - 1} \ , \tag{5.51}$$

where $\gamma = c_p/c_v = (c_v + R)/c_v$ is the ratio of specific heats (approximately 1.4 for air). Since the increase of internal energy is not only caused by the engine work output, but also by the "free" enthalpy of fresh air inducted by the engine, the term coefficient of performance is chosen rather than efficiency. This definition corresponds to that used for heat pumps.

Figure 5.29 shows typical values of the COP of the pneumatic pump mode as a function of p_{mi} (i.e., indicated torque). For a given tank pressure, only a limited range of brake torques can be applied in an efficient way. Higher torques would be possible, but then the additional work would not lead to a higher enthalpy flow to the tank, but rather would be lost by throttling, back-flow, or heat transfer. Depending on the temperature of the engine, it might thus be preferable to use the conventional brakes to provide this additional braking torque. Highest levels of COP can be achieved when the tank is empty.

Pneumatic Motor Mode

The pneumatic motor mode enables a fast start of the engine and even the launch of the vehicle. High and smooth torque transients can be provided running the motor in two-stroke mode. The pneumatic motor mode enables fast stop–start, stop-and-go operations without causing any emissions and launching the vehicle without using the clutch.

The load control in this mode is effected by varying the charge valve closing (CVC) and exhaust valve closing (EVC) simultaneously, as shown in Fig. 5.30. The efficiency-optimal timing is discussed in [242].

Figure. 5.31 shows the COP of the pneumatic motor mode as a function of tank pressure and p_{mi}. In contrast to the pump mode operation, the range of feasible torques is much larger. Since the pneumatic motor is running in two-stroke mode, a very high torque can be applied on the crankshaft. An average charge-discharge efficiency of $\approx 70\%$ can be assumed, neglecting mechanical friction.

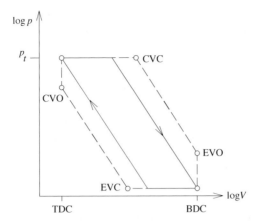

Fig. 5.30. Double logarithmic p–V diagram of the pneumatic motor mode. Nomenclature as in Fig. 5.26.

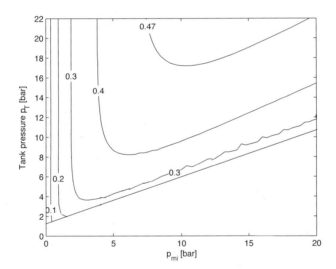

Fig. 5.31. Iso-COP lines for the pneumatic motor mode.

6

Fuel-Cell Propulsion Systems

This chapter contains three sections. The first section briefly describes the application of fuel-cell systems as stand-alone energy sources for powertrains or, in combination with a storage system, as fuel-cell hybrid powertrains. The second section introduces some thermodynamic and electrochemical models of fuel cells, as well as some fluid dynamic models of the complete fuel-cell system. The last section introduces the on-board production of hydrogen through fuel reforming and presents some system-level models of methanol reformers.

The main objective of this chapter is to introduce models that are useful in the context of energy management. Readers interested in the low-level control of fuel-cell systems are referred to [203].

6.1 Fuel-Cell Electric Vehicles and Fuel-Cell Hybrid Vehicles

Fuel cells are electrochemical devices that convert chemical energy directly into useful electrical energy. In contrast to internal combustion engines, there is no intermediate conversion into thermal energy and from that to mechanical energy. The efficiency of a fuel cell is thus not limited by the Carnot efficiency, and its work output theoretically can reach values that are higher than the lower heating value of the fuel.

Fuel cells deliver electrical energy, i.e., pure exergy, without any combustion products. This is the reason why, along with purely electric vehicles, fuel-cell vehicles are classified as zero-emission vehicles. As shown in Chap. 1, the specific energy of hydrogen as a fuel is substantially better than that of electrochemical batteries. Therefore, fuel-cell vehicles seem to be able to combine the best features of EVs and of ICE-based vehicles, namely zero local emissions, high efficiency, and a reasonable range. However, the operation of the on-board auxiliaries and all the conversion steps required to obtain and store the fuel (see Chap. 1) strongly affect the performance of fuel-cell systems. These shortcomings thus must be carefully taken into account when comparisons with other propulsion systems are made.

In principle, fuel cells can convert all fluid oxidable substances. However, from a technical point of view, only hydrogen, natural gas, and methanol are currently of practical use. Other liquid fossil fuels, such as ethanol and gasoline, can be used as energy carriers that are converted ("reformed") on-board into hydrogen. Hydrogen has the decisive advantage that the reaction product is pure water. Table 6.1 presents a comparison of the most important data for hydrogen and gasoline.

Table 6.1. Comparison of fuel data for hydrogen and gasoline.

	Hydrogen	Gasoline
Molecular weight	2.016	107
Boiling point (K)	13.308	310–478
Vapor density at normal conditions (g/m^3)	84	4400
Liquid density (kg/m^3)	70.8	700
Higher heating value (MJ/kg)	141.86	48
Lower heating value (MJ/kg)	119.93	43.5
Gas constant (J/kg/K)	4124	78.0
Flammability limit in air (vol. %)	4.0–75	1.0–7.6
Detonation limit in air (vol. %)	18.3–59.0	1.1–3.3
Auto-ignition temperature (K)	858	501–744
Adiabatic flame temperature in air (K)	2318	2470
Flame front speed in air (cm/s)	≈ 300	37–43

The integration of a fuel cell in a propulsion system seems simple in principle, and various prototypes have already been developed [280]. Since the fuel cell delivers energy in electrical form, the final power train is reduced to that of an electric vehicle (fuel-cell electric vehicles, FCEV). However, the resulting system is rather complex due to the multiple energy storage systems and load-levellers, various auxiliary devices, etc. that are required.

The combined problems of on-board hydrogen storage and of the lack of a hydrogen refuelling infrastructure so far have represented an impediment to the wide-scale adoption of FCEVs. On-board fuel processors that generate hydrogen from on-board hydrocarbons (methanol, gasoline, diesel, or natural gas) have therefore been proposed as an alternative hydrogen source for FCEVs. Although the energy density is much higher than for hydrogen storage, these systems are rather complex and introduce additional efficiency losses related to the various stages of fuel preparation, reforming, and hydrogen cleaning. Moreover, they produce CO_2 emissions and a poor dynamic response that makes control a difficult task.

The drivability and the power performance of an FCEV may be improved by the addition of a short-term storage system, resulting in a fuel-cell hybrid vehicle (FCHEV) [208]. A promising option consists of using supercapacitors, the alternative being represented by electrochemical batteries. Supercapacitors (see Sect. 4.5) have an extremely high power density and a higher efficiency than batteries for energy charge and discharge. They are used to cover power peaks, typically during accelerations, and for regenerative braking, while the fuel cell is operated almost stationarily. There are many advantages associated with this operation. First, the prime mover can be downsized with respect to the peak power and thus it can be a smaller and cheaper unit. Secondly, the stationary operation increases not only the efficiency of the fuel cell but also its lifetime.

6.1.1 Concepts Realized

In the 1990s and 2000s many prototypes of passenger cars and other types of vehicles equipped with a fuel cell have been demonstrated by research institutions and major car manufacturers. Some of these concepts, already realized in small series, e.g., for car sharing, can be considered to be at a pre-commercial stage. Concerning the propulsion architecture, purely fuel-cell vehicles must be distinguished from fuel-cell hybrid-electric vehicles. After pioneering attempts in the preceding decades, the development of FCEVs received a renewed impulse in the mid 1990s. Often the same manufacturer explored various hydrogen storage scenarios, i.e., gaseous, liquid, metal hydride, methanol, or gasoline reforming, in successive prototypes (e.g., DaimlerChrysler Necar series [61], Ford Focus FCV/FC5 [85], Opel HydroGen3 series [181]), although commercial or pre-commercial projects mostly employ compressed gaseous hydrogen (DaimlerChrysler F-Cell [60], GM's Hy-Wire [93] and Citaro Bus [58]). A recent concept uses sodium borohydride as a hydrogen storage system (DaimlerChrysler Natrium [59]).

Fuel-cell hybrids with a secondary battery as an energy storage system have been similarly investigated with various solutions for hydrogen storage (Toyota FCHV series [249], Ford Focus FCV Hybrid [84], Nissan X-Trail FCV [176], Daihatsu FC-EV series [56]). More recently, hybrid fuel-cell systems using a supercapacitor have been introduced. From being the subject of research and development projects [213, 144], this solution has become an approach adopted by several car manufacturers (Mazda Demio FCEV [165], Honda FCX [114]).

6.2 Fuel Cells

Hydrogen Storage Systems

As mentioned above, one of the main problems for the development of fuel cells as prime movers in passenger cars arises from the usage of hydrogen as a fuel. The lower energy density of hydrogen in comparison with gasoline makes its storage a difficult task. Various hydrogen storage technologies are the subject of research and testing [255]. Example technologies are pressure vessels, cryogenic accumulators, and metal sponges (chemical adsorption). These technologies lead to a specific energy which is significantly lower than that of gasoline. However, in consideration of the higher system efficiency of fuel-cell power sources, the range of such vehicles is comparable to that of ICE-based passenger cars. Additional disadvantages are generally higher cost and weight, aside from a more complicated fuel management.

Storage under pressures of up to 350 bar is achieved in conventional pressurized vessels. High-pressure tanks must be periodically tested and inspected to ensure their safety. However, this technology is widely developed, efficiently controllable, and relatively inexpensive. Therefore most of the current fuel-cell vehicle applications use compressed gaseous hydrogen. The specific energy of the fuel stored in a pressurized vessel is

$$\frac{E_{ht}}{m_{ht}} \approx \frac{E_{ht}}{m_m} = H_h \cdot \rho_h \cdot \gamma_{ht} , \tag{6.1}$$

where m_{ht} is the mass of the fully charged hydrogen tank, practically coincident with the mass of the vessel m_m, H_h is the lower heating value (120 MJ/kg or 33.3 kWh/kg) of hydrogen, ρ_h its density (depending on the pressure in the vessel), and γ_{ht} is the storage capacity of the vessel, i.e., the volume stored per unit mass. The energy density E_{ht}/V_{ht} is given by the product $H_h \cdot \rho_h$.

Hydrogen can be liquefied, but only at extremely low temperatures. Liquid hydrogen typically has to be stored at 20 K or –253°C. The storage tanks are insulated to preserve that low temperature and reinforced to store the liquid hydrogen under pressure. The energy density is still evaluated with (6.1). It is higher than with pressurized vessels, but energy losses of about 1% of the lower heating value are typical during vehicle operation, aside from those occurring in the liquefaction and compression process (typically, 30% of the lower heating value). Moreover, the cryogenic hydrogen must be heated before supplying the fuel cell, which causes additional losses and possibly further problems, especially in transient operation. However, the cryogenic storage has already found some application in fuel-cell vehicles (DaimlerChrysler Necar 4 [61]).

Metal hydride storage systems (metal sponges) represent a relatively new technology, although they have already been used in various fuel-cell vehicle prototypes (Toyota FCHV-3 [249], Mazda Demio [165]). The structure of metal hydrides causes hydrogen molecules to be decomposed and hydrogen atoms to be incorporated in the interstices of specific combinations of metallic alloys. During vehicle operation, these hydrogen atoms are liberated through the addition of heat. The energy density is comparable to that of liquid hydrogen. However, the overall specific energy drops to lower values due to the additional weight of the metal hydride material. The main advantage of metal hydride storage is a relatively simple and safe fuel handling and delivery. This technology avoids the risks due to high pressure or low temperature and, due to chemical bonds, it liberates only very little hydrogen in case of an accident. A serious drawback is the higher cost that originates on the one hand from the material and on the other hand from the complicated thermal management. The life of a metal hydride storage system is directly related to the purity of the hydrogen. The specific energy of metal hydride storage systems is calculated differently from (6.1), i.e.,

$$\frac{E_{ht}}{m_{ht}} \approx \frac{E_{ht}}{m_m} = H_h \cdot \xi_{ht} \, , \tag{6.2}$$

where ξ_{ht} is the mass fraction of hydrogen stored in the system. This value is typically 1–2%, though some alloys (e.g., MgH_2) are capable of storing up to 8% hydrogen, but only at high temperatures. The energy density of a metal hydride system is obtained from the specific energy by dividing it by γ_{ht}.

Carbon nanotubes are microscopic tubes of carbon with a diameter of approximately 2 nm that store hydrogen within microscopic pores in the tube structure. Similar to metal hydrides in their mechanism for storing and releasing hydrogen, carbon nanotubes are expected to be capable of a storage efficiency of 4–8%. However, this hydrogen storage capacity is still in the research and development stage.

Another promising technology is represented by glass microspheres, which are tiny hollow glass spheres warmed and filled by being immersed in high-pressure hydrogen gas. The spheres are then cooled, locking the hydrogen inside. A subsequent increase in temperature releases the hydrogen trapped in the spheres. Microspheres have the potential to be very safe, resistant to contamination, and able to store hydrogen at a low pressure, thus increasing the margin of safety. However, this technology is still at a very early stage of development.

Another technology that has found some application (DaimlerChrysler Natrium [59]) consists of storing hydrogen in sodium borohydride ($NaBH_4$). When this chemical is combined with a specific catalyst, liquid borax and pure hydrogen gas are produced. The former subsequently can be recycled back into sodium borohydride. This technology is rather expensive due to the costs of the catalyst (ruthenium) and of the processes to produce sodium borohydride and to recycle borax. Moreover, the energy losses associated with the several conversion steps are substantial.

Table 6.2 lists typical[1] values of the storage parameters of current storage technologies together with reference values for conventional gasoline storage. Also shown are the technical targets set by the US Department of Energy (DOE) for the year 2015 [256]. Note that these data do not include the conversion efficiencies as was the case in Fig. 1.7.

---

[1] The research efforts in this field are currently very active such that a reasonable and updated estimation of the average values is very difficult.

Table 6.2. Typical storage parameters of current storage technologies; [a]: for a steel tank, aluminium 1.5 l/kg, composite 3–4 l/kg, [b]: see [253], [c]: the mass of gasoline is not negligible when compared with that of the tank.

	γ_{ht} (l/kg)	ρ_h (kg/m^3)	E_{ht}/m_{ht} (kWh/kg)	E_{ht}/V_{ht} (kWh/l)	ξ_{ht} (%)
Pressure vessel	1^a	15	0.5	0.5	
Cryogenic storage	1.7	71	4.0	2.4	
Metal hydride storage^b	0.3	60	0.6	2.0	1.8
Gasoline	1.2^c	750	10.8	8.8	
DOE target, 2015			3.0	2.7	4.5

Types of Fuel Cells

The classification of fuel cells follows the type of electrolyte. The main types are listed in Table 6.3.

Table 6.3. Types of fuel cells. See below for the definition of the fuel cell efficiency η_{fc}.

Type	Electrolyte	η_{fc}(%)	ϑ (°C)	Use
AFC	Alkaline (NaOH, KOH)	50–65	80–250	aerospace
PEM	Ionic membrane (Nafion)	50–60	40–100	automotive
PAFC	Phosphoric acid (H_3PO_4)	35–45	160–220	power
MCFC	Molten carbonate ($KLiCO_3$)	40–60	600–650	power
SOFC	Solid oxides (ZrO_2, Y_2O_3)	50	850–1000	power

Alkaline fuel cells (AFC) use an aqueous solution of alkaline (e.g., potassium or sodium) hydroxide soaked in a matrix as the electrolyte. The cathode reaction is faster than in other electrolytes, which means a higher performance. In fact, AFCs yield the highest electrochemical efficiency levels, up to 65%. Their operating temperature ranges from 80 to 250°C, although newer, low-temperature designs can operate below 80°C. They typically have an output of 300 to 5000 W. This technology is widely developed since it was initially used in aerospace applications as far back as the 1960s, to provide not only the power but also the drinking water for the astronauts. Although some experimental vehicles (e.g., ZEVCO London hybrid taxi, 1998, Lada Antel, 2001) were powered by AFCs, this technology is not considered to be suitable yet for automotive applications. On the one hand, the caustic electrolyte is highly corrosive and thus high standards are demanded from the material and the safety technology. On the other hand, this type of cell is very sensitive to contaminations in the supply gases, thus requiring very pure hydrogen to be used. Moreover, until recently AFCs were too costly for commercial applications.

Proton-exchange membrane fuel cells (PEM) use a thin layer of solid organic polymer[2] as the electrolyte. This ion-conductive membrane is coated on both sides with highly dispersed metal alloy particles (mostly platinum, an expensive material) that are active catalysts. The PEM fuel cell basically requires hydrogen and oxygen as reactants, though the oxidant may also be ambient air, and these gases must be humidified to prevent membrane dehydration. Hydrogen must be as pure as possible, since CO poisons platinum catalysts (up to 100 ppm of CO are tolerated). Methane and methanol reforming is possible only for low loads. Because of the limitations imposed by the thermal properties of the membrane, PEM fuel cells operate at relatively low temperatures of about 80°C, which permits a quick start-up. Thanks to this fact and to other advantages such as higher power density and higher safety with the solid electrolyte, PEM fuel cells are particularly suitable for automotive applications. Further improvements are required in system efficiency, the goal being an efficiency level of 60%.

Direct methanol fuel cells (DMFC) are similar to the PEM fuel cells in that their electrolyte is also a polymer membrane. However, in the DMFC the anode catalyst itself draws the hydrogen from the liquid methanol, eliminating the need for a fuel reformer. That is quite an advantage in the automotive area where the storage or generation of hydrogen is one of the main obstacles for the introduction of fuel cells. Another field of application is in the very small power range, e.g., laptops. There are principal problems, including the lower electrochemical activity of the methanol as compared with hydrogen, giving rise to lower cell voltages and efficiency levels. Efficiencies of only about 40% may be expected from the DMFC, at a typical temperature of operation of 50–100°C. Also, DMFCs use expensive platinum as a catalyst and, since methanol is miscible in water, some of it is liable to cross the water-saturated membrane and cause corrosion and exhaust gas problems on the cathode side.

Phosphoric-acid fuel cells (PAFC) use liquid phosphoric acid soaked in a matrix as the electrolyte. This type of fuel cell is the most commercially developed and is used particularly for power generation. Already various manufacturers are represented on the market with complete power plants that generate from 100 to 1000 kW. The efficiency of PAFCs is roughly 40%. The fact that they can use impure hydrogen as a fuel allows the possibility of reforming methane or alcohol fuels. Nevertheless, due to the higher operating temperatures of 160–220°C and the associated warm-up times, for automotive applications this type is suitable only for large vehicles such as buses.[3] Moreover, PAFCs use expensive platinum as a catalyst and their current and power density is small compared with other types of fuel cells.

Molten-carbonate fuel cells (MCFC) use a liquid solution of lithium, sodium, and/or potassium carbonates, soaked in a matrix as the electrolyte. This cell operates at a temperature of about 650°C, which is required in or-

---

[2] The most commonly used material is Nafion by DuPont.

[3] In 1994 a prototype PAFC bus was demonstrated by Georgetown University.

der to achieve a sufficiently high conductivity of the electrolyte. The higher operating temperature provides the opportunity for achieving higher overall system efficiencies (up to 60%) and greater flexibility in the use of available fuels and inexpensive catalysts, although it imposes constraints on choosing materials suitable for a long lifetime. As of this writing, MCFCs have been operated with various fuels in power plants ranging from 10 kW to 2 MW. The necessity for large amounts of ancillary equipment would render a small operation, such as an automotive application, uneconomic.

Solid-oxide fuel cells (SOFC) use solid, nonporous metal oxide electrolytes. The metal electrolyte normally used in manufacturing SOFCs is stabilized zirconia. This cell operates at a temperature of about 1000°C, allowing internal reforming and/or producing high-quality heat for cogeneration or bottoming cycles. Thus this type is used in large power plants of up to 100 kW, where it reaches efficiency levels of 60% or even as high as 80% for the combined cycle. However, high temperatures limit the use of SOFCs to stationary operation and impose severe requirements on the materials used. On the other hand, SOFCs do not need expensive electrode materials. Moreover, various fuels can be used, from pure hydrogen to methane to carbon monoxide. Some developers are testing SOFC auxiliary power units for automotive applications.

Electrochemistry of Hydrogen Fuel Cells

In electrochemical cells, the reaction consists of two semi-reactions, which take place in two spatially separated sections. These two zones are connected by an electrolyte that conducts positive ions but not electrons. The electrons that are released by the semi-reaction of oxidation can arrive at the reduction electrode (cathode) only through an external electric circuit. This process yields an electric current, which is the useful output of the cell. The normal direction of the external current is from the reduction side (cathode, positive electrode) to the oxidation side (anode, negative electrode). In Fig. 6.1 a simple PEM fuel cell is depicted. It consists of a particular membrane that does not conduct electrons but in rather lets ions pass. This membrane, impermeable for neutral gas, serves as an electrolyte. At both sides of the membrane porous electrodes are mounted. The electrodes allow for the gas diffusion, and they accomplish a triple contact gas–electrolyte–electrode. As described below, at the anode and the cathode sides of the membrane hydrogen and oxygen are supplied, respectively.

At the anode, molecular hydrogen is in equilibrium with simple protons and electrons. The anode reaction is written as

$$H_2 \rightarrow 2H^+ + 2e^- \ . \tag{6.3}$$

Under standard operating conditions, the dissociation rate is small, such that most of the hydrogen is present in the form of electrically neutral molecules. The equilibrium can be modified with a change in the boundary conditions,

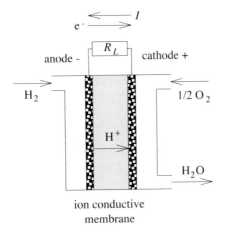

ion conductive
membrane

Fig. 6.1. Principle of operation of a fuel cell.

e.g. the adoption of a catalyst or an increase in temperature. Since the H^+ ions are formed at the anode, a concentration gradient is established between the two ends of the membrane, generating an ion diffusion toward the cathode. Such an H^+ ion current transports a positive charge from the anode to the cathode. As a consequence, a difference of potential arises across the membrane, with the anode as a negative electrode and the cathode as a positive electrode. Since the free electrons cannot follow the electric field through the membrane, they flow through the external circuit.

The electron current dissipates energy across an external resistance. This energy is generated by the chemical reaction that takes place at the cathode, where electrons (from the external path), protons (through the membrane), and oxygen from an external source are combined to yield water. The cathode reaction is

$$2H^+ + 2e^- + \frac{1}{2}O_2 \rightarrow H_2O . \tag{6.4}$$

The water product is at a lower energetic state than the original combination of protons, electrons, and oxygen molecules. The difference is the energy delivered by the fuel cell. The overall reaction is the combination of (6.3) and (6.4),

$$H_2 + \frac{1}{2}O_2 \rightarrow H_2O . \tag{6.5}$$

Equations (6.3)–(6.5) only show the chemical transformations. But together with chemical species the reactions also involve energy. In contrast to a combustion process, the energy generated is not accumulated as thermal energy, but it is directly converted into exergy in the form of electric energy. The limits of such a conversion are described in terms of "free energy" (or Gibbs' potential), as the following sections will show more clearly.

Thermodynamics of Hydrogen Fuel Cells

When a mixture of O_2 and H_2 burns, for every kmole of H_2O a defined quantity of heat known as *formation energy* is liberated. If the reaction takes place at constant pressure, this heat equals the decrease of enthalpy H and is often referred to as "heating value,"

$$Q_H = -\Delta H = -\left(H_{products} - H_{reactants}\right) . \tag{6.6}$$

The heating value is a function of the temperature and the pressure at which the reaction takes place. For a reaction such as that of (6.5), Q_H also depends on the state of water in the products. If water occurs in the form of vapor, the lower heating value is obtained. Vice versa, condensated water frees an additional amount of energy that leads to the higher heating value. Since the enthalpy of all pure substances in their natural state is zero by definition, the enthalpies of H_2 and O_2 are zero, thus $H_{reactants} = 0$ in (6.6). The enthalpy of water thus corresponds to the heating value of the reaction. For vapor water at 1 atmosphere and 298.15 K (reference temperature and pressure, RTP), $\Delta H_H = -241.8\,\mathrm{MJ/kmol}$. For liquid water at RTP, $\Delta H_l = -285.9\,\mathrm{MJ/kmol}$. The difference corresponds to the enthalpy of the vaporization of water.

If the system where the reaction takes place was an isolated system, all the fuel heating value in principle could be converted into useful electrical work. In fact, in a system without any heat exchange with the surrounding ambient, the work not related to variations of volume equals the variations of enthalpy,

$$W_{id} = -\Delta H = Q_H . \tag{6.7}$$

However, it is theoretically impossible to collect all the work W_{id} from a fuel cell. One limiting condition playing a role similar to that of the Carnot efficiency for thermal systems arises when the entropy is taken into account.

Besides its internal energy each substance is also characterized by a certain entropy level that depends on the particular thermodynamic state. In a closed system and in the ideal case of a reversible reaction, the heat dissipated to the surrounding ambient equals the entropy variation,

$$Q_S = -\vartheta \cdot \Delta S , \tag{6.8}$$

where Q_S is positive if released and negative if absorbed. In the general case, (6.8) is transformed into the well-known Clausius inequality. For the substances involved in an H_2–O_2 fuel cell, the entropy values are listed in Table 6.4 [73]. If 1 kmol of H_2O is formed, 1 kmol of H_2 and 0.5 kmol of O_2 disappear with their entropy. In the balance, 44.4 kJ/(kmol K) are missing. Thus the fuel cell releases the heat $Q_S = 298 \cdot 44.4 = 13.2\,\mathrm{MJ/kmol}$ to the ambient.

The fact that in association with the entropy variations there is always a certain amount of heat liberated limits the useful work available. In fact, for

Table 6.4. Thermodynamic data for hydrogen fuel cells at RTP.

H_2 (gaseous)	$H = 0\,MJ/kmol$	$S = 130.6\,kJ/(kmol\,K)$
O_2 (gaseous)	$H = 0\,MJ/kmol$	$S = 205.0\,kJ/(kmol\,K)$
H_2O (vapor)	$H = -241.8\,MJ/kmol$	$S = 188.7\,kJ/(kmol\,K)$

a reversible reaction at a constant temperature ϑ, a simple energy balance yields

$$-W_{rev} - Q_S = \Delta H \rightarrow W_{rev} = -\Delta H + \Delta(\vartheta \cdot S) = -\Delta G , \qquad (6.9)$$

where G is the state function known as "free energy"

$$G = H - \vartheta \cdot S . \qquad (6.10)$$

The quantity W_{rev} in (6.9) is the maximum electrical work that can be obtained from a fuel cell operating at constant pressure and temperature. When non-reversible processes are taken into account, the value of (6.9) is further diminished by a quantity corresponding to the non-reversible entropy sources.

Equation (6.9) can also be written as

$$W_{rev} = Q_H - Q_S , \qquad (6.11)$$

clearly showing that not all the heating value can be converted into useful work. Thus the highest possible cell efficiency for an ideal electrochemical cell ("electrochemical Carnot efficiency") is given by the ratio between the maximum work available in the case of a reversible process and the heating value of the single energy carriers,

$$\eta_{id} = \frac{-\Delta G}{-\Delta H} = 1 - \frac{\vartheta \cdot \Delta S}{\Delta H} = 1 - \frac{Q_S}{Q_H} . \qquad (6.12)$$

Table 6.5 shows a collection of theoretical efficiency levels for various fuels that may be used in fuel cells.

The work available per unit quantity of reactants with the cell voltage is calculated as follows. Each kmol of hydrogen contains $N_0 = 6.022 \cdot 10^{26}$ molecules. Equation (6.3) shows that for each molecule of hydrogen two electrons circulate in the external circuit. In general, the useful work W can be expressed in terms of cell voltage U_{fc} as

$$W = n_e \cdot q \cdot N_0 \cdot U_{fc} = n_e \cdot F \cdot U_{fc} , \qquad (6.13)$$

where n_e is the number of free electrons for every kmol of hydrogen, with $q = 1.6 \cdot 10^{-19}\,C$ being the charge of an electron and $F = q \cdot N_0 = 96.48 \cdot 10^6\,C/kmol$ being the Faraday constant. Assuming that the cells work in a reversible way, (6.13) can be written as

$$W_{rev} = n_e \cdot q \cdot N_0 \cdot U_{rev} = n_e \cdot F \cdot U_{rev} = -\Delta G , \qquad (6.14)$$

Table 6.5. Reversible voltage and efficiency for various fuel cell types; [a]: vapor water.

Fuel	Reaction	U_{rev} (V)	η_{id}
Hydrogen	$H_2 + 0.5\ O_2 \rightarrow H_2O$	1.18	0.94^a
Methane	$CH_4 + 2\ O_2 \rightarrow CO_2 + 2\ H_2O$	1.06	0.92
Methanol	$CH_3OH + 0.5\ O_2 \rightarrow CO_2 + 2\ H_2O$	1.21	0.97
Propane	$C_3H_8 + 5\ O_2 \rightarrow CO_2 + 2\ H_2O$	1.09	0.95
Carbon monoxide	$CO + 0.5\ O_2 \rightarrow CO_2$	1.07	0.91
Formaldehyde	$CH_2O + O_2 \rightarrow CO_2 + H_2O$	1.35	0.93
Formic acid	$HCOOH + 0.5\ O_2 \rightarrow CO_2 + H_2O$	1.48	1.06
Carbon	$C + 0.5\ O_2 \rightarrow CO$	0.71	1.24
	$C + O_2 \rightarrow CO_2$	1.02	1.00
Ammonia	$NH_3 + 0.75\ O_2 \rightarrow 0.5\ N_2 + 1.5\ H_2O$	1.17	0.88

where U_{rev} is the reversible cell voltage. Solving (6.14) for U_{rev}, the expression for the cell voltage obtained is

$$U_{rev} = -\frac{\Delta G}{n_e \cdot F}. \qquad (6.15)$$

If water is in its vapor form, $W_{rev} = 228.6\,\text{MJ/kmol}$ and the cell voltage is $U_{rev} = 1.185\,\text{V}$ (see Table 6.5). If in contrast the fuel cell produces liquid water, the higher heating value leads to $W_{rev} = 237.2\,\text{MJ/kmol}$ and thus the ideal open-circuit voltage is 1.231 V.

Besides that of the reversible voltage U_{rev}, an important role for further considerations is played by the "caloric voltage" U_{id}. This voltage is defined by

$$U_{id} = -\frac{\Delta H}{n_e \cdot F} \qquad (6.16)$$

and it measures the voltage (impossible to reach) that would be provided by a total conversion of enthalpy into electrical energy. From (6.12), $U_{rev} = \eta_{id} \cdot U_{id}$ follows.

Under load ($I_{fc}(t) > 0$), real cells deliver a voltage $U_{fc}(t)$, which is lower than U_{rev}. The power lost in the form of heat is evaluated as [173]

$$P_l(t) = I_{fc}(t) \cdot (U_{id} - U_{fc}(t)) \qquad (6.17)$$

This heat has to be removed by means of a dedicated cooling device (see Sect. 6.2.1).

Fuel-Cell Systems

For typical automotive applications, the voltage yielded by a single cell is too low. Higher voltages can be obtained by arranging several cells in series. This combination, often referred to as a "stack", is depicted in Fig. 6.2.

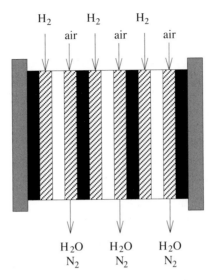

Fig. 6.2. Typical arrangement of a PEM fuel-cell stack. White fields represent the membrane–electrode arrangement (MEA), gray fields are the flow fields of reactant gases, black fields are bipolar plates.

The core of a single PEM cell consists of the membrane, the catalyst layers, and the porous electrodes (often referred to as "diffusion layers" or "backing layers"). Together, these layers form the membrane–electrode arrangement, MEA. Single cells are separated by flow fields that supply reactant gas to each electrode. Flow fields are arranged in bipolar plates as parallel flow channels (in coflow or counterflow form), a serpentine channel (one long channel with many passes over the diffusion layer), and interdigitated channels that force the flow through the diffusion layer. Bipolar plates also have to conduct electricity to the external circuit and allow for water flow to remove heat generated by the reaction. An alternative possibility is the use of porous plates [28].

A fuel-cell stack needs to be integrated with other components to form a fuel-cell system able to power a vehicle. Figure 6.3 shows a typical fuel-cell system with such components. These are grouped in four subsystems or circuits: (i) hydrogen circuit, (ii) air circuit, (iii) coolant circuit, and (iv) humidifier circuit. Most fuel cells use deionized water as a coolant, so that the subsystems (iii) and (iv) can be combined in a single water circuit [53]. The description in the following refers to a system with hydrogen storage. For systems fed by a fuel reformer, see Sect. 6.3.

Hydrogen is supplied to the fuel cell anode from its storage system, e.g., a pressurized tank, through a regulating valve that adjusts the pressure to the fuel cell level. A pressure tap from the air circuit serves as the reference pressure for the regulator. In some applications the hydrogen flow is humidified

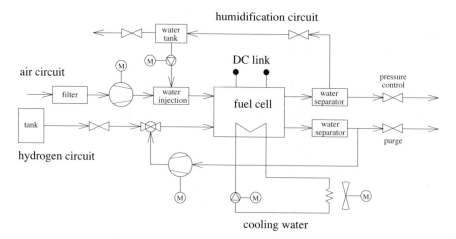

Fig. 6.3. Typical fuel-cell system, with distinct circuits for the humidification and the cooling.

(see below) in a chamber with a water injector. At the outlet of the fuel cell, hydrogen is recirculated by a pump to form a closed loop. This prevents the hydrogen from being consumed when the fuel cell is not absorbing current. In such a case the tank can be isolated by closing a shutoff valve located at the exit of the tank. A manual shutoff valve is also present in the recirculation loop to purge hydrogen during a shut-down.

The air supply system provides the fuel cell with clean air at a high relative humidity (>80%). A motor-driven volumetric (screw, scroll, rotary piston) or dynamic (centrifugal) compressor raises the air pressure typically by 70 kPa [271, 179]. The air is humidified (see below) and then fed to the fuel cell cathode. The exhaust air flow from the fuel cell outlet is regulated by a valve.

Due to the high flow rates and water carrying capacity of the reactants, current PEM fuel cells are quite sensitive to humidity. Humidification is for example accomplished by a controlled water injection system. Liquid water accumulates inside the humidity chamber as the humidified air passes through it and cools down. A reservoir collects this condensate and supplies it to the injection system and the cooling circuit. The exhaust stream from the fuel cell contains air, water droplets, and steam. To keep the water level in the deionized water tank even, the water in the air exhaust stream must be captured. This is accomplished by using an air–water separator in the exhaust stream [179].

The coolant circuit provides cooling for the fuel cell and some auxiliaries, such as the motor that drives the compressor. The coolant circuit includes a radiator, often of the multi-stage type, a radiator fan, a coolant pump, a water reservoir, and often a deionizer. If deionized water is used, the coolant circuit is strongly integrated with the humidifier circuit, since both use the same water reservoir. Unlike most ICE-based vehicles, fuel-cell vehicles have an electrically driven coolant pump whose speed can be varied independently of the operating conditions of the prime mover. Strategies of "intelligent cooling" can be used to minimize power consumption [215].

6.2.1 Quasistatic Modeling of Fuel Cells

The causality representation of a fuel cell in quasistatic simulations is sketched in Fig. 6.4. The input variable is the terminal power $P_2(t)$. The output variable is the hydrogen consumption $\overset{*}{m}_h(t)$.

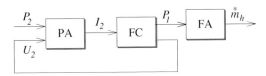

Fig. 6.4. Fuel cells: causality representation for quasistatic modeling.

First, the behavior of a single cell is discussed, then the arrangement of multiple cells in a "stack" and the related system. In the following, PEM fuel cell behavior is assumed where not explicitly stated otherwise. Fundamental fuel cell models that describe multi-phase, multi-dimensional flows in the electrodes, membrane, and catalyst layers, are not discussed here. The reader can find additional information in the bibliographies of the cited works, which deal with system-level, lumped-parameter modeling of PEM fuel cells.

Cell Voltage

The behavior of a single cell is characterized in terms of cell voltage $U_{fc}(t)$ and current density, $i_{fc}(t)$, which is defined as the cell current per active area,

$$i_{fc}(t) = \frac{I_{fc}(t)}{A_{fc}} . \tag{6.18}$$

Figure 6.5 shows a typical polarization curve of a fuel cell, i.e., its static dependency between $U_{fc}(t)$ and $i_{fc}(t)$. The curve is depicted for given operating parameters such as partial pressure, humidity, and temperature. Increasing operating pressure or air humidity would lift the curve up [98, 271, 159]. The effect of stack temperature may be more difficult to predict [211].

The cell voltage is given by the equilibrium potential diminished by irreversible losses, often called overvoltages or polarizations. These may be due to three sources: (i) activation polarization, (ii) ohmic polarization, (iii) concentration polarization,

$$U_{fc}(t) = U_{rev} - U_{act}(t) - U_{ohm}(t) - U_{conc}(t) . \tag{6.19}$$

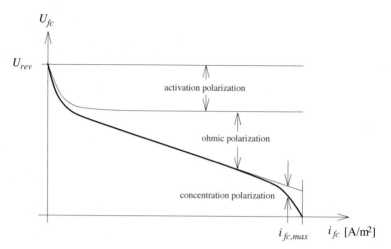

Fig. 6.5. Polarization curve of a hydrogen fuel cell with major loss contributions.

The reversible cell potential U_{rev} is a function of cell temperature and partial pressure of reactants and products, but not of cell current. The potential at standard conditions depends only on the reaction stoichiometry and is given in the previous section. The classical Nernst equation provides a correction for any temperature and pressure values different from the standard values [73]. Sometimes it is necessary to modify the theoretical coefficients of the Nernst equation to take into account other phenomena, e.g., the liquid water flooding the cathode [211].

The activation polarization $U_{act}(t)$ is a result of the energy required to initiate the reaction, which depends on the type of catalyst used. A better catalyst causes lower activation losses. The limiting reaction is that occurring at the cathode, which is inherently slower than the anode reaction. Activation losses increase as current density increases. The relation between $U_{act}(t)$ and $i_{fc}(t)$ is described by the semi-empirical Tafel equation [73, 159, 3, 163],

$$U_{act}(t) = c_0 + c_1 \cdot \ln(i_{fc}(t)) , \qquad (6.20)$$

where c_0 and c_1 are temperature-depending coefficients (c_0 also depends on reactant partial pressure). However, such a model is not valid for very small current densities, i.e., where the influence of the activation polarization is dominant. Therefore, a convenient empirical approximation of the Tafel equation can be derived [98] as

$$U_{act}(t) = c_0 \cdot \left(1 - e^{-c_1 \cdot i_{fc}(t)}\right) , \qquad (6.21)$$

where c_0 and c_1 depend on the partial pressure of the reactants and on the cell temperature. Equation (6.21) describes the fact that beyond a certain current density the activation polarization can be considered as a constant added to ohmic losses (see Fig. 6.5).

The ohmic losses $U_{ohm}(t)$ are due to the resistance to the flow of ions in the membrane and in the catalyst layers and of electrons through the electrodes, the former contributions being dominant. Assuming that both membrane and electrode behavior may be described by Ohm's law, the ohmic losses are expressed in terms of an overall resistance R_{fc} as

$$U_{ohm}(t) = i_{fc}(t) \cdot \tilde{R}_{fc} \; , \tag{6.22}$$

where $\tilde{R}_{fc} = R_{fc} \cdot A_{fc}$. The ohmic resistance R_{fc} includes electronic, membrane (ionic), and contact resistance contributions. Usually, only the dominant membrane resistance is taken into account, which is related to the membrane conductivity [202]. This resistance depends strongly on the cell temperature and the membrane humidity. A minor nonlinear influence of current density may be included as well [159, 3], though it is often small enough to be negligible [163].

The concentration polarization $U_{conc}(t)$ results from the change in concentration of the reactants at the electrodes as they are consumed in the reaction. This loss becomes important only at high current densities. Its dependency on $i_{fc}(t)$ may be described [98] by the law

$$U_{conc}(t) = c_2 \cdot i_{fc}^{c_3}(t) \; , \tag{6.23}$$

where c_2 and c_3 are complex functions of the temperature and the partial pressure of the reactants. A different formulation [146] expresses the concentration polarization as

$$U_{conc}(t) = c_2 \cdot e^{c_3 \cdot i_{fc}(t)} \; . \tag{6.24}$$

The measured polarization curve of a 100-cell stack is shown in Fig. 6.6. In the operating range of useful current densities, the figure clearly shows the effects of the activation losses and of the ohmic losses. With the exclusion of very low current densities $(i_{fc}(t) < 0.1\,\mathrm{A/cm^2})$, the curve can be conveniently linearized [98, 16], i.e., fitted by the equation

$$U_{fc}(t) = U_{oc} - R_{fc} \cdot I_{fc}(t) \; . \tag{6.25}$$

In this equation, U_{oc} is the voltage at which the linearized curve crosses the y axis with $i_{fc} = 0$ (no-current state). Thus U_{oc} should not be confused with the reversible voltage U_{rev} that is the voltage value corresponding to the nonlinear curve at $i_{fc} = 0$. The "resistance" R_{fc} is constant for a given type of cell and at constant operating conditions (pressure, temperature, humidity). Equation (6.25) is equivalent to the electrical circuit depicted in Fig. 6.7, which will be used in the following to describe fuel cells.

Fuel-Cell System

The voltage of a stack is obtained by multiplying the cell voltage by the number N of single cells in series,

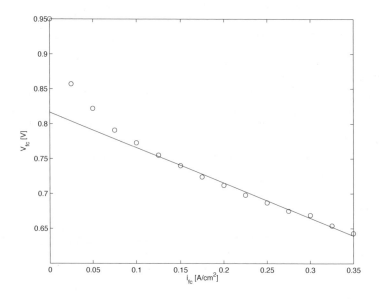

Fig. 6.6. Linear fitting of the polarization curve of a PEM fuel cell. Experimental data (circles) are from measurements taken on a stack of $N = 100$ cells with an active area of $A_{fc} = 200\,\text{cm}^2$ [211]. Operating conditions: $\vartheta_{fc} = 60°\text{C}$, $p_{ca,in} = 2.0\,\text{bar}$, $p_{an,in} = 2.2\,\text{bar}$, $\lambda_a = 2.2$. Voltage values: $U_{rev} = 0.95\,\text{V}$, $U_{oc} = 0.82\,\text{V}$.

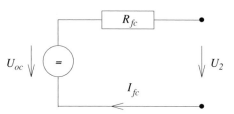

Fig. 6.7. Equivalent circuit of a fuel cell.

$$U_2(t) = N \cdot U_{fc}(t) \,, \tag{6.26}$$

while the stack current equals the cell current $I_{fc}(t)$. The parallel arrangement of several stacks is also possible. In the model introduced above it corresponds to an increase of the cell surface, since the output current increases.

The output power of a fuel-cell stack is

$$P_{st}(t) = I_{fc}(t) \cdot U_{fc}(t) \cdot N \,. \tag{6.27}$$

This power must cover the load demand $P_2(t)$ and the requirements of the auxiliaries $P_{aux}(t)$ (Fig. 6.3),

$$P_2(t) = P_{st}(t) - P_{aux}(t) \,. \tag{6.28}$$

The auxiliary power $P_{aux}(t)$ is the sum of various contributions,

$$P_{aux}(t) = P_0 + P_{em}(t) + P_{ahp}(t) + P_{hp}(t) + P_{cl}(t) + P_{cf}(t) , \qquad (6.29)$$

where $P_{em}(t)$ is the power of the compressor motor, $P_{hp}(t)$ is the power of the hydrogen circulation pump, $P_{ahp}(t)$ is the power of the humidifier water circulation pump, $P_{cl}(t)$ is the power of the coolant pump, and $P_{cf}(t)$ is the power of the cooling fan motor. The term P_0 is the value of the bias power that covers the linkage current necessary to keep a minimum flow of reactants throughout the whole operating range of the cell in order to keep it from shutting down. All the other power contributions will be described in the following sections.

Hydrogen Circuit

A sketch of the hydrogen circuit is shown in Fig. 6.8. Four branches can be distinguished: a supply branch regulated by a supply valve with a hydrogen mass flow rate $\overset{*}{m}_{h,c}(t)$, an inlet branch with a mass flow rate $\overset{*}{m}_{h,in}(t)$, a circulation branch with a mass flow rate $\overset{*}{m}_{h,out}(t)$, and another one with a reacting mass flow rate $\overset{*}{m}_{h,r}(t)$. The gas pressure at the node is $p_{an,in}(t)$ and at the outlet side of the cell is $p_{an,out}(t)$.

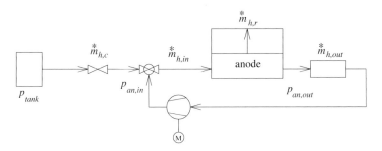

Fig. 6.8. Relevant variables in the hydrogen circuit.

The electric current intensity $I_{fc}(t)$ is given by the product of the quantity flow rate of the electrons, $\overset{*}{n}_e(t) = \overset{*}{n}_h(t) \cdot n_e$, the electron charge q, and the Avogadro constant N_0,

$$I_{fc}(t) = \overset{*}{n}_h(t) \cdot n_e \cdot q \cdot N_0 = \overset{*}{n}_h(t) \cdot n_e \cdot F , \qquad (6.30)$$

where F is the Faraday constant and $n_e = 2$ for hydrogen.

From the molar flow rate of hydrogen, $\overset{*}{n}_h(t) = I_{fc}(t)/(n_e \cdot F)$, the hydrogen mass flow rate $\overset{*}{m}_{h,r}(t) = M_h \cdot \overset{*}{n}_h(t)$ follows, where M_h is the hydrogen molar mass. By substitution, the relationship between the hydrogen mass flow rate and the fuel cell current for N cells is obtained as

$$\overset{*}{m}_{h,r}(t) = \frac{N \cdot I_{fc}(t) \cdot M_h}{n_e \cdot F} \; . \tag{6.31}$$

To obtain a uniform feeding of hydrogen on the cell surface, particularly in transient operation, the system must be fed with a hydrogen excess quantified by the variable $\lambda_h(t)$. The actual hydrogen entering the fuel cell is usually expressed as

$$\overset{*}{m}_{h,in}(t) = \lambda_h(t) \cdot \frac{N \cdot I_{fc}(t) \cdot M_h}{n_e \cdot F} = \lambda_h(t) \cdot \kappa_h \cdot I_{fc}(t) \; . \tag{6.32}$$

The value of $\lambda_h(t)$ is regulated by adjusting the opening of the hydrogen supply valve, since the mass flow rate $\overset{*}{m}_{h,in}(t)$ is the sum of a quantity recirculated $\overset{*}{m}_{h,rec}(t)$ and a quantity extracted from the hydrogen tank,

$$\overset{*}{m}_{h,in}(t) = \overset{*}{m}_{h,rec}(t) + \overset{*}{m}_{h,c}(t) \; . \tag{6.33}$$

The mass flow rate recirculated may differ from the anode output for a quantity that is periodically purged,[4]

$$\overset{*}{m}_{h,rec}(t) = \overset{*}{m}_{h,out}(t) - \overset{*}{m}_{h,pur}(t) \; . \tag{6.34}$$

The output mass flow rate is in turn given by

$$\overset{*}{m}_{h,out}(t) = \overset{*}{m}_{h,in}(t) - \overset{*}{m}_{h,r}(t) = (\lambda_h(t) - 1) \cdot \kappa_h \cdot I_{fc}(t) \; , \tag{6.35}$$

so that if there is no purge the mass flow rate extracted must equal the mass flow rate consumed in the cell reaction.

The hydrogen mass flow rate is related to the pressure difference between the cell inlet and outlet. Such a relationship depends on the type of flow fields that distribute the reactant gas to the electrodes. Graphite or carbon composite plates with small grooves (dimensions of ≈ 0.5 mm) for gas flow are extensively used in commercial fuel-cell stacks. For these flow fields, assuming laminar gas flow with continuous, uniform subtraction of mass due to the cell reaction, it is possible to write [98, 246]

$$\overset{*}{m}_{h,in}(t) = K_h \cdot (p_{an,in}(t) - p_{an,out}(t)) + \frac{1}{2} \cdot \overset{*}{m}_{h,r}(t) \; , \tag{6.36}$$

that is, to assume a linear dependency between mass flow rate and pressure drop. For other types of flow fields, e.g., porous plates, a quadratic dependency described by Darcy's law arises [246, 112].

The mass extracted from the tank flow rate is related to the hydrogen tank pressure and the cell pressure through the opening of the supply control valve. Such a dependency can be simplified as

---

[4] Purging prevents the accumulation of impurities in the hydrogen feed and N_2 accumulation due to diffusion from the cathode and, if done properly [214], removes excess water.

$$p_{an,in}(t) = p_{tank} - \xi_h(t) \cdot \overset{*}{m}{}_{h,c}^2(t) . \tag{6.37}$$

The circulation mass flow rate $\overset{*}{m}_{h,rec}(t)$ as well as the pressure levels $p_{an,out}(t)$ and $p_{an,in}(t)$ are related to the operation of the hydrogen circulation pump. The performance of the hydrogen pump is expressed by quasistatic characteristic curves, which represent the dependency

$$f_{hp}\left(\omega_{hp}(t), \overset{*}{m}_{h,rec}(t), \Pi_{hp}(t)\right) = 0 , \tag{6.38}$$

where $\Pi_{hp}(t) = p_{an,in}(t)/p_{an,out}(t)$ is the pump pressure ratio. The power absorbed by the motor-driven hydrogen pump is calculated as

$$P_{hp}(t) = \overset{*}{m}_{h,rec}(t) \cdot c_{p,h} \cdot \vartheta_{fc} \cdot \left(\Pi_{hp}(t)^{\frac{\gamma_h-1}{\gamma_h}} - 1\right) \cdot \frac{1}{\eta_{hp} \cdot \eta_{em}} , \tag{6.39}$$

where η_{em} is the efficiency of the drive motor and η_{hp} is the efficiency of the pump. Equation (6.39) can be approximated for small pressure drops as

$$P_{hp}(t) \approx \frac{\overset{*}{m}_{h,rec}(t) \cdot (p_{an,in}(t) - p_{an,out}(t))}{\rho_h \cdot \eta_{hp} \cdot \eta_{em}} , \tag{6.40}$$

where ρ_h is the average hydrogen mass density.

The system represented by equations (6.31)–(6.39) is a system of nine equations with the thirteen unknowns $\overset{*}{m}_{h,r}(t)$, $\overset{*}{m}_{h,in}(t)$, $\overset{*}{m}_{h,out}(t)$, $\overset{*}{m}_{h,rec}(t)$, $\overset{*}{m}_{h,pur}(t)$, $\overset{*}{m}_{h,c}(t)$, $I_{fc}(t)$, $p_{an,in}(t)$, $p_{an,out}(t)$, $\lambda_h(t)$, $\xi_h(t)$, $P_{hp}(t)$, and $\omega_{hp}(t)$. Since the cell current $I_{fc}(t)$ is the independent variable, three other variables must be specified for the system equations to be solved. In quasistatic simulations, system outputs are prescribed and control variables are calculated therefrom.

The hydrogen circuit control system is usually designed to keep the anode pressure $p_{an,in}(t)$ and the hydrogen circulation ratio $\lambda_h(t)$ at prescribed values. In particular, the anode pressure is prescribed as a function of the cathode pressure $p_{ca,in}(t)$, usually with a constant pressure difference (e.g., 0.2 bar) between the two sides. To meet the control requirements, the control system acts on the power of the hydrogen pump motor $P_{hp}(t)$, on the supply valve opening $\xi_h(t)$, and, if necessary, on the purge mass flow rate $\overset{*}{m}_{h,pur}(t)$.

Based on (6.36) and (6.32), the pressure drop $p_{an,in}(t) - p_{an,out}(t)$ is proportional to the reacting mass $\overset{*}{m}_{h,r}(t)$. As given by (6.35) the hydrogen circulating mass flow rate $\overset{*}{m}_{h,out}(t)$ is also proportional to $\overset{*}{m}_{h,r}(t)$. The pump power calculated with the approximation of (6.40) may thus be expressed as

$$P_{hp}(t) = \kappa_{hp} \cdot I_{fc}^2(t) . \tag{6.41}$$

Air Circuit

The oxygen mass flow rate that reacts with hydrogen is calculated from the cell electrochemistry. For each mole of hydrogen converted, 0.5 mol of oxygen is needed. Thus the molar flow rate of air that is strictly necessary for the reaction is given by

$$\overset{*}{n}_o(t) = \frac{1}{2} \cdot \overset{*}{n}_h(t) \ . \tag{6.42}$$

The reacting air mass flow rate thus can be calculated for the whole stack as a function of the fuel cell current,

$$\overset{*}{m}_{o,r}(t) = \frac{N \cdot I_{fc}(t) \cdot M_o}{2 \cdot n_e \cdot F} \ , \tag{6.43}$$

where M_o is the molar mass of oxygen.

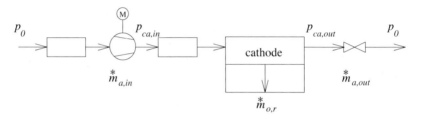

Fig. 6.9. Relevant variables in the air circuit.

For a stack that operates with excess air quantified by the variable $\lambda_a(t) > 1$, the effective air mass flow rate needed is given by

$$\overset{*}{m}_{a,in}(t) = \lambda_a(t) \cdot \frac{N \cdot I_{fc}(t) \cdot M_o}{2 \cdot n_e \cdot F} \frac{100}{21} = \lambda_a(t) \cdot \kappa_a \cdot I_{fc}(t) \ , \tag{6.44}$$

assuming the mass fraction of oxygen in dry inlet air to be 21%. The relationship between mass flow rate and pressure drop in the air circuit (see Fig. 6.9) is similar to (6.36) for laminar flow in the cathode channel,

$$\overset{*}{m}_{a,in}(t) = K_a \cdot (p_{ca,in}(t) - p_{ca,out}(t)) + \frac{1}{2} \cdot \overset{*}{m}_{o,r}(t) \ . \tag{6.45}$$

This relationship can be properly adapted to represent a nozzle-type behavior (see below).

The mass flow rate $\overset{*}{m}_{a,in}(t)$ and the supply manifold pressure $p_{ca,in}(t)$ are related to the air compressor operation. In fact, the performance of the air compressor is expressed by quasistatic characteristic curves, which represent the dependency

$$f_{cp}\left(\omega_{cp}(t), \overset{*}{m}_{a,in}(t), \Pi_{cp}(t)\right) = 0 \ , \tag{6.46}$$

where $\Pi_{cp}(t) = p_{ca,in}(t)/p_0$ is the compressor pressure ratio. Various approaches are available in the literature to fit steady-state compressor data and to derive the analytical expression of (6.46) [246]. The power absorbed by a motor-driven air compressor is calculated as

$$P_{cp}(t) = \overset{*}{m}_{a,in}(t) \cdot c_{p,a} \cdot \vartheta_a \cdot \left(\Pi_{cp}(t)^{\frac{\gamma_a-1}{\gamma_a}} - 1\right) \cdot \frac{1}{\eta_{cp}} , \qquad (6.47)$$

where ϑ_a is the supply air temperature and η_{cp} is the compressor efficiency.

At the output side of the cell, there are residual air mass flow rates, given by

$$\overset{*}{m}_{a,out}(t) = \overset{*}{m}_{a,in}(t) - \overset{*}{m}_{o,r}(t) . \qquad (6.48)$$

In the case where the mass flow $\overset{*}{m}_{a,out}(t)$ is discharged to the ambient, it is related to the cathode outlet pressure through the law that describes the pressure losses in the circuit,

$$p_{ca,out}(t) - p_0 = \xi_a(t) \cdot \overset{*}{m}^2_{a,out}(t) . \qquad (6.49)$$

In the presence of an expander to recuperate part of the enthalpy contained in the exhaust air, (6.49) is substituted by the dependency expressed by the expander characteristic curves,

$$f_{ex}\left(\omega_{cp}(t), \overset{*}{m}_{a,out}(t), \Pi_{ex}(t)\right) = 0 , \qquad (6.50)$$

where $\Pi_{ex}(t) = p_{ca,out}(t)/p_0$ is the expansion ratio. The power recuperated through the expander is given by

$$P_{ex}(t) = \overset{*}{m}_{a,out}(t) \cdot c_{p,a} \cdot \vartheta_{fc} \cdot \left(1 - \Pi_{ex}(t)^{-\frac{\gamma_a-1}{\gamma_a}}\right) \cdot \eta_{ex} , \qquad (6.51)$$

where η_{ex} is the expander efficiency. Unfortunately, expanders often have a rather small efficiency. Moreover, the available enthalpy is small as well. For these reasons, expanders are rarely installed, and controllable outlet valves (acting on $\xi_a(t)$) are used instead. Notice that the composition of the outlet air is different from that of the inlet air, since oxygen is consumed in the fuel cell. In some cases, it may be necessary to take this effect into account by properly modifying the specific heat $c_{p,a}$.

At steady state, a relationship exists between the compressor power, the expander power (if any), and any additional power provided by an external source, e.g., an electric motor,

$$-P_{cp}(t) + P_{ex}(t) + P_{em}(t) \cdot \eta_{em} = 0 , \qquad (6.52)$$

where η_{em} is the efficiency of the electric motor.

The system represented by (6.43)–(6.52)[5] is a system of nine equations with the twelve unknowns $\omega_{cp}(t)$, $p_{ca,in}(t)$, $p_{ca,out}(t)$, $P_{cp}(t)$, $P_{ex}(t)$, $\overset{*}{m}_{a,in}(t)$,

---

[5] With (6.49) in alternative to (6.50).

$\overset{*}{m}_{a,out}(t)$, $\overset{*}{m}_{o,r}(t)$, $\lambda_a(t)$, $I_{fc}(t)$, $\xi_a(t)$, and $P_{em}(t)$. Since the cell current $I_{fc}(t)$ is the independent variable, two other variables must be specified for the system equations to be solved. In quasistatic simulations, system outputs are prescribed and control variables are calculated therefrom.

The air circuit control system is usually designed such as to keep the excess air $\lambda_a(t)$ and the cathode pressure $p_{ca,in}(t)$ – and thus $\Pi_{cp}(t)$ – at prescribed constant values, acting on the purge valve opening $\xi_a(t)$ and on the compressor motor power $P_{em}(t)$. Based on (6.52) and (6.47), with constant motor efficiency η_{em} and omitting the possibility of an expander, it is possible to express $P_{em}(t)$ as

$$P_{em}(t) = \kappa_{cp} \cdot I_{fc}(t) . \tag{6.53}$$

Water Circuit

To prevent the dehydration of the fuel cell membrane, the air flow and in some cases also the hydrogen flow are humidified by the injection of water. The humidifier can act also as an air cooler, to reduce the temperature of the air entering the cell. Even if the hydrogen is not humidified, there still may be some water content at the anode outlet, due to the water diffusion through the fuel cell membrane. In the following, the presence of liquid water flooding the electrodes will not be taken into consideration for simplicity, though its influence is treated in the literature [211].

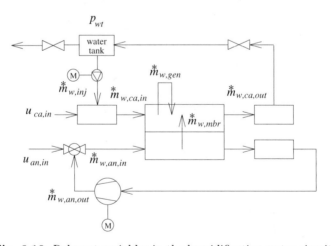

Fig. 6.10. Relevant variables in the humidification water circuit.

The water mass entering the cathode is calculated as the sum (see Fig. 6.10)

$$\overset{*}{m}_{w,ca,in}(t) = u_0 \cdot \overset{*}{m}_{a,in}(t) + \overset{*}{m}_{w,inj}(t) , \tag{6.54}$$

where u_0 is the humidity mass fraction in the ambient air and $\overset{*}{m}_{w,inj}(t)$ is the injected water mass flow rate. This is regulated by the water pump to achieve a specified humidity $u_{ca,in}(t)$ at the cathode inlet. This humidity is given by

$$u_{ca,in}(t) = \frac{\overset{*}{m}_{w,ca,in}(t)}{\overset{*}{m}_{a,in}(t)} . \tag{6.55}$$

At the cathode outlet, the water mass flow rate is increased by the water production in the reaction and by the water flow rate across the membrane,

$$\overset{*}{m}_{w,ca,out}(t) = \overset{*}{m}_{w,ca,in}(t) + \overset{*}{m}_{w,gen}(t) + \overset{*}{m}_{w,mbr}(t) . \tag{6.56}$$

The humidity at the cathode outlet thus is

$$u_{ca,out}(t) = \frac{\overset{*}{m}_{w,ca,out}(t)}{\overset{*}{m}_{a,out}(t)} . \tag{6.57}$$

The water production rate is a consequence of the cell reaction in (6.4) and it is evaluated as

$$\overset{*}{m}_{w,gen}(t) = \overset{*}{m}_{h,r}(t) \cdot \frac{M_w}{M_h} . \tag{6.58}$$

The water mass flow rate across the membrane is due to various phenomena [211]. First, the electro-osmotic drag phenomenon is responsible for the water molecules dragged across the membrane from anode to cathode by the hydrogen ions. Second, the gradient of the water concentration across the membrane causes a back diffusion of water from cathode to anode. Another source of back diffusion is the presence of a gradient in the partial pressure of water across the membrane. Since the membrane is usually very thin (approximately 50 μm) the gradients can be approximated as constant. Neglecting the third term, which depends on the concentration of water in the membrane [202], the water flux is evaluated as

$$\overset{*}{m}_{w,mbr} = M_w \cdot N \cdot \left(n_d \cdot \frac{I_{fc}(t)}{F} - A_{fc} \cdot D_w \cdot \frac{\phi_{ca}(t) - \phi_{an}(t)}{\delta_{mbr}} \right) , \tag{6.59}$$

where δ_{mbr} is the membrane thickness, n_d is the electro-osmotic drag coefficient, and D_w is the diffusion coefficient, the latter two being functions of the membrane water content and of the stack temperature [211]. The membrane water content (related to concentration of water in the membrane) in turn can be expressed as a function of $\phi_{ca}(t)$ and $\phi_{an}(t)$, which are the relative humidity of the cathode and of the anode, respectively. The relationship between the u's (absolute humidity) and the ϕ's (relative humidity) is generally

$$u = \frac{M_w}{M_a} \cdot \frac{\phi \cdot p_{sat}(\vartheta)}{p - \phi \cdot p_{sat}(\vartheta)} . \tag{6.60}$$

It is possible to approximate $u_{ca}(t)$ as an average of the values $u_{ca,in}(t)$ and $u_{ca,out}(t)$, from which $\phi_{ca}(t)$ to be used in (6.59) follows. The evaluation of $\phi_{an}(t)$ for the anode side is analogous.

The mass flow $\overset{*}{m}_{w,ca,out}(t)$ is purged and recuperated in the air/water separator. An additional mass flow to compensate inevitable losses is provided by a water pump. The power for the air humidifier water pump is given by

$$P_{ahp}(t) = \frac{\overset{*}{m}_{w,inj}(t) \cdot \Delta p_{ahp}(t)}{\rho_w \cdot \eta_{ahp}} = \kappa_{ahp} \cdot \overset{*}{m}_{w,inj}(t) , \qquad (6.61)$$

where $\Delta p_{ahp}(t) = p_{ca,in}(t) - p_{wt}$ is the pressure rise of the air humidifier pump, p_{wt} is the constant pressure in the water tank, and η_{ahp} is the efficiency of the pump.

If there is no hydrogen humidifier, at the anode inlet only the vapor in the supply hydrogen is present. The water mass flow rate entering the anode is given by

$$\overset{*}{m}_{w,an,in}(t) = \overset{*}{m}_{h,c}(t) \cdot u_{an,c} + \overset{*}{m}_{w,an,out}(t) , \qquad (6.62)$$

with $u_{an,c}$ usually assumed to correspond to a relative humidity of 100%. At the anode outlet, the water flow rate across the membrane is lost. Thus the remaining flow rate, assuming no additional water purge, is given by

$$\overset{*}{m}_{w,an,out}(t) = \overset{*}{m}_{w,an,in}(t) - \overset{*}{m}_{w,mbr}(t) . \qquad (6.63)$$

The humidity levels at the anode inlet and outlet stages, useful to evaluate the quantity $\overset{*}{m}_{w,mbr}(t)$, are

$$u_{an,in}(t) = \frac{\overset{*}{m}_{w,an,in}(t)}{\overset{*}{m}_{h,in}(t)} , \qquad (6.64)$$

$$u_{an,out}(t) = \frac{\overset{*}{m}_{w,an,out}(t)}{\overset{*}{m}_{h,out}(t)} . \qquad (6.65)$$

The system of (6.54)–(6.65) is a system of eleven equations with the twelve unknowns $\overset{*}{m}_{w,ca,in}(t)$, $\overset{*}{m}_{w,ca,out}(t)$, $\overset{*}{m}_{w,inj}(t)$, $\overset{*}{m}_{w,gen}(t)$, $\overset{*}{m}_{w,mbr}(t)$, $u_{ca,in}(t)$, $u_{ca,out}(t)$, $P_{ahp}(t)$, $\overset{*}{m}_{w,an,out}(t)$, $\overset{*}{m}_{w,an,in}(t)$, $u_{an,in}(t)$, and $u_{an,out}(t)$. Thus, one of the variables must be specified for the system to be solved. In quasistatic simulations, system outputs are prescribed and control variables are calculated therefrom.

The water circuit control system is designed such as to keep the humidity of the inlet air $u_{ca,in}(t)$ at a prescribed constant value, acting on the water injected mass flow rate $\overset{*}{m}_{w,inj}(t)$. Based on (6.55) and (6.61), $\overset{*}{m}_{w,inj}(t)$ must be proportional to the inlet air mass flow rate. This is in turn related to the cell current, so that it is possible to express $P_{ahp}(t)$ as

$$P_{ahp}(t) = \kappa_{ahp} \cdot I_{fc}(t) . \qquad (6.66)$$

Coolant Circuit

The power for the coolant pump is evaluated as

$$P_{cl}(t) = \frac{\overset{*}{m}_{cl}(t) \cdot \Delta p_{cl}}{\rho_{cl} \cdot \eta_{cl}} . \tag{6.67}$$

The coolant mass flow rate is calculated as a function of the heat to be removed, as

$$\overset{*}{m}_{cl}(t) = \frac{P_{l,st}(t)}{c_{p,cl} \cdot \Delta \vartheta_{cl}} , \tag{6.68}$$

where $c_{p,cl}$ and $\Delta \vartheta_{cl}$ are the specific heat and the temperature rise of the coolant.

The heat power to be removed by the coolant, $P_{l,st}(t)$, is practically coincident with the amount of heat generated by the cells, since exhaust gas contains only a very limited amount of enthalpy. Thus the evaluation of $P_{l,st}(t)$ for a stack may be derived from (6.17) for a single cell,

$$P_{l,st}(t) = (U_{id} - U_{fc}(t)) \cdot N \cdot I_{fc}(t) . \tag{6.69}$$

The temperature difference in (6.68) cannot exceed an admissible value which is a characteristic value of the cooling system. Equation (6.67) may thus be written as $P_{cl}(t) = \kappa_{cl} \cdot P_{l,st}(t)$.

The thermal power must be removed from the water with the help of cooling fans, usually capable of an on/off operation, such as in internal combustion engines. The power required for the cooling fan $P_{cf}(t)$ is proportional to the mass flow rate of the cooling air, by an expression similar to (6.47). In turn, the cooling air mass flow rate is proportional to the coolant mass flow rate, so that it is possible to write $P_{cf}(t) = \kappa'_{cf} \cdot \overset{*}{m}_{cl}(t) = \kappa_{cf} \cdot P_{l,st}(t)$.

The coolant control system regulates the coolant mass flow rate and the cooling fan operation in order to remove the heat generated by the fuel cell. If (6.69) is combined with (6.25), it is possible to express the power requirement of the cooling system as

$$P_{cl}(t) + P_{cf}(t) = \kappa_{cl,1} \cdot I_{fc}(t) + \kappa_{cl,2} \cdot I_{fc}^2(t) . \tag{6.70}$$

Overall Model

Now the total auxiliary power can be evaluated as a function of the fuel cell current. Using the simplified expressions derived above, i.e., (6.41), (6.53), (6.66), and (6.70), equation (6.29) may be written as

$$P_{aux}(t) = P_0 + \kappa_{hp} \cdot I_{fc}^2(t) + \kappa_{ahp} \cdot I_{fc}(t) + \kappa_{cp} \cdot I_{fc}(t) + \kappa_{cl,1} \cdot I_{fc}(t) +$$
$$+ \kappa_{cl,2} \cdot I_{fc}^2(t) = P_0 + \kappa_1 \cdot I_{fc}(t) + \kappa_2 \cdot I_{fc}^2(t) . $$
$$\tag{6.71}$$

On the other hand, the stack output power $P_{st}(t)$ has a quadratic dependency on $I_{fc}(t)$, as described by (6.25) and (6.27). Thus, the result of (6.71) agrees with semi-empirical data that suggest a linear dependency between auxiliary power and stack output power [248]. For a larger output power the dependency may be less than linear, thus a nonlinear approximation should be used, e.g., exponential [231]. Other formulations were derived leading to a quadratic dependency between $P_{aux}(t)$ and $P_{st}(t)$ [16].

In a first approximation, (6.71) can be further simplified using a linearized accessory power [99],

$$P_{aux}(t) = P_0 + N \cdot \kappa_{aux} \cdot I_{fc}(t) , \tag{6.72}$$

which explicitly takes into account multiple cells.

Equation (6.72) can be combined with (6.28), (6.27), and (6.25), or generally with (6.19), to obtain the quadratic expression for $I_{fc}(t)$

$$N \cdot R_{fc} \cdot I_{fc}^2(t) - (N \cdot U_{oc} - N \cdot \kappa_{aux}) \cdot I_{fc}(t) + P_2(t) + P_0 = 0 , \tag{6.73}$$

from which the final equation for the cell current may be written as [99]

$$I_{fc}(t) = \frac{N \cdot (U_{oc} - \kappa_{aux}) - \sqrt{N^2 \cdot (U_{oc} - \kappa_{aux})^2 - 4 \cdot N \cdot R_{fc} \cdot (P_2(t) + P_0)}}{2 \cdot N \cdot R_{fc}} . \tag{6.74}$$

Once the cell current is known, the hydrogen consumption can be calculated using (6.33),

$$\overset{*}{m}_{h,c}(t) = N \cdot \frac{I_{fc}(t) \cdot M_h}{n_e \cdot F} = k_h \cdot I_{fc}(t) . \tag{6.75}$$

Fuel Cell Efficiency

Besides the electrochemical efficiency $\eta_{id} = U_{rev}/U_{id}$, various other efficiencies can be defined for fuel-cell stacks. The voltage efficiency for a single cell is defined as

$$\eta_V(I_{fc}) = \frac{U_{fc}(I_{fc})}{U_{rev}} . \tag{6.76}$$

In the affine approximation (6.25), the voltage efficiency also is an affine function of the cell current. It decreases linearly as $I_{fc}(t)$ increases, showing its higher values at low loads. Another possible efficiency is the current or Faradaic efficiency that compares the effective current with the current theoretically delivered,

$$\eta_I(\lambda_h) = \frac{I_{fc}}{I_{th}} = \frac{1}{\lambda_h} . \tag{6.77}$$

The global efficiency of a fuel cell $\eta_{fc}(I_{fc}) = \eta_V(I_{fc}) \cdot \eta_I$ has the same dependency as $\eta_V(I_{fc})$ on the cell current, since η_I typically is a constant.

The system efficiency takes into account also the auxiliary power. The effective power delivered is compared with the theoretically deliverable power,

$$\eta_{st}(I_{fc}) = \frac{P_2(I_{fc})}{N \cdot U_{id} \cdot I_{fc}} \;, \tag{6.78}$$

thus the total stack efficiency is $\eta_{st,tot}(I_{fc}) = \eta_{st}(I_{fc}) \cdot \eta_I$. From (6.73) and (6.12), an expression for the efficiency may be derived

$$\eta_{st}(I_{fc}) = \eta_{id} \cdot \frac{U_{oc}}{U_{rev}} \cdot \left(1 - \frac{R_{fc} \cdot I_{fc}}{U_{oc}} - \frac{P_0}{U_{oc} \cdot I_{fc} \cdot N} - \frac{\kappa_{aux}}{U_{oc}}\right) \;. \tag{6.79}$$

Obviously, this expression has a maximum for η_{st} at $I_{fc,max} = \sqrt{P_0/(N \cdot R_{fc})}$.

Figure 6.11 illustrates the dependency of various power terms and efficiency values on the fuel cell current.

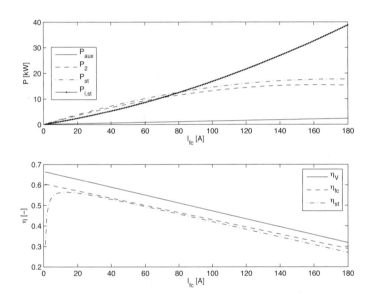

Fig. 6.11. Fuel cell power (top) and efficiency (bottom) as a function of fuel cell current. Fuel cell data: $N = 250$, $U_{rev} = 1.23\,\mathrm{V}$, $U_{oc} = 0.82\,\mathrm{V}$, $A_{fc} = 200\,\mathrm{cm}^2$, $R_{fc} = 0.0024\,\Omega$, $\lambda_h = 1.1$, $P_0 = 100\,\mathrm{W}$, $\kappa_{aux} = 0.05\,\mathrm{V}$.

6.2.2 Dynamic Modeling of Fuel Cells

The physical causality representation of a fuel cell is sketched in Fig. 6.12. The model input variable is the stack current $I_2(t) = P_2(t)/U_2(t)$. The model output variables are the stack voltage $U_2(t)$ and the fuel consumption $\overset{*}{m}_{h,c}(t)$.

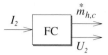

Fig. 6.12. Fuel cells: physical causality for dynamic modeling.

The dynamic model of a PEM fuel cell that will be discussed here is a system-level, lumped-parameter model based on the ideas presented in the previous section. The dynamic effects include in principle electrochemical, fluid dynamic, and thermal effects. However, the order of magnitude of the relevant time constants of such processes is quite different [98],

- Hydrogen and air manifolds: $O(10^{-1}$ s)
- Membrane water content: $O(10^{0}$ s)
- Control system: $O(10^{0}$ s)
- Stack temperature: $O(10^{2}$ s)

where $O(\cdot)$ denotes the order of magnitude.

The fastest transient phenomena are the electrochemical ones. These dynamics are due to charge double layers at the membrane/electrode interfaces. The ion/electron charge separation at these interfaces creates a charge storage that can be described by a double-layer capacitance C_{dl} [185]

$$\frac{d}{dt}U_{act}(t) = \frac{i_{fc}(t)}{C_{dl}} - \frac{U_{act}(t)}{R_{act} \cdot C_{dl}} ,$$ (6.80)

where R_{act} is defined as the ratio of the steady-state activation polarization, given by (6.20) or (6.21) and the current density $i_{fc}(t)$. The equivalent circuit of a cell is modified accordingly, as depicted in Fig. 6.13. For simplicity, the concentration polarization is not considered. However, the resulting time constants $R_{act} \cdot C_{dl}$ are so small that this class of dynamic effects may be neglected without causing any substantial errors.

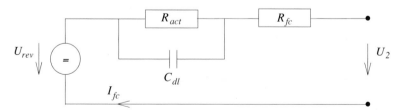

Fig. 6.13. Equivalent dynamic electric circuit for a fuel cell.

The fluid dynamic transient effects taken into consideration are usually of the capacitive type, i.e., they describe the variation of gas pressure in a reservoir as a consequence of mass flows entering and leaving the reservoir.

The models available in the literature differ in the number of pressure levels introduced.

The most immediate extension of the quasistatic equations presented in the previous section consists of keeping the same pressure levels, i.e., $p_{ca,in}(t)$ and $p_{ca,out}(t)$ for the cathode channel, $p_{an,in}(t)$ and $p_{an,out}(t)$ for the anode channel, but letting them vary as a function of the mass flow rates $\overset{*}{m}_{a,in}(t)$, $\overset{*}{m}_{a,out}(t)$, $\overset{*}{m}_{h,in}(t)$, $\overset{*}{m}_{h,out}(t)$, etc. Assuming that the supply manifold temperature is kept close to the operating stack temperature ϑ_{st} by means of a cooling device,[6] the state equation for $p_{ca,in}(t)$ is

$$\frac{d}{dt}p_{ca,in}(t) = \frac{R_a \cdot \vartheta_{st}}{V_{a,sm}} \cdot \left(\overset{*}{m}_{a,cp}(t) - \overset{*}{m}_{a,in}(t) \right) , \qquad (6.81)$$

where R_a is the air gas constant and $V_{a,sm}$ is the volume of the air supply manifold. Analogously, the state equation for $p_{ca,out}(t)$ is

$$\frac{d}{dt}p_{ca,out}(t) = \frac{R_a \cdot \vartheta_{st}}{V_{a,rm}} \cdot \left(\overset{*}{m}_{a,out}(t) - \overset{*}{m}_{a,rm}(t) \right) , \qquad (6.82)$$

where $V_{a,rm}$ denotes the return manifold volume.

Equations (6.81)–(6.82) introduce two additional variables, namely $\overset{*}{m}_{a,cp}(t)$ and $\overset{*}{m}_{a,rm}(t)$, representing the air mass flow rate delivered by the compressor and the one discharged through the control valve or the expander. These quantities replace $\overset{*}{m}_{a,in}(t)$ and $\overset{*}{m}_{a,out}(t)$ in (6.46)–(6.47) and (6.49)–(6.51), respectively.

Substituting its quasistatic counterpart (6.52), an additional state equation can be written for the compressor speed as

$$\frac{d}{dt}\omega_{cp}(t) = \frac{P_{em}(t) \cdot \eta_{em} - P_{cp}(t) + P_{ex}(t)}{\Theta_{cp} \cdot \omega_{cp}(t)} . \qquad (6.83)$$

Inertial effects in the connecting pipes, particularly at the compressor output [211], also may be taken into account in a lumped-parameter fashion, using the approach of the quasi-propagatory model [52]. This model describes the mass flow rates in the pipes as new state variables, evaluated as a function of the pressure levels at the manifolds using equations of the type

$$\frac{d}{dt}\overset{*}{m}(t) = \frac{\kappa \cdot \Delta p - \overset{*}{m}(t)}{\tau} , \qquad (6.84)$$

where $\kappa \cdot \Delta p$ is the steady-state mass flow rate and τ is the time constant of the process that depends primarily on the pressure levels and on the pipe lengths [52].

Compared with the quasistatic case, if inertial effects are not modeled there are two more variables, namely $\overset{*}{m}_{a,cp}(t)$, and $\overset{*}{m}_{a,rm}(t)$, balanced by two

---

[6] The humidifier can perform this function.

additional equations. The cell current $I_{fc}(t)$ is still the independent variable, while $P_{em}(t)$ is the main control variable of the air circuit. An additional control input that may contribute to the regulation of the cathode pressure is $\xi_a(t)$, although in most applications this valve opening is kept constant. Consequently, all the remaining quantities are determined, in particular the excess air $\lambda_a(t)$ by (6.44).

A similar modeling approach can be used for the anode channel. The state equation for $p_{an,in}(t)$ is

$$\frac{d}{dt}p_{an,in}(t) = \frac{R_h \cdot \vartheta_{st}}{V_{h,sm}} \cdot \left(\overset{*}{m}_{h,c}(t) - \overset{*}{m}_{h,in}(t) - \overset{*}{m}_{h,rec}(t) \right) , \qquad (6.85)$$

with obvious meaning of the variables. The state equation for $p_{an,out}(t)$ similarly is

$$\frac{d}{dt}p_{an,out}(t) = \frac{R_h \cdot \vartheta_{st}}{V_{h,rm}} \cdot \left(\overset{*}{m}_{h,out}(t) - \overset{*}{m}_{h,rec}(t) - \overset{*}{m}_{h,pur}(t) \right) . \qquad (6.86)$$

As in the quasistatic case, the number of variables exceeds the number of relationships by four. The cell current $I_{fc}(t)$ is still the independent variable, while $P_{hp}(t)$, $\xi_h(t)$, and $\overset{*}{m}_{h,pur}(t)$ are the control variables of the hydrogen circuit. Therefore, all the remaining quantities are determined, in particular the excess hydrogen $\lambda_h(t)$ by (6.32) and the hydrogen consumption $\overset{*}{m}_{h,c}(t)$ by (6.33).

As for the water circuit, usually no dynamic effects are introduced in the manifolds. Therefore, as in the quasistatic case, the number of variables exceeds the number of relationships available by one. Since the control variable is the injected mass flow rate $\overset{*}{m}_{w,inj}(t)$, all the remaining quantities are determined, in particular the inlet air humidity $u_{ca,in}(t)$ by (6.55).

A different method of lumping the capacitive properties may lead to recognizing distinct pressure levels for the anode and the cathode channels, $p_{an}(t)$ and $p_{ca}(t)$, respectively [281, 185, 202, 204]. Mass flow rates entering and leaving the channels are evaluated according to some "nozzle-type" equation as a function of the pressure difference between the channels and the supply and return manifolds introduced above, which are characterized by their own pressure levels as described earlier.

This model structure allows a generalization of the equations above to the non-isothermal case. The energy conservation law must be invoked together with the mass conservation law to derive at least two additional equations (one for the cathode, one for the anode) at the two additional variable temperature levels $\vartheta_{an}(t)$ and $\vartheta_{ca}(t)$. Energy fluxes to be considered are the enthalpy flows associated with mass fluxes and heat exchanged with the solid walls facing the gas flows. An additional state equation may be written for the solid body temperature $\vartheta_{st}(t)$, which may vary due to heat exchanged with fluids, heat produced by the electrochemical reaction, or converted electric power [281, 185, 5].

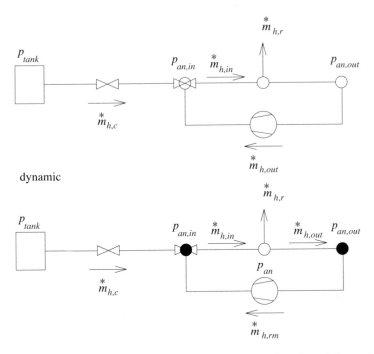

quasistatic

dynamic

Fig. 6.14. Fluid dynamics modeling of anode circuit, quasistatic and dynamic cases.

Another model refinement possible consists of introducing two more state variables representing the vapor water masses $m_{w,ca}(t)$ and $m_{w,an}(t)$ in the cathode and in the anode, respectively. The corresponding state equations express the conservation of water mass in the electrode channels – thus replacing (6.56) and (6.63) – including the terms due to membrane transport and reaction. This allows a diract evaluation of the absolute humidity $u_{ca}(t)$, $u_{an}(t)$ to be used, e.g., in (6.59) [202, 204].

In summary, in dynamic simulation $U_2(t)$ is calculated using (6.19), possibly combined with (6.80) and (6.26). The hydrogen consumption rate is calculated by solving the coupled system of differential equations for the anode, the cathode, and the water circuit, taking the various control variables as inputs ($P_{hp}(t)$, $\xi_h(t)$, $\dot{m}_{w,inj}(t)$, $P_{em}(t)$, and $\xi_a(t)$).

6.3 Reformers

In view of the problems concerning the hydrogen storage, the use of different energy carriers that are easier to handle is a significant and still open research task. The main interest of current research efforts for automotive applications[7] is concentrated on liquid hydrocarbons, although they require an additional on-board process to extract from the supply fuel the hydrogen to operate the fuel cell ("reforming"). The advantages are that (i) there is no need for special storage systems, (ii) until the production and distribution of hydrogen is better established, the existing infrastructure for fossil fuels can be used, and (iii) the consumer acceptance is surely higher. The refueling operation does not change for the user from how it is today. This would clearly favor the adoption of this new technology. Disadvantages are that (i) the resulting vehicle is not a zero-emission vehicle (CO_2 emissions), (ii) higher complexity and costs, (iii) lower system efficiency since the reformer block requires energy, (iv) a lower life span due to the impurities in the reformer gas, and especially (v) poor response times, which makes the use of reformers critical during transient operation. These major drawbacks limit the application of fuel reforming in vehicles, although some prototypes of fuel-cell vehicles have adopted this technology (DaimlerChrysler Necar 5 [61]). The use of reforming-based fuel-cell systems as small, stationary auxiliary power units for trucks and camper vans, where efficiency and response time is not an issue, seems to be more promising.

Among liquid hydrocarbons, gasoline, diesel, and methanol are the most common reforming fuels for vehicles. The advantages of methanol (methyl alcohol), CH_3OH, are that (i) methanol can be obtained from various renewable resources (e.g., biomass), (ii) the conversion of natural gas into methanol allows the use of remote natural gas sources, (iii) due to its simpler molecular structure, it is technically easier to reform, which simplifies the hydrogen production and yields a very high H_2/CO_2 ratio for liquid fuel, and (iv) there exist already prototypes of fuel cells (DMFCs) that allow for a direct electrochemical conversion of methanol and thus render an upstream on-board reformer superfluous. However, methanol needs a dedicated distribution infrastructure and corrosion-resistant refuelling and storage equipment, it is poisonous if swallowed, and it burns with an invisible flame. Methanol is also water soluble, which makes it more dangerous.

For the on-board production of hydrogen from methanol there are basically three methods, the "steam reforming," the partial oxidation (POx), and the methanol scission. Steam reforming is generally used for methanol reforming. Together with carbon dioxide, hydrogen is produced from methanol and water vapor. The overall reaction can be written as

$$CH_3OH + H_2O \rightarrow CO_2 + 3H_2, \quad \Delta h_R = 58.4\,\text{kJ/mol} . \tag{6.87}$$

In real reactions another product is carbon monoxide CO. The CO formation is due to the direct decomposition of methanol. Its concentration is affected by the water-gas shift reaction

$$CH_3OH \rightarrow CO + 2H_2, \quad \Delta h_R = 97.8\,\text{kJ/mol} , \tag{6.88}$$

---

[7] There is also a deep interest in methane reforming, however, not for mobile applications.

$$CO + H_2O \rightarrow CO_2 + H_2 \quad \Delta h_R = -39.4\,\text{kJ/mol}\,. \tag{6.89}$$

However, even this model is a very simplified approximation of the reality. Reforming is a complex mechanism with many side reactions.

With the partial oxidation method the methanol is directly oxidized with the aid of oxygen. The overall reaction can be written as

$$CH_3OH + 1/2O_2 \rightarrow CO_2 + 2H_2, \quad \Delta h_R = -193\,\text{kJ/mol}\,, \tag{6.90}$$

and it can be regarded as the result of the two partial reactions

$$CH_3OH + 1/2O_2 \rightarrow CO + H_2 + H_2O, \quad \Delta h_R = -153.6\,\text{kJ/mol}\,, \tag{6.91}$$

$$CO + H_2O \rightarrow CO_2 + H_2, \quad \Delta h_R = -39.4\,\text{kJ/mol}\,, \tag{6.92}$$

with the possible formation of formaldehyde as another byproduct. Although partial oxidation allows an exothermal reaction, for proper methanol–oxygen ratios, the quality of the exhaust gas is not suitable for low-temperature fuel cells. Lower hydrogen concentrations, a higher carbon monoxide content, and an incomplete methanol conversion make a complex aftertreatment necessary. The same problems arise also with the methanol scission method, in which basically a thermal cracking takes place. Combinations of steam reforming and POx are also studied, since in this way autothermal reformers can be obtained.

Figure 6.15 shows a schematic of the methanol steam reforming process. Steam reforming is typically carried out over a catalyst bed containing oxides of copper, zinc, and aluminum ($CuO/ZnO/Al_2O_3$). With this catalyst, reformer operation is limited at low temperature by the formation of water condensate on the catalyst and at high temperature by sintering of the catalyst.

Since the methanol reforming reaction (6.87) is endothermic, external heat must be supplied to the reformer. Usually heat is transferred directly to the reformer reactants though, to avoid excess temperatures, certain solutions have been proposed with an intermediate heat transfer fluid (oil) heated in a separate heat exchanger. Heat is mostly produced in a burner by combustion or catalytic oxidation of excess hydrogen leaving the fuel cell anode, or by combustion of methanol extracted from the main feedstock. Of course the former solution is preferable, since the combustion of methanol is not pollution-free. Besides the recuperation of heat from the anode outlet ($\overset{*}{m}_{h,pur}$ of Sect. 6.2), other thermal integrations with the fuel-cell system are possible, including recovery of low-temperature heat from the stack cooling system [2]. Additional heat is required to preheat, vaporize, and superheat the reactant steam–methanol mixture fed to the reformer at its operating temperature. This heat can be recovered from the anode outlet as well, or from the reformer outlet, which is at a temperature normally higher than the operating temperature of a (PEM) fuel cell.

Reformate gas produced by the reformer has a small content of CO (typi-cally, 2% by volume [2]), which, being a severe poison to the platinum catalyst used in the fuel cell, must be eliminated. Various methods to clean CO from hydrogen exist, including selective oxidation on a catalyst (platinum on alu-mina) bed.

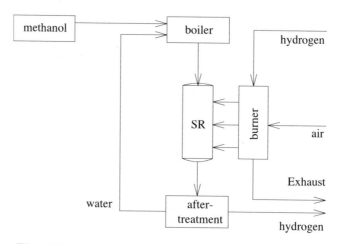

Fig. 6.15. Schematic of methanol steam reforming process.

Autothermal reforming is essentially a combination of POx and steam reforming, consisting of adding air (thus oxygen) to the steam reforming reac-tion. Steam reforming is an endothermal reaction, while POx is an exothermal reaction. Under adiabatic conditions (insulated reformer) the equilibrium tem-perature is given by the relative ratios of steam and air fed into the reformer.

In autothermal reformers, the burner is only responsible for the preheating, vaporization, and superheating of methanol. The construction of the reactor itself is similar to that in steam reforming systems, but inherently less com-plicated. However, an additional complexity is due to the handling of reactant air, for which a compressor is required [247].

6.3.1 Quasistatic Modeling of Fuel Reformers

The causality representation of a fuel reformer in quasistatic simulations is sketched in Fig. 6.16. The model input variable is the hydrogen mass flow rate $\overset{*}{m}_{h,c}$. The model output variable is the methanol consumption rate $\overset{*}{m}_m$.

Conversion Ratio

A simple model of a methanol steam reformer can be derived [4, 205] consider-ing the reactions taking place in a tube (plug-flow reactor) filled with a porous

Fig. 6.16. Fuel reformers: causality representation for quasistatic modeling.

catalyst bed. Input flows to the catalyst tube are pre-vaporized methanol and steam. An isothermal process without pressure losses is assumed here. More complex, one-dimensional models accounting for detailed heat transfer, diffusion of species, and friction, are available in literature both for steam reformers [174] and for autothermal reformers [44].

For the range of conditions of interest to vehicle applications, the water-gas shift reaction (6.89) is usually neglected without a substantial loss of accuracy [4]. Experimental work has shown that the reaction rate of the reforming reaction is linear with the concentration of methanol, while the reaction rate of the gas-shift reaction is affected only slightly by the concentration of methanol or water, and thus it can be regarded as a constant [4]. Introducing the reaction rate constants k_1 for (6.87) and k_2 for (6.88), which are both functions of temperature and pressure (and k_1 also of the steam-to-methanol ratio of the feed gas σ), the rate equations for methanol and hydrogen inside the reformer (see Fig. 6.17) are

$$d\overset{*}{n}_m = -(k_1 \cdot C_m(x) + k_2) \cdot dm_c = -\overset{*}{n}_m(0) \cdot dx \;, \qquad (6.93)$$

$$d\overset{*}{n}_h = (3 \cdot k_1 \cdot C_m(x) + 2 \cdot k_2) \cdot dm_c \;, \qquad (6.94)$$

where x is the fraction of methanol converted (methanol conversion ratio), $m_c(x)$ is the mass of the catalytic bed, and $C_m(x)$ is the molar concentration of methanol. The latter is a function of x and of σ and can be calculated considering that for each mole of methanol that reacts, the total number of moles increases by two. Hence, at constant pressure and temperature the concentration of methanol is given by [205]

$$C_m(x) = (1-x) \cdot \frac{1+\sigma}{1+\sigma+2\cdot x} \cdot C_m(0) \;, \qquad (6.95)$$

where $C_m(0)$ is the initial concentration of methanol (see below) and σ is the steam-to-methanol ratio of the feed gas (typically ranging from 0.67 to 1.5). The integration of (6.93) yields the catalyst mass that is necessary to achieve a certain conversion ratio x,

$$m_c(x) = \int_0^x \frac{\overset{*}{n}_m(0)}{k_1 \cdot C_m(\xi) + k_2} \, d\xi \;, \qquad (6.96)$$

which, after substitution of (6.95), is solved as

$$m_c(x) = \overset{*}{n}_m(0) \cdot \left(c_1 \cdot \ln \frac{c_2}{c_2 - c_3 \cdot x} - c_4 \cdot x \right) \;, \qquad (6.97)$$

where $c_1 = (U \cdot c_3 + 2 \cdot c_2)/c_3^2$, $c_2 = U \cdot (k_1 \cdot C_m(0) + k_2)$, $c_3 = U \cdot k_1 \cdot C_m(0) - 2 \cdot k_2$, $c_4 = 2/c_3$, and $U = 1 + \sigma$. Now, (6.94) is integrated, yielding

$$\overset{*}{n}_h(x) = \overset{*}{n}_m(0) \cdot \int_0^x \frac{3 \cdot k_1 \cdot C_m(\xi) + 2 \cdot k_2}{k_1 \cdot C_m(\xi) + k_2} \, d\xi = 3 \cdot x \cdot \overset{*}{n}_m(0) - k_2 \cdot m_c(x) . \quad (6.98)$$

The latter equation describes the fact that the hydrogen output molar rate is a fraction x of the theoretical value $3 \cdot \overset{*}{n}_m(0)$, diminished by the molar rate $k_2 \cdot m_c(x)$ of the CO production.[8]

Now, for a given reformer, m_c is given and $\overset{*}{n}_h(x)$ is calculated from the hydrogen mass flow rate required, as

$$\overset{*}{n}_h(x) = \frac{\overset{*}{m}_{h,c}}{M_h} , \quad (6.99)$$

where M_h is the molar mass of hydrogen. Therefore, the system of highly nonlinear coupled equations (6.97)–(6.98), which has to be solved iteratively, yields the corresponding values of x and $\overset{*}{n}_m(0)$. Notice that $C_m(0)$ is also a function of the unknown quantity $\overset{*}{n}_m(0)$,

$$C_m(0) = \frac{\overset{*}{n}_m(0)}{1 + U} \cdot \frac{p_{ref}}{R \cdot \vartheta_{ref}} , \quad (6.100)$$

where R, p_{ref} and ϑ_{ref} are the universal gas constant, the pressure, and the temperature of the reformer. Typical values of p_{ref} range from 1 to 3 bar and of ϑ_{ref} from 430 to 570 K.

Finally, the methanol mass flow rate consumed is evaluated as

$$\overset{*}{m}_m = \overset{*}{n}_m(0) \cdot M_m , \quad (6.101)$$

where M_m is the molar mass of methanol ($M_m = 32$ kg/kmol).

Fuel Processing Efficiency

A clear definition of a (steam) reformer efficiency that is independent of the fuel cell operation is complicated by the critical feedback loop, in which the anode exhaust is burned to partially satisfy the heat requirements for the steam reforming reaction. If the hydrogen excess ratio (or utilization factor) λ_h is taken as a measure of the hydrogen exhaust burned, then the related losses are taken into account in the fuel cell utilization factor η_I (6.77) that can now be rewritten as

---

[8] This model does not consider the hydrogen lost during CO removal. If, for instance, selective oxidation is used, typically about the same number of moles of hydrogen as of CO are lost.

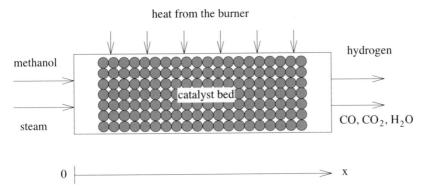

Fig. 6.17. Schematic of a catalyst bed steam reformer tube.

$$\eta_I(t) = \frac{\overset{*}{m}_{h,r}(t)}{\overset{*}{m}_{h,c}(t)} = \frac{1}{\lambda_h(t)} \ . \tag{6.102}$$

The reformer efficiency is defined as the ratio of chemical output power to total input power, which is the sum of the methanol chemical power and the fraction of the external power required[9] that is not recuperated within the fuel-cell system. This fraction is usually rather complicated to estimate, the calculation involving various chemical, thermal, and fluid dynamic aspects. In a first approximation, it is convenient to assume that the total external power is proportional to the methanol mass flow rate,

$$P_{ref}(t) = \kappa_{ref} \cdot \overset{*}{m}_m(t) \ . \tag{6.103}$$

Now, if a fraction μ of $P_{ref}(t)$ is recuperated from the anode outlet, from the cooling system, or within loops internal to the reformer circuits, the reformer efficiency is

$$\eta_{ref}(t) = \frac{\overset{*}{m}_{h,c}(t) \cdot H_h}{\overset{*}{m}_m(t) \cdot H_m + (1 - \mu) \cdot \kappa_{ref} \cdot \overset{*}{m}_m(t)} \ , \tag{6.104}$$

with H_h and H_m being the lower heating values of hydrogen ($120\,\text{MJ/kg}$) and methanol ($19.9\,\text{MJ/kg}$), respectively. Of course, if $\mu = 1$, the reformer efficiency is simply a ratio of the energy flows across the reformer. This may be the case of POx reformers, where the exhaust anode energy is not necessary for the catalytic reformer since the reaction is exothermal, although it may be useful, e.g., to vaporize the reactants. Typical values for η_{ref} are 0.62 for steam reforming and 0.69 for POx [180].

Using the conversion ratio x as in (6.98), the reformer efficiency becomes

---

[9] That is the sum of the heat power required for the steam reforming process and the power required to drive the auxiliaries.

$$\eta_{ref}(t) = \frac{H_h \cdot M_h}{M_m} \cdot \frac{3 \cdot x - k_2 \left(c_1 \cdot \ln \dfrac{c_2}{c_2 - c_3 \cdot x} - c_4 \cdot x \right)}{H_m + (1 - \mu) \cdot \kappa_{ref}}. \qquad (6.105)$$

The overall efficiency of the fuel-cell system with reforming is finally evaluated as

$$\eta_{fcr} = \eta_{st} \cdot \eta_I \cdot \eta_{ref}. \qquad (6.106)$$

6.3.2 Dynamic Modeling of Fuel Reformers

The physical causality of a fuel reformer is the reverse of that sketched in Fig. 6.16. The input side is represented by the control variables of the various circuits, while the output variables are the hydrogen mass flow rate $\overset{*}{m}_{h,c}$ and the methanol mass flow rate $\overset{*}{m}_m$. Notice that this changes the physical causality of the dynamic model of the fuel cell alone as sketched in Fig. 6.12. The fuel cell submodel now is not controlled on the hydrogen side.[10] Thus it receives the hydrogen mass flow rate as an input variable.

A simple dynamic model of a fuel processor system uses the conversion ratio defined in the previous section and describes the reformer dynamics, including the vaporizer/superheater and the gas clean-up stage, by assuming a second-order behavior [107],

$$\tau^2 \cdot \frac{d^2}{dt^2} \overset{*}{n}_h(t) + 2 \cdot \tau \cdot \frac{d}{dt} \overset{*}{n}_h(t) + \overset{*}{n}_h(t) = 3 \cdot x \cdot \frac{\overset{*}{m}_m(t)}{M_m} - k_2 \cdot m_c(x), \qquad (6.107)$$

where τ is the time constant of the process and x is a static function of $\overset{*}{n}_m$ and m_c, as described in the previous section. This model describes the response to an input positive step of methanol flow as a critically damped (essentially, exponential) increase over time. The response to a decrease in methanol flow is also modeled with a second-order differential equation. However, due to the fact that the decrease of hydrogen flow requires no heat, these dynamics are much faster than the positive step. Typical values of τ are 2 s for step-up transients and 0.4 s for step-down transients [107].

---

[10] The recirculation loop is usually deactivated and the control valve is moved to the methanol side.

7

Supervisory Control Algorithms

In all types of hybrid vehicles, a supervisory controller must determine how to operate the single power paths, in order to satisfy the power demand of the drive line in the most convenient way. The main objective of that optimization is the reduction of the overall energy use, usually in the presence of various constraints due to driveability requirements and the characteristics of the components.

Based on the review article [225], this chapter describes the theoretical concepts of various types of control strategies for parallel hybrid-electric vehicles. Appendix I contains examples of the application of these ideas. Similar approaches that have been investigated also for series hybrids [272, 19, 40], combined hybrids [209, 51], and fuel-cell hybrids [184, 213] are not treated in this book.

7.1 Introduction

A parallel hybrid-electric vehicle can be operated in any of the modes summarized in Table 7.1. Besides the power split ratio u (see Chap. 4 for its definition), additional control variables are the clutch status and the engine status. Both are Boolean, clutch engaged ($B_c = 1$) or disengaged ($B_c = 0$), engine on ($B_e = 1$) or off ($B_e = 0$). Both zero-emission (ZEV) and regenerative braking modes ($u = 1$) can be operated in principle either with the engine shut down and disengaged or shut down but still engaged. The other modes (ICE, power assist, battery recharge) all require the engine to be on and engaged. In the trivial case of no power demand, these values are always zero, of course ($u = 0$, $B_e = 0$, $B_c = 0$).

In relation to the torque and speed values required at the drive line the supervisory controller determines at each instant the operating mode to be adopted and the value of the ratio $u(t)$. In all practical control strategies, the engine is shut down when the torque at the wheels is negative or zero, i.e.,

Table 7.1. Control parameters for different parallel hybrid operating modes.

	Mode	u	B_e	B_c
1	ICE	0	1	1
2a	ZEV	1	0	0
2b	ZEV	1	0	1
5a	Regenerative braking	1	0	0
5b	Regenerative braking	1	0	1
3	Power assist	$\in (0,1)$	1	1
4	Recharge	< 0	1	1

when the vehicle is coasting or braking. The control strategies in the literature differ for the choice of $u(t)$ and of $B_e(t)$ when the power required is positive.

Control strategies may be classified according to their dependency on the knowledge of future situations. Non-causal controllers require the detailed knowledge of the future driving conditions. This knowledge is available when the vehicle is operated along regulatory drive cycles, or for public transportation vehicles that have prescribed driving profiles. In all other cases, driving profiles are not predictable in advance, at least not in the sense that the exact speed and altitude profiles as a function of time would be known a priori. In these cases, causal controllers must be used.

A second classification can be made among heuristic, optimal, and sub-optimal controllers. The first class of controllers represents the state of the art in most prototypes and mass-production hybrids. Optimal controllers are inherently non-causal, although the next sections will show how to substantially reduce the amount of information required. Sub-optimal controllers are often causal.

7.2 Heuristic Control Strategies

Heuristic controllers are based on Boolean or fuzzy rules involving various vehicular variables. A typical heuristic approach, sometimes called "electric assist" strategy [272, 273, 150, 207], is based on the torque demand and on the vehicle speed:

- below a certain vehicle speed the motor is used alone ($u = 1$);
- above this speed threshold and below the maximum engine torque at the current engine speed, the engine alone is used ($u = 0$);
- however, if the battery state of charge is too low, the engine is forced to deliver excess torque to recharge the battery ($u < 0$);
- if the state of charge is too high, the motor is used alone ($u = 1$); and
- above the engine maximum torque at the current engine speed, the motor is used to assist the engine ($0 < u < 1$).

In the baseline "electric assist" strategy the key control parameter is the speed threshold at which the choice is made between motor or engine operation (see Fig. 7.1). Other strategies may define different thresholds based on different combinations of vehicular variables. One possibility is to operate the engine only above a specified fraction of the maximum torque of the engine at the current engine speed [272]. Another strategy consists of using an acceleration threshold at which the choice is made between the quasi-stationary, ICE-based ($u = 0$) mode, and the transient, electrical ($u = 1$) mode [35]. Also power thresholds are often used. They offer the advantage of being immediately comparable with the power limits of the prime movers [151, 33, 269, 29, 80]. More complex combinations of power demand, speed, and possibly other variables may also be used [14].

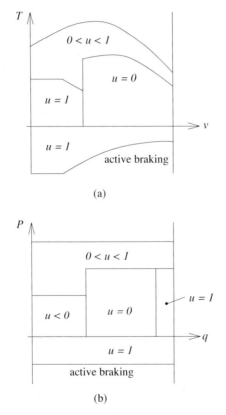

Fig. 7.1. Typical heuristic power management control for parallel hybrids, in terms of motor torque and vehicle speed (a) or motor power and state of charge of the battery (b).

When both prime movers are on, the value of u is mainly determined by the battery state of charge. In the baseline "electric assist" strategy, the engine delivers excess torque only when the state of charge reaches a specified lower bound. In contrast, the strategy sometimes referred to as "balanced electric assist" [272] continuously modulates the power split ratio to keep the state of charge at a constant level.

In other control systems, the value of u varies continuously as a function of two to four vehicular parameters, based on a "fuzzy" set of rules. Vehicular variable sets may include vehicle speed and engine speed [142], state of charge, power demand and motor speed [223, 218], temperature, and vehicle acceleration and speed, state of charge [88], torque demand and state of charge [11, 24], vehicle acceleration and engine speed [148].

The main advantage of heuristic controllers is that they are intuitive to conceive and rather easy to implement. If properly tuned, they can provide good results in terms of fuel consumption reduction and charge sustainability. Unfortunately, the behavior of heuristic controllers strongly depends upon the choice of the thresholds involved, which actually can vary substantially with the driving conditions [227]. The resulting limited robustness of heuristic controllers, in addition to the tuning effort required, motivates the development of model-based controllers that optimize the power flows.

7.3 Optimal Control Strategies

7.3.1 Optimal Behavior

The main objective of the energy-management controller is to minimize fuel consumption along a route. Obviously, it is not necessary to minimize the fuel mass-flow rate at each instant of time, but rather the total fuel consumed during a driving mission.

Possible missions are single or multiple repetitions of the governmental test-drive cycles, see Chap. 2. Alternatively, missions can be driving patterns recorded on typical routes. During operation, an HEV energy-management controller can explicitly use all of the available information about the mission. The mission information is either provided by the driver or identified implicitly by the control algorithm.

The energy-management controller must respect various hard and soft constraints. For instance, the battery must never be depleted below a specified threshold, while the torque provided by the engine is limited.

Performance Index

As illustrated in Figure 7.2, the simplest performance index $J = m_f(t_f)$ is the fuel mass m_f consumed over a mission of duration t_f. Hence, J can be written as [62, 20, 172, 236, 143, 245, 268, 276, 140, 282]

$$J = \int_0^{t_f} \overset{*}{m}_f(t, u(t))\, dt. \tag{7.1}$$

Pollutant emissions can also be included in the performance index J by considering the more general expression

$$J = \int_0^{t_f} L(t, u(t))\, dt, \tag{7.2}$$

where $L(\cdot)$ is the cost function. The emission rates of the regulated pollutants can be included in the performance index (7.1) by introducing a weighting factor for each pollutant species [125, 31, 153]. However, if the ICE is a spark-ignited engine operated with stoichiometric air/fuel ratios, its pollutant emissions can usually be reduced to negligible levels using a three-way catalytic converter. Accordingly, the pollutant emission is not considered as part of the optimization problem, although in practice "duty-cycle" (on/off operation) or engine shutoff at idle can cause problems due to excessive pollutant emissions caused by engine or catalyst cooling.

Drivability issues are sometimes included in the optimality criterion. For example, the cost function in [284] includes an anti-jerk term, which consists of the engine acceleration squared, multiplied by an arbitrary weighting factor. Smoothness and driver-acceptance considerations are included in [268] among the local constraints discussed below.

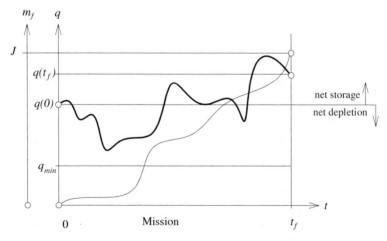

Fig. 7.2. Typical trajectories of the state variable $q(t)$ and consumed fuel mass $m_f(t)$ (bold) along a mission.

Integral Constraints

Obviously, the drive mode that minimizes the performance index (7.1) corresponds to a purely electrical strategy in which all of the traction power is provided by the battery. However, if the energy recovered by regenerative braking is not sufficient to sustain the battery charge, this choice can leave the battery completely discharged at the end of the mission.

The sustenance of the energy-storage system is required for the vehicle certification process. Since only small deviations from the nominal value of the state of charge (SoC) are permitted at the end of tests to assess vehicle energy consumption, energy-management controllers must ensure small SoC variations over drive cycles.

In principle, the sustenance constraint can be taken into account in two different ways, namely, as a soft constraint, that is, by penalizing deviations from the initial value of the energy stored at the end of the mission, or as a hard constraint, by requiring that the energy stored at the end of the mission equal the value at the start of the mission.

To represent constraints on the final SoC $q(t_f)$, a penalty function $\phi(q(t_f))$ is added to the performance index (7.2) to obtain a charge-sustaining performance index of the form

$$J = \phi(q(t_f)) + \int_0^{t_f} L(t, u(t)) \, dt. \tag{7.3}$$

A hard constraint, in which $q(t_f)$ must exactly match the initial value $q(0)$, is often explicitly assumed [62, 20, 172, 236, 143, 245, 268, 140, 282].

Soft constraints can be added as functions of the difference $q(t_f) - q(0)$. In [152] the quadratic penalty function $\phi(q(t_f)) = \alpha \cdot (q(t_f) - q(0))^2$ is used, where α is a positive weighting factor. In [153], the term $\alpha \cdot (q(t) - q(0))^2$ is included in the cost function.

A quadratic penalty function tends to penalize deviations from the target SoC, regardless of the sign of the deviation. In contrast, a linear penalty function of the type

$$\phi(q(t_f)) = w \cdot (q(0) - q(t_f)), \tag{7.4}$$

where w is a positive constant, penalizes battery use, while favoring the energy stored in the battery as a means for saving fuel in the future. Since the penalty function (7.4) can be expressed as

$$\phi(q(t_f)) = w \cdot \int_0^{t_f} \dot{q} \, dt, \tag{7.5}$$

the variable $w \cdot \dot{q}(t)$ can be added to the cost function of (7.3) to yield the performance index

$$J = \int_0^{t_f} (L(t, u(t)) + w \cdot \dot{q}(t)) \, dt. \tag{7.6}$$

The parameter w is often chosen arbitrarily [125, 284, 276], while in the regulatory standard SAE J1711 [216], $w = 38$ kWh per gallon of gasoline. Other physically meaningful definitions of w are discussed below.

The piecewise-linear penalty function

$$\phi(q(t_f)) = \begin{cases} w_{dis} \cdot (q(0) - q(t_f)), & q(t_f) > q(0), \\ w_{chg} \cdot (q(0) - q(t_f)), & q(t_f) < q(0), \end{cases} \qquad (7.7)$$

is at the core of equivalent-consumption minimization strategies (ECMS) [183], which represent real-time implementations of optimal control algorithms. Their formulation is presented in the next section.

Local Constraints

Local constraints can also be imposed on the state and control variables. These constraints mostly concern physical operation limits, notably the maximum engine torque and speed, the motor power, or the battery state of charge. Constraints on the control variables are imposed in [268] to enhance smoothness and driver acceptance.

7.3.2 Optimization Methods

This section presents various approaches to evaluating optimal control laws. These approaches are grouped into three subclasses, namely, static optimization methods, numerical dynamic optimization methods, and closed-form dynamic optimization methods.

Static Optimization

Since a mission usually lasts hundreds to thousands of seconds, while, at each time t, multiple values of $u(t)$ must be evaluated, finding the optimal control law by inspecting all possible solutions requires excessive computational and memory resources. The simplified approach described in [20] for a series HEV can easily be extended to parallel HEVs. This approach does not require detailed knowledge of the actual power demand at the wheels $P_m(t)$, but only its average and root mean square values. The charge sustenance is guaranteed only when duty-cycle operations are performed. For continuous operation of the primary energy source, charge sustenance must be achieved using an additional slow, integrative controller, which makes the controller inherently suboptimal. Dynamic optimization techniques, as presented in the next section, avoid this drawback.

Numerical Optimization Methods

Dynamic programming is commonly used for optimization over a given time period [172, 284, 152, 153, 13]. This method can be used to minimize the performance index (7.1) in the presence of a hard or a soft constraint on the terminal value of the SoC.

Dynamic programming requires gridding of the state and time variables, and thus the optimal trajectory is calculated only for discretized values of time and SoC. Consequently, the integral (7.3) and the state dynamics

$$\dot{q}(t) = f(t, q(t), u(t)) \tag{7.8}$$

are replaced by discrete counterparts. A useful property of all dynamic programming algorithms is that their computational burden increases linearly with the final time t_f. Since the computational burden increases exponentially with the number of state variables, however, reasonably long missions can be analyzed only if the number of state variables is small. Conveniently, (7.8) is a scalar equation.

The cost-to-go function $\Gamma(t, q)$ is the cost over the optimal trajectory passing through the point (t, q) in the time-state space, up to the terminal time t_f, as shown in Fig. 7.3. Based on this definition, the value of $J = \Gamma(0, q(0))$. To evaluate $\Gamma(t, q)$, the computation proceeds with a time-discretization step Δt backward from time $t_f - \Delta t$ to time $t = 0$ [30, 27]. The cost-to-go function is then evaluated from the recursion

$$\Gamma(t, q(t)) = \min_{u} \left\{ \Gamma\left(t + \Delta t, q(t) + \dot{q}(t, q(t), u(t)) \cdot \Delta t\right) + L(t, u(t)) \cdot \Delta t \right\}. \tag{7.9}$$

The initial condition for (7.9) is imposed at time t_f by

$$\Gamma(t_f, q(t_f)) = \phi(q(t_f)). \tag{7.10}$$

The feedback function

$$U(t, q(t)) = \arg\min_{u} \left\{ \Gamma\left(t + \Delta t, q(t) + \dot{q}(t, q(t), u(t)) \cdot \Delta t\right) + L(t, u(t)) \cdot \Delta t \right\} \tag{7.11}$$

represents the control strategy to be stored for real-time operation.

Due to the discretization of the state, the values of q are either interpolated or approximated by the nearest available values on the grid. In the latter case, rounding can artificially increase or decrease the battery energy calculated over the optimal trajectory. The energy artificially introduced or deleted by rounding determines whether the adopted state-space discretization is acceptable or the number of grid points must be increased.

Improved algorithms with reduced computational time are available. They will be discussed in detail in Appendix III. For example, the iterative dynamic programming algorithm in [13] is based on the adaptation of the state space.

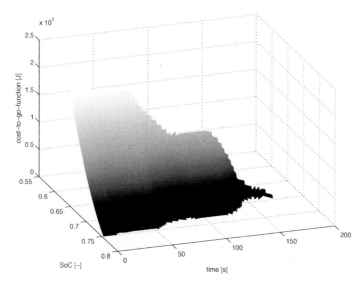

Fig. 7.3. Illustrative cost-to-go function Γ given by (7.9). The data are calculated for the operation of an A-class HEV in the ECE cycle, for a terminal time of 196 s, a target terminal SoC of 0.7, a time step of 1 s, and an SoC discretization of 0.01% of full charge. The target terminal SoC can be reached starting from an initial SoC lower than 0.753 and greater than 0.648.

At each iteration, the state space is selected as a small fraction of the entire space, centered around the optimal trajectory evaluated in the previous iteration. Another approach, used in [282], reduces the computing time by splitting the mission into a series of time sections and solving an optimization problem for each of those sections. This approach generally produces a suboptimal solution.

Approximations of the original optimization problem can substantially reduce the computational burden. For instance, if the cost function is linearized with respect to the control variable, a standard linear programming method, such as the simplex algorithm, can be used [236, 245]. Alternatively, quadratic programming [143, 282] requires a quadratic cost function with linear constraints. Of course, all of these simplifications yield suboptimal solutions.

Analytical Optimization Methods

Direct numerical optimization methods require substantial amounts of computational time. One approach that often permits a reduction of the computational effort is based on the minimum principle [62, 236, 140]. This method introduces a Hamiltonian function to be minimized at each time, that is,

$$u(t) = \arg \min_v \{H(t, v, \mu(t))\}, \tag{7.12}$$

where

$$H(t, q, u, \mu) = L(t, u) + \mu \cdot f(t, q, u). \tag{7.13}$$

In this formulation, t is a continuous variable, and the dynamics of the SoC are given by (7.8). The parameter $\mu(t)$, which corresponds to the adjoint state in classical optimal control theory, is described by the Euler-Lagrange equation

$$\dot{\mu}(t) = -\frac{\partial}{\partial q} f(t, q(t), u(t)). \tag{7.14}$$

The approximation

$$\dot{q}(t) \approx \widetilde{f}(t, u(t)) \tag{7.15}$$

of (7.8) is introduced in [62, 236], where the influence of the SoC $q(t)$ on the internal battery parameters, such as open-circuit voltage and internal resistance, is neglected. Consequently, (7.14) becomes $\dot{\mu} \approx 0$. This assumption may be not valid in, e.g., hydraulic hybrids [79], but it is reasonable for hybrid electric HEVs, where only large deviations of the SoC can cause substantial variations of the internal battery parameters. With this assumption, the adjoint state $\mu(t)$ remains approximately constant along the optimal trajectory. The optimization problem is thus reduced to searching for a constant parameter μ_0 that approximates $\mu(t)$ for a given mission.

This optimization problem is straightforward. For every time t the Hamiltonian must be minimized with respect to the control variable $u(t)$. A further simplification is possible when the Hamiltonian can be expressed as an explicit function of the control variable. This approach requires an explicit description of $L(t, u(t))$ and the SoC deviation rate $\dot{q}(t)$, which can be obtained using simple, but accurate modeling tools [210].

In some special applications, the Hamiltonian turns out to be an affine function of the control variable [231]. In this case, the minimum principle states that the optimal control variable is found at its extreme values, depending on the sign of the switching function $\partial H / \partial u$. When the switching function is zero, determining the optimal trajectory requires additional information, typically the second derivative of the Hamiltonian with respect to the variable u.

Equivalent-Consumption Minimization Strategies

The value of the adjoint state $\mu(t)$ depends primarily on the choice of $\phi(q(t_f))$. In the case of a hard constraint, μ_0 guarantees the fulfillment of the constraint. This value must be determined iteratively, using numerical methods.

In the case of a linear soft constraint such as (7.4), the value of μ_0 is [62]

$$\mu_0 = \frac{\partial \phi}{\partial q(t_f)} = -w . \tag{7.16}$$

If the soft constraint on the final SoC is of the piecewise-linear type (7.7), then the value of the constant adjoint state must be consistent with the final sign of the SoC deviation [230], in particular,

$$\mu_0 = \frac{\partial \phi}{\partial q(t_f)} = \begin{cases} -w_{dis}, & q(t_f) > q(0) , \\ -w_{chg}, & q(t_f) < q(0) . \end{cases} \tag{7.17}$$

A meaningful expression for the ECMS is obtained when both terms in (7.13) are reduced to power terms, namely,

$$H(t, s(t), u(t)) = P_f(t, u(t)) + s(t) \cdot P_e(t, u(t)) . \tag{7.18}$$

In this equation, $P_f(t, u(t)) = H_{LHV} \cdot \overset{*}{m}_f(t, u(t))$ is the fuel power (with H_{LHV} being the lower heating value of the fuel) and $P_e(t, u(t)) = -\dot{q}(t, u(t)) \cdot V_b \cdot Q_{max}$ is the battery power. The *equivalence factor*

$$s(t) = -\mu(t) \cdot \frac{H_{LHV}}{V_b \cdot Q_{max}} \tag{7.19}$$

represents a nondimensional scaling of the adjoint state. The equivalence factor thus converts battery power to an equivalent fuel power that must be added to the actual fuel power to attain a charge-sustaining control strategy [230, 172]. Under the assumption (7.15), (7.18) can be reduced to

$$H(t, u(t)) = P_f(t, u(t)) + s_0 \cdot P_e(t, u(t)) , \tag{7.20}$$

where s_0 is a constant equivalence factor. The ECMS approach is also referred to as a *cost-based strategy* [31], *real-time control strategy* [128], or *online optimization strategy* [140].

Alternative definitions of Hamiltonian-like functions, often derived on a purely heuristic basis, can be reduced to (7.18). In [140] the Hamiltonian to be minimized is

$$H(t, u(t), \alpha(t)) = P_{diss}(t, u(t)) + \alpha(t) \cdot P_e(t, u(t)) + K_{cond}(u(t)). \tag{7.21}$$

Similarly to (7.13), the main idea of (7.21) is to minimize the overall power dissipation $P_{diss}(t, u(t)) = P_f(t, u(t)) + P_e(t, u(t)) - P_m(t)$. This minimization does not guarantee SoC sustenance. Therefore, the weighted correction $\alpha(t) \cdot P_e(t, u(t))$ is appended to the cost function to penalize the SoC deviations. Moreover, a penalty term $K_{cond}(u(t))$ is added to prevent frequent on/off switching of the engine, which leads to additional energy losses and engine wear. Obviously $P_m(t)$ does not depend on $u(t)$ and, therefore, the Hamiltonian function (7.21) is equivalent to (7.18) except for $K_{cond}(u(t))$.

The nondimensional performance measure of [125] is heuristically designed to locally minimize the fuel-consumption rate and maximize the efficiency of the electrical path. Thus this approach cannot be reduced to a form similar to (7.18).

In ECMS, the optimization problem is shifted to the evaluation of the equivalence factor $s(t)$. The simplest approach is to assume that a constant value of s is approximately valid for every type of driving condition. In [276], such a value is implicitly assumed to be equal to unity, since the Hamiltonian is simply the sum of the fuel power and battery power. In [31], the objective function is not explicitly shown, but is assumed to be a combination of higher order polynomials based on fuel-conversion efficiency and battery SoC. In general, the equivalence factor s depends on the driving conditions along the mission.

Equivalence Factors

In general, the function $\phi(q(t_f))$ depends on the conversions occurring within the system that transform fuel energy to electrical energy and vice versa. In certain special cases, it is possible to evaluate $\phi(q(t_f))$ precisely. For example, in a purely electric vehicle the fuel equivalent of a given battery energy use can be evaluated easily. It equals the fuel necessary to reload the battery by the same amount of energy, which can be calculated from the well-to-tank (plug) efficiency of the electric grid.

In an autonomous HEV, the electric storage system (a battery, a supercapacitor, etc.) can only be recharged via regenerative braking, or by the fuel converter (an engine, a fuel cell, etc.), without any external device. Therefore, $\phi(q(t_f))$ cannot be evaluated as easily as for an electric vehicle. One case for which it is rather simple to derive expressions for the fuel equivalent is the case of constant efficiencies both of the electrical path and of the thermal path (see Fig. 7.4a).

In this case, the fuel equivalent $\phi(E_e(t_f))$ of a positive amount of battery energy used in the mission $E_e(t_f) = \int_0^{t_f} P_e(\tau)\,d\tau$, i.e., provided by the storage system, is the fuel energy necessary to reload the same amount. Figure 7.4b shows that $\phi(E_e(t_f)) = E_e(t_f)/(\eta_e \cdot \eta_f)$. The fuel equivalent therefore is a linear function of $E_e(t_f)$. The proportionality coefficient, or *equivalence factor*, for this case is calculated as

$$s_{dis} = \frac{1}{\eta_e \cdot \eta_f} . \qquad (7.22)$$

The fuel equivalent $\phi(E_e(t_f))$ of a negative amount of energy $E_e(t_f)$, i.e., one that recharges the storage system, is the fuel energy that can be saved by using $E_e(t_f)$ in the future. Figure 7.4c shows that in this case $\zeta(E_e(t_f))$ equals $E_e(t_f) \cdot \eta_e/\eta_f$. Again, the fuel equivalent is a linear function of $E_e(t_f)$. The equivalence factor for this case is calculated as

$$s_{chg} = \frac{\eta_e}{\eta_f} . \qquad (7.23)$$

In general, $s_{chg} < s_{dis}$ holds. The values of s_{dis} and s_{chg} are equal only in the case in which there are no losses in the electrical path, i.e., $\eta_e = 1$.

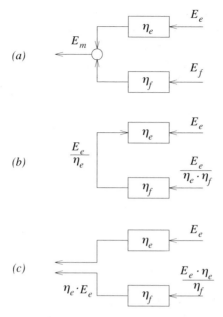

Fig. 7.4. Fuel equivalent of the electrical energy for constant efficiencies of the parallel paths (a), for positive (b), and for negative (c) electrical energy.

The general case of variable path efficiencies has been analyzed in several publications. Some authors used constant, average values for the fuel and electrical efficiencies, further distinguishing between $\bar{\eta}_e^{(d)}$ in the discharge phase and $\bar{\eta}_e^{(c)}$ in the charge phase [283, 182, 183, 140, 37]. The equivalence factors are thus

$$s_{dis} = \frac{1}{\bar{\eta}_e^{(d)} \bar{\eta}_f} \,, \tag{7.24}$$

$$s_{chg} = \frac{\bar{\eta}_e^{(c)}}{\bar{\eta}_f} \,. \tag{7.25}$$

However, this approach is strongly dependent on the way the average efficiencies are defined, and it often requires heuristic corrections to avoid excessive SOC excursions [37].

A more consistent analysis has shown [226] that s_{chg} and s_{dis} can be evaluated purely from energy considerations, without any assumption on the path efficiencies. The procedure requires collecting data on the electrical energy use $E_e(t_f)$ and the fuel energy use $E_f(t_f)$ over a mission of duration t_f, obtained with similarly structured control policies. A possible choice is to use various constant values of the control variable, in the range $u \in [-u_l, u_r]$ given by the upper and lower bounds for the SOC that are admissible during the system operation. Figure 7.5 illustrates such a procedure.

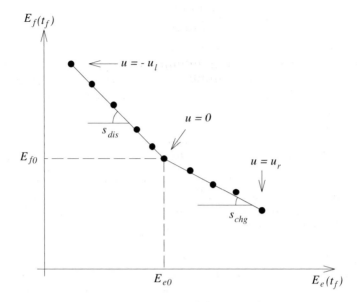

Fig. 7.5. Typical dependency between $E_f(t_f)$ and $E_e(t_f)$ for a given vehicle configuration and for a given drive cycle.

In the pure thermal case $u = 0$ the fuel energy used, E_{f0}, is also the energy that would be used to drive the cycle if no electrical path were present. The electrical energy use in the pure thermal case, E_{e0}, is not zero, due to regenerative braking power (which provides a negative contribution) and idle losses in the electrical path (i.e., the losses when the power at the output stage of the electrical path is zero, which add a positive contribution to E_{e0}). For many common scenarios, it has been observed that the pure thermal case separates the curve $E_f(t_f) = f(E_e(t_f))$ in two branches, which are nearly linear in the range of interest. The slopes of these lines that fit the data are the two equivalence factors s_{dis} and s_{chg}.

The linear form of the curve $E_f(t_f) = f(E_e(t_f))$ is observed even if the efficiencies of the parallel paths vary depending on the operating point. This may be explained by an averaging effect that is due to the large number of operating points included in a drive cycle. A straightforward but tedious analysis [229] shows that the average efficiencies of the thermal and electrical paths over the drive cycle can be computed from the equivalence factors, but not vice versa as in (7.24)–(7.25). The equivalence factors can be effectively used to combine the fuel consumption and the state of charge variations in an equivalent specific fuel consumption conveniently expressed, for instance, in liter/100 km.

7.3.3 Real-time Implementation

All of the optimization techniques discussed above require knowledge about future driving conditions. This fact makes their implementation in real-time controllers a challenging task. This section lists the level of information required by various control strategies and how such information can be achieved during real-time operation.

Predictive Control

The highest level of information is available when the complete mission is known at the outset. When this is the case, an optimization procedure, such as dynamic programming, can be applied. For public transportation vehicles along fixed routes, where the mission is known in advance, such an approach may be viable. The resulting feedback function $U(t, q(t))$ given by (7.11) can then be implemented in the powertrain control systems.

For passenger cars, the mission is usually unknown at the outset, and the estimation of future driving conditions must be made online. The combination of such an estimation with the application of dynamic programming follows the model-predictive control (MPC) paradigm [143, 13], which requires estimation of the power demand $P_m(t)$ on a prediction horizon of duration t_f. Dynamic programming is then used to calculate the optimal control law, which is applied for a shorter control horizon $t_c < t_f$. In [13] the power demand is estimated using speed limit, curve radius, and road slope, as a function of the distance along the route. These data are obtained by combining onboard and GPS navigation. The vehicle speed is calculated using a dynamic model of the vehicle as a function of the target speed, which is constrained by the speed limit, maximum safe speed, especially in curves, and maximum speed allowed by traffic conditions (a parameter that may be available in future applications using radar sensors).

Prediction of the adjoint state

In strategies derived from optimal control theory, such as ECMS, uncertainty about future driving conditions is transferred to uncertainty on the correct (optimal) value of the constant adjoint state approximation μ_0. The advantage with respect to predictive control is that only one parameter must be determined instead of a power demand $P_m(t)$ as a function of time. Various methods are available for the online estimation of μ_0. These methods, which are described below, can be classified into three approaches depending on the information used, namely, past driving conditions, past and present driving conditions, and past, present and future driving conditions.

Past Driving Conditions

One technique for estimating μ_0 is based on ideas borrowed from pattern recognition [125]. Optimal values of μ_0 are pre-calculated offline for a set of representative driving patterns, which are composed of urban, expressway, and suburban driving patterns. Up to 24 characteristic parameters, such as average velocity, standstill time, and total time, can be chosen to characterize driving patterns. During real-time operation, a neural network periodically decides which representative driving pattern is closest to the current driving pattern. Then, the energy-management controller switches to the corresponding value of the parameter μ_0.

Past and Present Driving Conditions

Pattern recognition methods use information only about past driving conditions. Alternative controllers evaluate μ_0 continuously by adapting μ_0 to the current driving conditions [143, 128, 123] or simply to the current value of the SoC [140].

In some implementations, the equivalence factor $s_0(t)$ is calculated as the partial derivative of the present fuel power with respect to the battery power, that is,

$$s_0(t) = -\frac{\partial P_f}{\partial P_e}(t). \tag{7.26}$$

In other words, by varying the control variable $u(t)$, it is possible to obtain a function $P_f(t) = \varphi(P_e(t))$ that is specific to the current driving conditions. The function $\varphi(\cdot)$, which has a negative slope, is a measure of the fuel cost of the *replacement* battery energy, which is the battery energy required in the future to compensate use at the present rate $P_e(t)$.

Consequently, the fuel equivalent of $P_e(t)$ is simply $\varphi(-P_e(t))$, while the corresponding Hamiltonian is

$$H(P_f(t), P_e(t)) = P_f(t) + \varphi(-P_e(t)), \tag{7.27}$$

which is consistent with the definition (7.26).

Such an approach assumes that similar operating conditions will exist in the future, that is, the replacement energy will "cost" the same amount of fuel energy as it does in the current driving conditions. In general, this assumption leads to trajectories that are neither fuel optimal nor charge sustaining. For this reason, [128] includes an SoC-control factor in the cost function. A modified adaptive strategy of the same type [123] assumes instead that the missing battery energy will be replaced when the ICE operates under more favorable conditions.

The strategy described in [140] mainly emphasizes the SoC-control factor. The weighting factor $\alpha(t)$ of (7.21), which plays the role of an equivalence factor, is chosen to be an affine function of the current SoC, that is,

$$\alpha(t) = \alpha_0 - \beta \cdot (q(t) - q(0)), \tag{7.28}$$

where α_0 and β are constants. In other words, when the battery is depleted, the value of α increases so that the use of fuel is favored. Some rules of thumb are given to assign proper values to the parameters α_0 and β. The former, in particular, is calculated using energy considerations that are similar to those underlying (7.26).

Past, Present, and Future Driving Conditions

Information about the present cannot guarantee optimality of the control action, while the charge cannot be sustained with such an approach. Therefore, future driving conditions must be predicted or estimated, although the power demand profile cannot be estimated in detail. In particular, ECMS-based strategies such as *telemetry ECMS* (T-ECMS) [230] or *adaptive ECMS* (A-ECMS) [172] do not consider the complete future power demand profile along the mission, but rather a few characteristic parameters to estimate the optimal value of μ_0. These two approaches are presented below.

A piece of information that is usually assumed to be available is the GPS-derived altitude profile of the route that the vehicle intends to follow. The altitude profile provides the road slope as a function of the distance covered. To transform the altitude profile into a slope function of time, the future vehicle speed profile must be estimated. Information on speed limits, which is often considered to be available, can be used for such an estimation. With regard to traffic conditions, future cars are expected to include radar and other sensors that can be used to obtain this information [230].

T-ECMS

The T-ECMS controller is based on (7.17). Assuming a piecewise-linear soft constraint for the final SoC, the value of the optimal adjoint state depends on the final sign of the SoC deviation. In real-time conditions, this sign is not known. Therefore, the equivalence factor $s(t)$ varies in time between two limit values, namely, s_{chg} and s_{dis}, according to a probability factor $p(t)$, that is,

$$s(t) = p(t) \cdot s_{dis} + (1 - p(t)) \cdot s_{chg}. \tag{7.29}$$

The probability $p(t)$ in turn is calculated as

$$p(t) = \frac{E_e^+(t)}{E_e^+(t) - E_e^-(t)}, \tag{7.30}$$

as a function of the two quantities E_e^+ and E_e^- (see Fig. 7.6), which represent the maximum positive and negative values of the electric energy use that can result at the end of the mission. A mission is defined here as a trait of the vehicle's route characterized by a given value $E_m(t_f) = \int_0^{t_f} P_m(\tau)\, d\tau$ of the required energy at the wheels ("energy horizon"), a quantity that is independent of the control law.

The estimation of E_e^+ and E_e^- strongly depends on the current value of the electrical energy use, $E_e(t) = \int_0^t P_e(\tau)\, d\tau$. In detail, the quantity $E_e^+(t)$ is given by the sum of three terms: (i) $E_e(t)$, (ii) the electrical energy that would be used for the traction with the system driven at a constant $u = u_r$ (see Fig. 7.6) from t till the end of the mission, and (iii) a negative term due to the "available energy," i.e., the electric energy that will be stored from t till the end of the mission. The term (ii) is calculated as $u_r \cdot (E_m(t_f) - E_m(t))/\bar\eta_e$. In fact, $E_m(t_f) - E_m(t)$ is the mechanical energy that still must be delivered before the end of the mission. When it is multiplied by u_r, the mechanical energy provided at the output stage of the electrical path is obtained. To derive the energy at the input stage, the average efficiency of the electrical path (7.24) is used. The term (iii) is evaluated assuming a constant ratio λ between the available energy as a function of time and $E_m(t)$. This parameter λ is calculated for various drive cycles as the ratio $E_{e0}/E_m(t_f)$. The final expression for $E_e^+(t)$ is

$$E_e^+(t) = E_e(t) + \frac{u_r \cdot (E_m(t_f) - E_m(t))}{\bar\eta_e} - \lambda \cdot (E_m(t_f) - E_m(t)) . \quad (7.31)$$

The (negative) quantity $E_e^-(t)$ is also given by three terms: (i) the current value of $E_e(t)$, (ii) the electrical energy that would be recharged during the traction with the system driven at constant $u = -u_l$ (see Fig. 7.6) from t till the end of the mission, and (iii) a negative term due to the free energy, evaluated as above since it is independent of the control law. The term (ii) is calculated as $-u_l \cdot \bar\eta_e \cdot (E_m(t_f) - E_m(t))$, applying the same considerations as above. Therefore, the final expression for $E_e^-(t)$ is

$$E_e^-(t) = E_e(t) - \bar\eta_e \cdot u_l \cdot (E_m(t_f) - E_m(t)) - \lambda \cdot (E_m(t_f) - E_m(t)) . \quad (7.32)$$

The resulting equation for $p(t)$ is

$$p(t) = \frac{u_r/\bar\eta_e - \lambda}{u_r/\bar\eta_e + \bar\eta_e \cdot u_l} + \frac{E_e(t)}{(u_r/\bar\eta_e + \bar\eta_e \cdot u_l) \cdot (E_m(t_f) - E_m(t))} , \quad (7.33)$$

with $p(t)$ limited between 0 and 1. For simplicity, (7.33) may be implemented with $u_r = u_l$ [226].

The flowchart of the ECMS is sketched in Fig. 7.7. At each time t with a time step Δt, the strategy consists of accomplishing the following steps:

- The vehicle speed and acceleration are measured. The torque and speed required at the wheels are evaluated from the model of the system.
- Tentative values of the control variables u are applied in the range from $u = -u_{max}$ (limited by the engine and generator power) to $u = 1$ (pure electrical mode), with a step Δu.

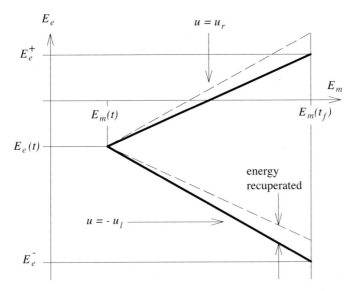

Fig. 7.6. Sketch of the quantities that lead to the evaluation of $E_e^-(t)$ and $E_e^+(t)$.

- For each tentative value of u, the model calculates the fuel and electrical energy use, and the cumulative quantities $E_e(t)$ and $E_m(t)$ are updated. The energy use $E_e(t)$ also can be related to the estimated SOC.
- Equation (7.33) is applied to evaluate $p(t)$ and then $s(t)$ is calculated from (7.29). This requires values for s_{dis}, s_{chg} and λ to be stored in memory. The energy horizon $E_m(t_f)$ is user-defined, and it determines the duration of the successive missions.
- For the tentative value u the Hamiltonian $H(t, u)$ is computed using the equation $H(t, u) = P_f(t, u) + s(t)P_e(t, u)$.
- The control value $u(t)$ is chosen as the tentative value which yields the minimal value of $H(t, u)$.

In the previous considerations leading to (7.33), the parameter λ, which is a quantity not depending on the control law, played a fundamental role. This parameter has a large variability and a large influence on the controller performance, thus it has to be estimated accurately. In contrast, the equivalence factors typically show a weaker influence, at least in their typical range of variability which is rather small. As a consequence of this fact, the T-ECMS keeps s_{dis} and s_{chg} constant, i.e., a pair of average values is conveniently selected to represent the vehicle and these values are used for every driving condition. The T-ECMS includes instead a sophisticated algorithm for the on-line estimation of λ, based on the information that is provided by an on-board telemetry system during a mission (Fig. 7.8). Every mission is assumed to have a defined point that is to be reached along a defined route. All the static features of the route, including the maximum speed allowed in its var-

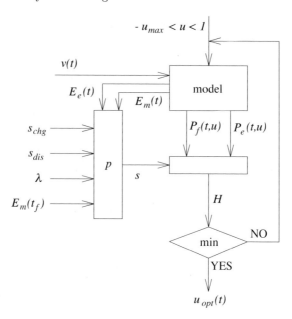

Fig. 7.7. Flowchart of the ECMS.

ious parts, the total distance to be covered, and the altitude profile are also assumed to be known.

If t_k is the time at which the k-th information I_k is available, $\hat{\lambda}_k$ is the related estimation of the parameter λ. The information I_k can be of two different types. If the presence of a moving or fixed obstacle is detected, I_k is a "stop" signal. In this case, the distance and the speed of the obstacle are assumed to be known also. When the obstacle is removed, the corresponding I_k is a "go" signal. In both cases, the estimation of λ derives from an estimation of the future velocity profile, $\hat{v}_k(t)$. The profile assumed is always the one that covers the rest of the mission in the minimum amount of time, and which fulfills the constraints concerning the maximum speed and the presence of obstacles. The estimated velocity profile may only consist of: (i) trajectories at constant speed, (ii) trajectories with maximum acceleration, and (iii) trajectories with maximum deceleration. Under certain assumptions, such profiles have been proven to be fuel optimal [102].

From the estimated velocity profile $\hat{v}_k(t)$, the mechanical energy delivered at the wheels $\hat{E}_m(t)$ may be calculated. Each portion of the profile is responsible for an energy contribution which is the sum of two terms. The former is due to altitude variations. The latter term depends on the other resistances, and it can be negative (energy being recuperated) or positive (energy being delivered). The contributions of the various portions of the estimated velocity profile to the electric energy recuperated are evaluated from the (negative) mechanical energy contributions and by using the model of the system.

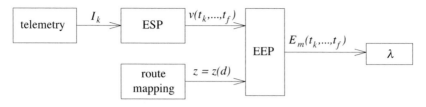

Fig. 7.8. Flowchart of the T-ECMS.

Further details on the operation of the T-ECMS are illustrated in the case study 8.4.

Time-Invariant Feedback Controllers

While the optimal and suboptimal strategies described above require global or local estimates, a simpler approach consists of storing the control algorithm in the form of a lookup table, providing the control variable $u(t)$ as a function of the current driving conditions and state variables. Generally, the feedback quantities that parameterize the control variable are the vehicle speed, power demand $P_m(t)$, and battery SoC. In place of vehicle speed, the wheel or engine speed is often used. Likewise, in place of power demand, it is possible to use the torque demand or vehicle acceleration, in the latter case neglecting the influence of road slope.

A substantial improvement with respect to heuristic controllers can be obtained using dynamic optimization to build a feedback map [284, 268, 152]. The optimal solution found with dynamic programming is statistically analyzed in terms of input and state variables, from which implementable rules are extracted to construct the control strategy. To limit the complexity of the feedback map, only two input variables are usually allowed. Examples include torque demand and SoC [152], power demand and SoC [284], and wheel speed and power demand [268]. Although this approach performs well in real hybrid vehicles, it is based on optimization with respect to a specific drive cycle and, in general, it is neither optimal nor charge-sustaining for other cycles. Moreover, the feedback solution obtained using dynamic programming cannot be implemented directly, and the rule-extraction process is not straightforward.

To overcome these drawbacks, the procedure of [152] is extended in [153] using stochastic dynamic programming. To obtain a time-invariant control strategy, an infinite-horizon optimization problem is solved. The feedback control law derived with stochastic dynamic programming is applicable to general driving conditions. Its validity is tested in [153] over regulatory as well as random drive cycles.

8

Appendix I – Case Studies

8.1 Case Study 1: Gear Ratio Optimization

This case study shows how the gear ratios of a manual gear box can be optimized to improve the fuel economy of a passenger car. This analysis is purely academic because it completely neglects all drivability issues and only focuses on the fuel economy of a vehicle that follows the MVEG–95 driving profile.

Nevertheless, it is instructive because it shows how a numerical parametric optimization problem can be defined and solved using a quasistatic problem formulation. The software tools used in this example are the QSS toolbox in conjunction with numerical optimization routines provided by Matlab/Simulink. This approach is quite powerful and can be used to solve non-trivial problems.

8.1.1 Introduction

The powertrain of the light-weight vehicle analyzed in Sects. 3.3.2 and 3.3.3 includes a standard five-speed manual gear box. The ratios of these five gears are chosen according to the approach introduced in Sect. 3.2.2 and satisfy the usual requirements with respect to acceleration performance, towing capability, etc.

These gear ratios do not yield the smallest possible fuel consumption when the vehicle is following the MVEG–95 test cycle and it is clear that there is a different set of gear ratios that improve the vehicle's fuel economy. However, it is not clear at the outset what gear ratios are optimal and — more importantly — what gains in fuel economy may be expected in the best case. These two questions can be answered using the approach shown below.

8.1.2 Software Structure

The vehicle model and its representation with the QSS toolbox have already been introduced in Sect. 3.3.3. The model shown in Fig. 3.11 is reused in this problem setup. As illustrated in Fig. 8.1, that model is embedded into a larger software structure that has a total of four hierarchy levels.

The top level (a Matlab .m file named `optimaster.m` in this example) initializes all system parameters and defines a first guess for the optimization variables u, i.e., for the five gear ratios. After that, a numerical optimization routine is called (`fminsearch.m` in this example). This routine, which is part of Matlab's optimization toolbox, calls a user-provided .m file (named `opti_fun.m` in this example) that computes the actual value of the objective function $L(u)$, i.e., the fuel consumption of the vehicle for the chosen set of gear ratios u. For that purpose, the vehicle model that has been programmed using the QSS toolbox is used (file `sys.mdl` in this example).[1]

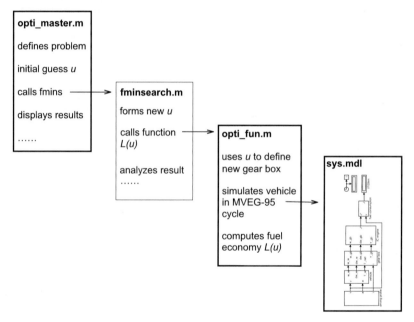

Fig. 8.1. Software structure used to find the fuel-optimal gear ratios of a standard IC engine/manual gear box powertrain. The blocks with thick frames are provided by the programmer, the block framed by a thin line is a subroutine provided by Matlab's optimization toolbox.

---

[1] Note that Matlab/Simulink encapsulates all variables within the corresponding software modules. The exchange of variables across modules can be accomplished with several methods. A convenient approach is to define the necessary variables on all levels to be global and, thus, accessible to all modules. However, this method must be used with caution in order to avoid using the same variable name for different objects.

All files necessary to solve this case study can be downloaded at the website http://www.imrt.ethz.ch/research/qss/. The programs are straightforward to understand. Some efforts have to be made in order to correctly handle situations in which infeasible gear ratios are proposed by the optimization routine.

8.1.3 Results

Obviously, each iteration started by the optimization routine requires one full MVEG–95 cycle to be simulated. Since many iterations are needed to find an optimum, short computation times become a key factor for a successful analysis. The computations whose results are shown below required approximately a total of 10 s CPU time on a 2 GHz power PC. This figure is acceptable and indicates that this method might be useful to solve more complex problems within a reasonable time.[2]

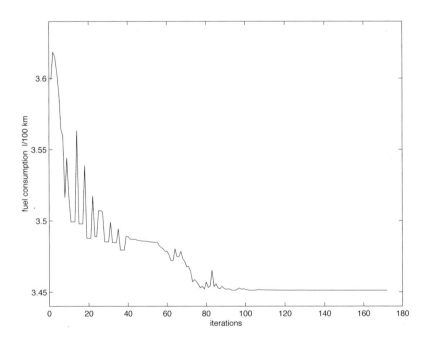

Fig. 8.2. Iterations of the fuel consumption (the non-feasible solutions have been smoothed out).

---

[2] Moreover, no attempts were made to optimize or compile the Matlab/Simulink program.

Figure 8.2 shows the evolution of the fuel consumption during one optimization run. The corresponding gear ratios are shown in Fig. 8.3. During the optimization, several non-feasible sets of gear ratios are proposed by the routine `fminsearch.m`. The fuel consumption of these non-feasible solutions is set to be higher than the initial fuel consumption. For clarity reasons, these outliers have been omitted in Fig. 8.2 and the fuel consumption of the previous feasible solution has been used instead.

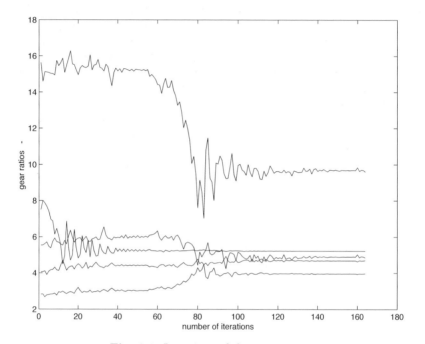

Fig. 8.3. Iterations of the gear ratios.

It goes without saying that the method used in this case study is not guaranteed to yield the best possible fuel consumption. As with all numerical optimization techniques, there is no guarantee that the optimum found by the algorithm is the global optimum. However, in this specific case, the optimization was started with several other initial guesses for the gear ratio, and all of these runs converged to the same set of gear ratios.

The main result of this case study is that there is little room for improvement by changing the gear ratios. As shown above, the expected gains in fuel economy (probably) are, even in the best case, less than 5%. This relatively small gain in fuel economy would not justify the poorer drivability following from the choice of fuel-optimal gear ratios.

8.2 Case Study 2: Dual-Clutch System - Gear Shifting

8.2.1 Introduction

This case study analyzes the problem of finding an optimal gear-shifting strategy for a vehicle equipped with a dual-clutch system. The vehicle model is illustrated in Fig. 8.4. Which gear $x \in \{1, 2, 3, 4, 5, 6\}$ is engaged when, substantially influences the total fuel consumption. Two gear shifting strategies are compared: (i) shifting the gears as proposed by the test cycle (the MVEG-95, in this case), or (ii) shifting the gears such that the smallest possible fuel consumption is realized. Of course, at the outset this optimal gear shifting strategy is not known. Deterministic dynamic programming (see Sect. 10) can be used to find it, provided the future driving profile is known. The resulting solution is not causal, but it will represent a benchmark for all possible causal control strategies.

The problem has one state variable x_k (the previous gear number) and one control input i_k (the desired future gear number), both with inherently discrete values. A DDP approach is feasible because the drive cycle, which the vehicle has to follow (the MVEG-95 cycle introduced in Fig. 2.6), is known a priori.

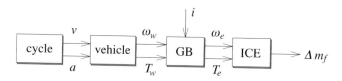

Fig. 8.4. QSS powertain model, gearbox GB includes a dual clutch system.

8.2.2 Model Description and Problem Formulation

The QSS model illustrated in Fig. 8.4 is equivalent to the following discrete model $f(x_k, i_k)$. The chosen time step is 1 s. The vehicle model contains the air drag force

$$F_a(v) = \frac{1}{2} \cdot \rho_a \cdot c_d \cdot A_f \cdot v^2 \tag{8.1}$$

the rolling friction force

$$F_r(v) = m_v \cdot g \cdot (c_{r0} + c_{r1} \cdot v^{c_{r2}}) \tag{8.2}$$

and the inertial force

$$F_i = (m_v + m_r) \cdot a \tag{8.3}$$

The torque required at the wheel axle is

$$T_w = (F_a + F_r + F_i) \cdot r_w \tag{8.4}$$

where r_w is the wheel radius. The rotational speed and acceleration of the wheel are

$$\omega_w = \frac{v}{r_w}, \qquad \dot{\omega}_w = \frac{a}{r_w} \tag{8.5}$$

The gearbox model includes a transformation of the torque and rotational speed required at the wheel to the torque and rotational speed at the engine

$$T_e = T_w/\gamma(x), \qquad \omega_e = \omega_w \cdot \gamma(x), \qquad \dot{\omega}_e = \dot{\omega}_w \cdot \gamma(x) \tag{8.6}$$

where $\gamma(x)$ is the gear ratio of gear x. The engine model is based on the Willans approximation introduced in Chap. 3. With this approach the fuel flow can be approximated by

$$\Delta m_f = \frac{\omega_e}{e(\omega_e) \cdot H_l} \cdot \left(T_e + \frac{p_{me0}(\omega_e) \cdot V_d}{4\pi} + \Theta_e \cdot \dot{\omega}_e \right) \cdot \Delta t \tag{8.7}$$

where $p_{me0}(\omega_e)$ is the engine friction pressure, $e(\omega_e)$ is the Willans efficiency, and Θ_e is the engine inertia. In the optimization discussed below, all numerical values correspond to a two liter naturally aspirated engine and a midsize vehicle. The model is now a simplified QSS-based model of a conventional vehicle with one state (the previous gear number) and one input (the desired new gear number).

The time needed to shift gears using the dual-clutch system and an automated gearbox is assumed to be much smaller than $\Delta t = 1$ s and therefore considered as instantaneous. Further it is assumed that the gearbox has limited possibilities to change gears, i.e., that there is a constraint on the possible next gears depending on the current gear. It is this constraint that makes DDP a suitable method to solve the problem. Typically, such a gearbox contains two shafts: the first shaft carries gears one, three, and five and the second shaft gears two, four, and six. The possible instantaneous gear shifting is limited to gears from one shaft to the other. Hence, running at the gear x the possible next gear must be in the set

$$I(x) = \begin{cases} \{2,4,6\} \text{ if } x \in \{1,3,5\} \\ \{1,3,5\} \text{ if } x \in \{2,4,6\} \end{cases}$$

Using the dynamic programming algorithm with the cost criteria

$$g_k(x_k, i_k) = \begin{cases} \Delta m_f & i_k \in I(x_k) \\ \infty & \text{otherwise} \end{cases} \tag{8.8}$$

it is possible to determine the optimal gear switching strategy for a given cycle that gives the minimum fuel consumption.

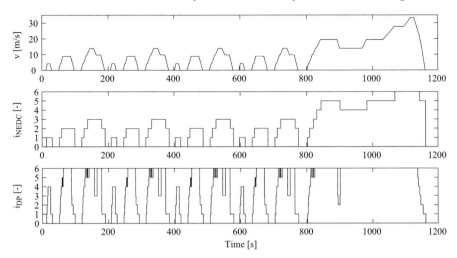

Fig. 8.5. Results DDP optimization of the dual clutch gear box problem. Velocity profile of the new European drive cycle NEDC (=MVEG-95) (top); standard MVEG-95 gear switching strategy (middle); and DDP optimal gear switching strategy (bottom).

8.2.3 Results

The resulting gear switching profile for the MVEG-95 is shown in Fig. 8.5. Figure 8.5 also shows the standard gear switching profile for that cycle. The average CO_2 emissions for the considered vehicle using the standard MVEG-95 gear switching strategy are approximately 200 g/km. With the dual clutch system and the optimized strategy this value is reduced to 172 g/km. It is clear that the result of the DDP problem is optimal only for the chosen test cycle. Moreover, in practice it is not possible to achieve this level of fuel economy because of the many constraints (driving comfort, energy use of the gear shifting device, etc.). Nevertheless, this result shows that there is a substantial potential to improve the fuel economy using dual-clutch systems.

8.3 Case Study 3: IC Engine and Flywheel Powertrain

This section presents an approach that can be used to improve the part-load fuel consumption of an SI engine system. The key idea is to avoid low-load conditions by operating a conventional IC engine in an on–off mode. The excess power produced by the firing engine is stored in a flywheel in the form of kinetic energy. During the engine-off phases this flywheel provides the power needed to propel the vehicle. A CVT with a wide gear ratio is necessary to kinematically decouple the engine from the vehicle.

From a mathematical point of view the interesting point in this example is the presence of state events that describe the transition of the clutch from slipping to stuck conditions. This is an example of a parameter optimization problem in which a forward system description must be used.

The problem analyzed in this case study was formulated and solved within the ETH Hybrid III project. General information on that project can be found in [66] and [275]. The work described in this section was first published in [104].

8.3.1 Introduction

The ETH Hybrid III project was a joint effort of several academic and industrial partners. Various aspects of hybrid vehicle design and optimization were analyzed during that project. All concepts proposed were experimentally verified on engine dynamometers and on proving grounds. The experimental vehicle used for that purpose is shown in Fig. 8.6.

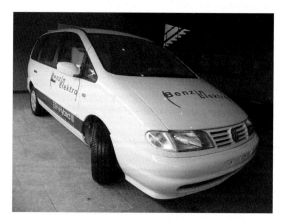

Fig. 8.6. The ETH Hybrid III vehicle.

This case study focusses on one particular aspect of the ETH Hybrid III design process. One of the key ideas analyzed and realized in this project was the development of a flywheel–CVT powertrain that permitted an efficient recuperation of the vehicle's kinetic energy while braking and an on–off operation of the IC engine during low-load phases. The CVT was realized in an "i^2" configuration[3] that yielded a very large gear ratio range (approximately $1:20$). Figure 8.7 shows a schematic representation of those parts of the ETH Hybrid III powertrain[4] that are relevant for the subsequent analysis.

The flywheel has a mass of approximately 50 kg and can store sufficient energy to accelerate the vehicle from rest to approximately 60 km/h (see Sect. 5.2 for more information on flywheels). The flywheel is mounted coaxially to the engine shaft and rotates in air at ambient conditions. Its gyroscopic influence on the vehicle is noticeable but does not pose any substantial stability problems.

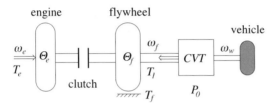

Fig. 8.7. Simplified schematic representation of the ETH Hybrid III powertrain structure.

For the sake of simplicity, it is assumed in this analysis that the vehicle is driving on a horizontal road with a constant velocity characterized by the constant wheel speed ω_w. Under this assumption the powertrain is operated in a periodic way as illustrated in Fig. 8.8.

At $t = 0$ one cycle starts with the IC engine off and the flywheel at its maximum speed $\bar{\omega}$. At $t = \tau_o$ the command to close the clutch is issued and the engine is accelerated from rest to $\underline{\omega}$. Since the point in time $t = \tau_c$ at which the engine speed reaches the flywheel speed is not known, its detection will be an important part of the solution presented below.

In the time interval $t \in [\tau_c, \tau_c + \vartheta)$ the engine is operated at the torque $T_{opt}(\omega_e)$ that yields the best fuel economy, i.e., almost at full load. Of course, the power produced in this phase exceeds the power consumed by the vehicle to overcome the driving resistances. Accordingly, the flywheel is accelerated

---

[3] With an appropriate system of external cog wheels and automatic clutches, the input and output of an "i^2" CVT can be interchanged. Therefore, the standard gear ratio range i can be used twice. Due to the fact that some overlapping is necessary at the switching point, the total gear ratio range is slightly smaller than the theoretical maximum of i^2.

[4] In addition to the IC engine and the flywheel, the powertrain included an electric motor in a parallel configuration as well. These three power sources justified the appendix "III" in the vehicle name.

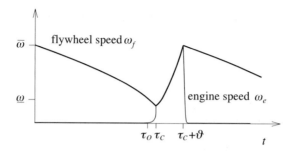

Fig. 8.8. Illustration of the flywheel and the engine speed as functions of time.

until it again reaches $\bar{\omega}$. At this point in time ($t = \tau_c + \vartheta$) the clutch is opened and the fuel is cut off such that the engine rapidly stops.

The optimization problem to be solved consists of finding for each constant vehicle speed ω_w those parameters $\underline{\omega}$ and $\bar{\omega}$ that minimize the total fuel consumption. The following two contradicting effects are the reason for the existence of an optimum: higher flywheel speeds produce smaller duty-cycles and, hence, better engine utilization, but higher flywheel speeds also produce larger friction losses. The optimal compromise that minimizes the fuel consumption is found using mathematical models of all relevant effects and numeric optimization techniques.

8.3.2 Modeling and Experimental Validation

The system to be optimized operates in two different configurations: "clutch open" with $\omega_f \neq \omega_e$ and "clutch closed" with $\omega_f = \omega_e$. Accordingly, it is described by two different differential equations. In the case "clutch open" this equation is

$$\Theta_f \cdot \frac{d}{dt}\omega_f(t) = -T_f(t) - \frac{P_0(\omega_w)}{\omega_f(t)} \, , \tag{8.9}$$

where $P_0(\omega_w)$ is the power consumed by the vehicle at the actual wheel (vehicle) speed ω_w. Following the assumptions mentioned above, the power P_0 is constant and depends on the wheel speed ω_w as follows

$$P_0(\omega_w) = p_1 \cdot \omega_w + p_3 \cdot \omega_w^3 \tag{8.10}$$

(the numerical values of all coefficients and parameters used in this case study are listed in Tables 8.1 and 8.2).

The torque $T_f(t)$ stands for all friction losses in the powertrain. It includes the aerodynamic losses of the rotating flywheel. In the ETH Hybrid III project it was possible to approximate these losses by

$$T_f(t) = k_0 + k_1 \cdot \omega_f(t) \, . \tag{8.11}$$

In the case "clutch closed" $(\omega_e(t) = \omega_f(t))$ the powertrain dynamics are described by

$$(\Theta_f + \Theta_e) \cdot \frac{d}{dt}\omega_f(t) = T_e(\omega_f) - T_f(t) - \frac{P_0}{\omega_f(t)} , \qquad (8.12)$$

where the variable $T_e(t)$ stands for the fuel-optimal engine torque. This function can be parametrized as

$$T_e(\omega_e) = c_0 + c_1 \cdot \omega_e + c_2 \cdot \omega_e^2 . \qquad (8.13)$$

The last missing element of the modeling process is an approximation of the fuel mass flow during the phase when the engine fires

$$\overset{*}{m}_f(\omega_e) = f_0 + f_1 \cdot \omega_e . \qquad (8.14)$$

Figure 8.9 shows the comparison of the predicted and the measured flywheel speeds during on on–off cycle. The deviations at lower speed are noticeable, but do not substantially influence the final result. This can be seen by comparing the predicted and the measured fuel consumption as illustrated in Fig. 8.12. Five different pairs of switching speeds are shown in that figure, including the optimal solution $\underline{\omega} = 114\,\text{rad/s}$ and $\overline{\omega} = 278\,\text{rad/s}$.

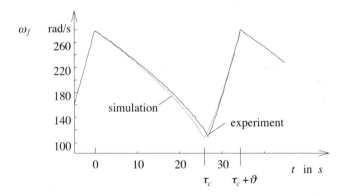

Fig. 8.9. Comparison of the predicted and measured flywheel speed during one full cycle.

8.3.3 Numerical Optimization

The optimization criterion that has to be minimized is the powertrain's fuel consumption per distance travelled in one cycle, i.e.,

$$J = \frac{\int_0^{\tau_c + \vartheta} \overset{*}{m}_f(\omega_e(t))\, dt}{\tau_c + \vartheta} . \qquad (8.15)$$

The two variables to be optimized are $\{\underline{\omega}, \bar{\omega}\}$. The fuel consumption $\overset{*}{m}_f$ and the times τ_c, ϑ depend on these quantities. Note that since the velocity of the vehicle is assumed to be constant, the distance it travels is proportional to the duration of one cycle.

The solution to the problem analyzed in this section can be found using numerical optimization techniques. Here, a fully numeric approach is presented. In [104] a semi-analytical approach is proposed in which the gradients of the objective function are approximated by analytic functions.[5]

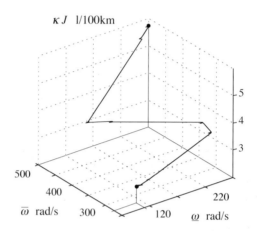

Fig. 8.10. Optimization process in the parameter space.

Figure 8.10 shows a typical optimization trajectory in the parameter space. Starting with a reasonable but arbitrary set of parameters $\{\underline{\omega}_0, \bar{\omega}_0\}$, the numerical minimization algorithm detects the steepest-descent directions by numerically approximating the gradients using finite differences. Following these directions, the optimal solution is approached in a reasonable amount of computing time.

A crucial point in such numerical optimizations is the correct handling of state events. In the problem analyzed in this case study the relevant state event is the switch between the system structure described by (8.9) and the structure described by (8.12). The time instant τ_c at which this event takes place is not known a priori but is determined by the evolution of the system state variables. This effect can cause problems in numerical optimizations.

In fact, numerical optimizations rely on approximations of the gradients of the objective functions by finite differences. In an idealized setting, the smaller the differences are, the better the gradients can be approximated. In real situations many errors limit the minimal differences that can be used. One

---

[5] Compared to the fully numerical solution, a speed-up factor of the order of 20 was observed in this particular example.

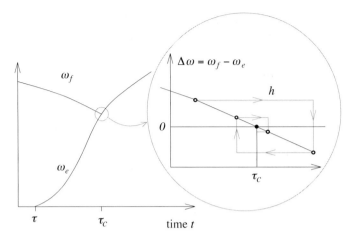

Fig. 8.11. Illustration of the iterations necessary to correctly detect the state event "clutch sticks".

of the most important sources of errors are too large step sizes used by the integration routines. While relatively large step sizes are acceptable whenever the system does not change its structure, close to a state event the integration steps must be adapted to localize the event within a predefined error bound that is compatible with the differences chosen to approximate the gradients. Figure 8.11 illustrates the main idea of such an iteration that has to take place each time a state event has been detected.

8.3.4 Results

The procedure outlined above was used to determine the fuel-optimal engine-on and engine-off speeds $\{\underline{\omega}, \bar{\omega}\}$ for various vehicle speeds. The results shown below are valid for the case $v = 50\,\text{km/h}$. Compared with a standard ICE-based powertrain, the fuel consumption was reduced by more than 50%. Extending the approach introduced above to the case of non-constant vehicle speeds and using the energy recuperation capabilities of the ETH Hybrid III prowertrain, the total fuel consumption in the city part of the MVEG–95 cycle could be reduced by almost 50%. A detailed description of all results of the ETH Hybrid III project can be found in [65].

Table 8.1 shows the model parameters and Table 8.2 lists the coefficients of the polynomials (8.10), (8.11), (8.13), and (8.14) used in the numerical optimization. The results of these calculations are shown in Fig. 8.12. Also shown in that figure are five measured data points. As can be seen in that figure, the measured data matches well the predicted values and — more importantly — the best fuel economy is obtained using very similar duty cycle parameters $\{\underline{\omega}, \bar{\omega}\}$.

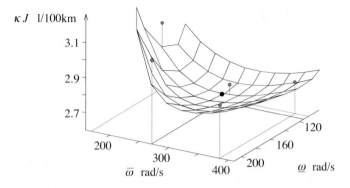

Fig. 8.12. Predicted (surface) and measured (dots) fuel consumption for five different parameter pairs $\{\underline{\omega}, \bar{\omega}\}$. Vehicle speed $v = 50\,\mathrm{km/h}$ (wheel speed $\omega_w = 51\,\mathrm{rad/s}$, driving power $3.2\,\mathrm{kW}$).

Table 8.1. Model parameters used in all calculations shown in this section.

wheel speed	$\omega_w = 51.3\,\mathrm{rad/s}$
engine inertia	$\Theta_e = 0.125\,\mathrm{kg\,m^2}$
flywheel inertia	$\Theta_f = 2.8\,\mathrm{kg\,m^2}$
dissipated power	$P_0 = 3.2\,\mathrm{kW}$
scaling factor	$\kappa = 9.37 \cdot 10^3\,\mathrm{1\,s/kg\,100\,km}$

Table 8.2. Coefficients of polynomials (8.10), (8.11), (8.13) and (8.14) (compatible units).

$p_1 = 4.0319 \cdot 10^1$	$p_3 = 8.3518 \cdot 10^{-3}$
$f_0 = -2.4796 \cdot 10^{-4}$	$f_1 = 7.3260 \cdot 10^{-6}$
$c_0 = 5.3375 \cdot 10^1$	$c_1 = 1.8222 \cdot 10^{-1}$
$c_2 = -2.4455 \cdot 10^{-4}$	
$k_0 = 1.0043$	$k_1 = 3.6707 \cdot 10^{-3}$

8.4 Case Study 4: Supervisory Control for a Parallel HEV

The problem discussed in this section is the supervisory control, i.e., the power split control of a parallel hybrid vehicle. This study is based on a quasistatic model of the system, which is validated with respect to experimental data of overall fuel consumption over regulatory drive cycles.

The availability of a validated model discloses the possibility of comparing the various control strategies treated in Chap. 7. In particular, the improvements of the sub-optimal controllers ECMS and T-ECMS with respect to a heuristic controller are assessed, with the performance calculated with the dynamic programming technique taken as a global optimum reference.

The problem was formulated and solved for the DaimlerChrysler Hyper, a prototypical parallel hybrid car based on the series-production Mercedes A-Class A 170 CDI. The work described in this section was first published in [226]. The description and the validation of the T-ECMS approach was first published in [230].

8.4.1 Introduction

The Hyper is a parallel hybrid vehicle with the two prime movers acting separately on the front and the rear axles. The thermal path consists of a front-wheel driven powertrain, with a 1700 cm^3 Diesel engine that yields 44 kW maximum power (66 kW in the series-production setup), and a 5-speed, automated manual gearbox. The electrical path includes a 6.5 Ah, 20 kW NiMH-battery pack and a motor/generator connected to the rear wheels via a second 5-speed gearbox. The two gearboxes are connected in such a way that they shift simultaneously.

The supervisory controller determines at each time how to split the mechanical power between the two parallel paths. In principle all of the ideas discussed in Chap. 7 could be applied to achieve minimum fuel consumption and a good self-sustainability of the battery during typical vehicle operation. The control results obtained with different strategies may be compared using a validated model of the system.

8.4.2 Modeling and Experimental Validation

The representation of the Hyper model with the QSS toolbox is shown in Fig. 8.13. The relevant model parameters are listed in Table 8.4.

The rolling resistance is calculated as a fifth-order polynomial function of the vehicle speed,

$$F_r(v) = g \cdot m_v \cdot \big(a_0 + a_1 \cdot v(t) + a_2 \cdot v^2(t) + \\ + a_3 \cdot v^3(t) + a_4 \cdot v^4(t) + a_5 \cdot v^5(t) \big) \ . \tag{8.16}$$

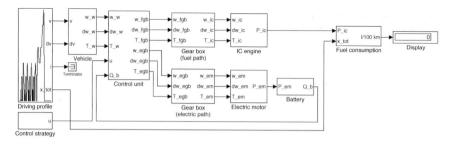

Fig. 8.13. QSS model of the Hyper parallel hybrid vehicle.

At the wheel axle, the basic relationship is the balance of torques

$$T_{fgb}(t) + T_{egb}(t) = T_w(t) .$$

The control variable is the torque split factor u that regulates the torque distribution between the parallel paths. As described in Sect. 4.7), this factor is defined as $u(t) = T_{egb}(t)/T_w(t)$. The value $u(t) = 1$ therefore means that all the (positive) torque needed at the wheels is provided by the electrical path, or that all the (negative) torque available at the wheels from regenerative braking is entirely absorbed by the electrical path. When $u(t) = 0$, it means that all the torque needed at the wheels is provided by the fuel path. When $T_w(t) = 0$, no control is needed and u remains undefined.

The fuel consumption of the engine is calculated as a tabulated function of engine speed and torque (engine map)

$$\overset{*}{m}_f(t) = f_{ic}\left(T_{ic}(t), \omega_{ic}(t)\right) . \tag{8.17}$$

The output power of the battery is calculated as a tabulated function of motor speed and torque

$$P_b(t) = f_{em}(T_{em}(t), \omega_{em}(t)) . \tag{8.18}$$

For the battery, the equivalent circuit model illustrated in Fig. 4.26 has been adopted, where the inner resistance R_b is a constant and the open circuit voltage $U_{b,oc}$ is a tabulated function of the state of charge.

The reliability of the model developed is demonstrated by comparing the fuel consumption for the conventional arrangement (i.e., without the hybrid equipment) with the data obtained from experiments and confirmed by the simulation tools used at DaimlerChrysler. The results of this comparison, which was made for three regulatory drive cycles and with the engine fully warmed-up or cold, is shown in Table 8.3.

8.4.3 Control Strategies

Three cumulative quantities are introduced for control purposes. They are the fuel energy use, $E_f(t) = \int_0^\tau H_l(\tau) \cdot \overset{*}{m}_f(\tau)\, d\tau$, the electrical energy use,

Table 8.3. Specific fuel consumption (liters/100 km) with the present model and as declared by DaimlerChrysler for various regulatory drive cycles and for warm as well as for cold start.

Cycle	Model	Warm	Cold
ECE	5.51	5.54	6.54
EUDC	3.91	3.92	4.04
MVEG–95	4.49	4.52	4.96

Table 8.4. Numerical values of the vehicle parameters.

a_0	rolling resistance coefficient	$8.80 \cdot 10^{-3}$
a_1	rolling resistance coefficient	$-6.42 \cdot 10^{-5}$
a_2	rolling resistance coefficient	$9.27 \cdot 10^{-6}$
a_3	rolling resistance coefficient	$-3.30 \cdot 10^{-7}$
a_4	rolling resistance coefficient	$6.68 \cdot 10^{-9}$
a_5	rolling resistance coefficient	$-4.46 \cdot 10^{-11}$
A_f	frontal area	$2.31 \, \mathrm{m}^2$
c_d	drag coefficient	0.32
$I_{b,max}$	battery maximum current	$\pm 200 \, \mathrm{A}$
m_v	vehicle total mass	$1680 \, \mathrm{kg}$
Q_0	battery charge capacity	$6.5 \, \mathrm{Ah}$
r_w	wheel radius	$0.29 \, \mathrm{m}$
R_b	battery inner resistance	$0.65 \, \Omega$
$\{\gamma_{egb}\}$	electric machine gear ratios	$\{12.38, 8.98, 6.45, 4.57, 3.32\}$
$\{\gamma_{fgb}\}$	engine gear ratio	$\{9.97, 5.86, 3.84, 2.68, 2.14\}$
η_{sm}	starter motor efficiency	0.6
Θ_{ic}	engine inertia	$0.195 \, \mathrm{kg} \, \mathrm{m}^2$
Θ_v	vehicle total inertia	$145 \, \mathrm{kg} \, \mathrm{m}^2$

$E_e(t) = \int_0^\tau I_b(\tau) \cdot U_{b,oc}(\tau) \, d\tau$, that is the variation (positive or negative) of the electrical energy stored, and the mechanical energy delivered at the wheels, $E_m(t) = \int_0^\tau T_w(\tau) \cdot \omega_w(\tau) \, d\tau$. The quantity H_l is the lower heating value of the fuel.

The control strategies whose performance will be compared are:

1. pure thermal operation, with $u(t) = 0$ except when (i) the required power at the wheels would be too high for the engine alone, and (ii) in case of regenerative braking, in which the regenerative power is collected with $u(t) = 1$;

2. the "parallel hybrid electric assist" strategy derived from the literature [128], according to which the engine is on when (i) the vehicle speed is higher than a set point, or (ii) if the vehicle is accelerating, or (iii) if the battery SOC is lower than a minimum value; when the engine is on, all the torque required at the wheels has to be provided by the engine path

($u = 0$), and an additional (positive or negative) torque is required that is proportional to the current SOC deviation; the minimum torque that the engine can provide is given by a specified fraction of the maximum torque at the current engine speed;

3. the global optimal control strategy, calculated off-line using dynamic programming techniques;
4. a pattern recognition technique, which estimates the parameter $s_0(t)$ from a comparison of the driving conditions in the near past with a set of reference patterns;
5. the ECMS as defined in Chap. 7; and
6. the T-ECMS also defined in Chap. 7.

The control performance considered in the following includes (i) the specific fuel consumption SFC (liters/100 km), (ii) the final battery SOC reached, starting from a value of 0.7, and (iii) the equivalent specific fuel consumption eSFC (liters/100 km). This latter figure is the sum of the fuel energy use and of the the electrical energy use weighted by the equivalence factor s_{dis} or s_{chg}. For the definition and the procedure to calculate these equivalence factors, see Chap. 7. For the Hyper and the MVEG–95 cycle, the values calculated are $s_{dis} = 2.7$, $s_{chg} = 1.6$. The corresponding plot $E_f(t_f) = f(E_e(t_f))$ is shown in Fig. 8.14. Other parameter values for the MVEG–95 cycle are $\lambda = 0.21$ and $E_m(t_f) = 4.98$ MJ.

The values of the mentioned parameters evaluated for other drive patterns are listed in Table 8.5. Clearly, λ has a wider variability, being allowed to take every value from zero (no energy recuperable) to infinity (no positive energy delivered). For the regulatory drive cycles (first five rows), which all have a zero road slope, the values of λ depend only on the relative importance of deceleration trajectories. Therefore, this parameter is higher for urban cycles such as the ECE or the Japanese 10–15. The variability of λ increases for drive patterns with altitude variations, such as the proprietary AMS cycle. In contrast, the two equivalence factors show a much smaller variability. Also shown in Table 8.5 is the parameter s_0 which is the constant equivalence factor that yields, when used in the policy given by (7.20), the globally optimum control law. This controller minimizes the fuel consumption and keeps the state of charge at the end of the mission equal to the initial value [229]. The values of s_0 are always between s_{chg} and s_{dis}.

8.4.4 Results

Table 8.6 summarizes the control performance evaluated for the MVEG–95 cycle with the control strategies mentioned in the previous section. The table clearly shows that, even without any particular control strategy, the hybridization process is beneficial for the fuel economy. In fact, the fuel consumption with the pure thermal mode is lower than the fuel consumption of the conventional arrangement. This reduction is due to two contributions: (i) an increase

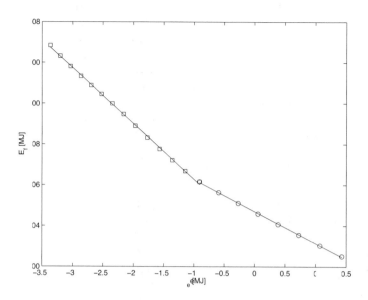

Fig. 8.14. Dependency between $E_f(t_f)$ and $E_e(t_f)$ for the MVEG–95 cycle, data calculated (circles) and linearly fitted (solid lines). The slopes of the fitting curves are the two equivalence factors. Values of $E_{e0} = -0.90\,\text{MJ}$, $E_{f0} = 16.16\,\text{MJ}$ (see Chap. 7 for their definition.

Table 8.5. Values of λ and of the equivalence factors for different drive patterns.

Cycle	λ	s_{dis}	s_{chg}	s_0
MVEG–95	0.21	2.7	1.6	2.5
ECE	0.35	2.9	1.4	2.6
EUDC	0.12	2.9	1.6	2.2
FTP75	0.17	2.9	1.7	2.4
JP 10–15	0.35	3.2	1.4	2.2
AMS	0.09	3.1	1.8	2.3
SMD/mean	52.3%	5.6%	13.3%	

of 1.92 MJ of fuel energy due to the additional mass, and (ii) a reduction of 2.70 MJ due to the suppression of the idle consumption by turning the engine off. An additional amount of 0.91 MJ stored in the battery due to regenerative braking can be used to reduce the fuel consumption even further.

The operation of the heuristic controller ("parallel hybrid electric assist") with the parameters as in [128] yields only a limited improvement with respect to the pure thermal mode. Much more significant is the reduction of fuel consumption obtained with the causal optimal controller with pattern recognition. The reduction of the value of SFC of about 20% with respect to the pure thermal mode is due to the relatively good agreement between the "true" value of s_0 and the values estimated during the drive cycle. A further reduction of the SFC of about 30% with respect to the pure thermal mode is obtained with the ECMS. This value is very close to the global optimum, calculated with the dynamic programming technique, and it corresponds to a zero deviation of the battery SOC. The control performance of the T-ECMS is reasonably close to the global optimum as well.

Table 8.6. Specific fuel consumption SFC (liters/100 km), final SOC (initial 0.7) and equivalent specific fuel consumption eSFC (liters/100 km) for the MVEG–95 cycle; ^{a}: the SFC in this case is lower than the global optimum because there is an associated depletion of SOC.

Strategy, MVEG–95	SFC	SOC	eSFC
Conventional	4.49 (100%)	–	–
Pure thermal hybrid	4.28 (95%)	0.83	3.91 (87%)
Electric assist	4.25 (95%)	0.74	4.12 (92%)
Global optimum	3.18 (71%)	0.70	3.18 (71%)
Pattern recognition	3.41 (76%)	0.75	3.27 (73%)
ECMS^a	3.13 (70%)	0.68	3.21 (71%)
T-ECMS	3.35 (75%)	0.74	3.25 (72%)

The traces of the battery SOC during the MVEG–95 obtained with the ECMS and the T-ECMS are shown in Fig. 8.15 together with the optimal trajectory calculated with the dynamic programming technique. The figure clearly shows that differences among the computed trajectories arise only in the second (i.e., extra-urban) part of the drive cycle.

Table 8.7 extends the of the various control strategies to more drive patterns, including five regulatory drive cycles, a proprietary test drive cycle, and five patterns with strong altitude variations. The last ones (VN1 to VN5) are partial records of 100 s each of a drive in a mountainous region. The control performance is summarized by the eSFC and the final state of charge.

For the regulatory drive cycles the performance of the T-ECMS is very close to the global optimum. In some cases (e.g., ECE) the T-ECMS performs even better than the ECMS. In fact, on one hand, the ECMS profits from the a priori knowledge of λ. On the other hand, this is only an average value, while the various values of $\hat{\lambda}_k$ used by the T-ECMS are more accurate estimations since they are based on the future driving conditions. Due to the fact that for each cycle a good approximation of the actual value of s_0 is always found in the reference set, the pattern recognition technique is very effective as well.

When the road grade is considered, the pattern recognition technique seems no longer to be able to sustain the battery state of charge at reasonable levels. Indeed, in some cases the battery is completely depleted or overcharged before the end of the mission. This is due to the fact that, in the presence

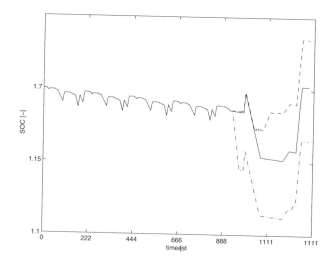

Fig. 8.15. ECMS – Battery state of charge for the MVEG–95: optimal trajectory (solid), with the ECMS (dashdot) and the T-ECMS (dashed).

of grade, even patterns that show a similar set of characteristic parameters could have significantly different values of s_0. Thus their estimation could be incorrect. This effect is not clearly visible in the flat terrain cycles, where the variations of s_0 are less prominent. In contrast, the T-ECMS seems to be able to sustain the charge even in the presence of large altitude variations. This is accomplished without reducing the capability of optimizing the fuel consumption. In fact, the values for the eSFC that are obtained with the T-ECMS are still close to the global optimum.

The performance of the ECMS strongly depends on the choice of the parameters λ, s_{dis}, s_{chg}. The influence of λ on the equivalent fuel consumption is shown in Fig. 8.16 for the ECE cycle. The cycle-averaged value $\lambda = 0.35$ actually yields a minimum value for the eSFC. For variations of λ in an interval of twice its standard mean deviation, variations of about 10% in the eSFC are calculated. For the MVEG–95 cycle, the same calculation yields a sensitivity of about 5%.

The two parameters s_{dis}, s_{chg} are responsible for weaker variations of the eSFC. In Fig. 8.16 the curves at given values of eSFC in the ECE cycle are plotted as a function of the two parameters (for $s_{dis} > s_{chg}$). The cycle values $(2.91, 1.41)$ yield an eSFC close to the minimum. For variations of s_{dis} and s_{chg} in an interval of twice their standard mean deviation, variations in the eSFC of about 2% and 1%, are calculated, respectively. Similar numbers have been obtained for the MVEG–95 cycle.

As described in Chap. 7, such an analysis confirms that the parameter λ has a large variability and a large influence on the controller performance.

Table 8.7. Equivalent specific fuel consumption (liter/100 km) and final state of charge (initial value: 0.70) for various drive patterns and control strategies.

		ECMS	T-ECMS	Pattern recog.	Global optimum
ECE	eSFC	2.93	2.86	2.85	2.82
	SOC	0.70	0.69	0.69	0.70
MVEG–95	eSFC	3.21	3.25	3.27	3.18
	SOC	0.68	0.74	0.75	0.70
EUDC	eSFC	3.77	3.85	3.87	3.77
	SOC	0.70	0.73	0.74	0.70
FTP	eSFC	3.38	3.39	3.39	3.37
	SOC	0.69	0.71	0.68	0.70
10–15	eSFC	2.97	2.97	3.15	2.90
	SOC	0.71	0.72	0.76	0.70
AMS	eSFC	4.35	4.33	–	4.30
	SOC	0.69	0.68	lim	0.70
VN1	eSFC	7.45	7.45	7.50	7.42
	SOC	0.68	0.65	0.47	0.70
VN2	eSFC	3.54	3.61	–	3.50
	SOC	0.70	0.67	lim	0.70
VN3	eSFC	6.12	6.13	6.24	6.11
	SOC	0.60	0.60	0.32	0.70
VN4	eSFC	0.79	0.68	0.84	0.68
	SOC	0.60	0.71	0.52	0.70
VN5	eSFC	3.31	3.25	3.27	3.18
	SOC	0.67	0.63	0.00	0.70

Thus, it has to be estimated accurately. In contrast, the equivalence factors have a weaker influence, at least in their typical range of variability which is rather small. It is as a consequence of this fact that the T-ECMS keeps s_{dis} and s_{chg} constant, i.e., a pair of average values is conveniently selected to represent the vehicle and is then used for every drive condition.[6]

An example of how the T-ECMS estimates the velocity profile $\hat{v}_k(t)$ during a typical urban drive cycle is shown in Fig. 8.17. The various plots refer to sequential values of t_k, $k = 1, \ldots, 8$. The instants t_k with the corresponding signals I_k are listed in Table 8.8. This set of information has been derived from the velocity profile of the ECE drive cycle. All the changes in the vehicle acceleration as scheduled in the drive cycle have been translated to a corresponding value I_k to simulate the real-time output of a telemetry system.

---

[6] For the system considered here, a suitable choice made after inspection of Table 8.5 was: $s_{dis} = 2.6$, $s_{chg} = 1.7$.

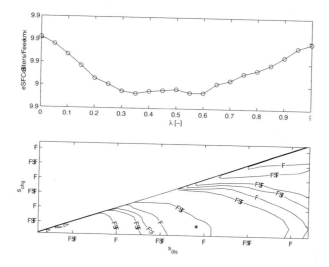

Fig. 8.16. Equivalent specific fuel consumption with the ECMS as a function of λ (top) and as a function of s_{dis} and s_{chg} (bottom), for the MVEG–95 cycle. The asterisk corresponds to the values listed in Table 8.5.

Figure 8.17 clearly shows that the more information is used, the closer the estimated velocity profile is to the original drive cycle.

Table 8.8. Scheduled information used to simulate the ECE cycle and estimated values of λ (EOM: end-of-mission).

k	t_k	I_k	d_o (m)	v_o (m/s)	$\hat{\lambda}_k$
1	11	GO	–	–	0.25
2	15	STOP	45.8	0	0.28
3	49	GO	–	–	0.26
4	61	STOP	266.6	0	0.41
5	117	GO	–	–	0.33
6	143	STOP	263.1	9.7	1.52
7	176	STOP	–	–	∞
8	180	EOM	–	–	–

The estimation of λ associated with the velocity profile of Fig. 8.17 is presented in Fig. 8.18. The successive estimates made at t_k, $k = 1, \ldots, 7$ are plotted as dots and shown with the "true" curve $\lambda(t)$. This is obtained by applying its definition at the various instants, with the energy terms evaluated from the current time to the end of the mission. The correspondence between the values of $\hat{\lambda}_k$ and $\lambda(t_k)$ is evident.

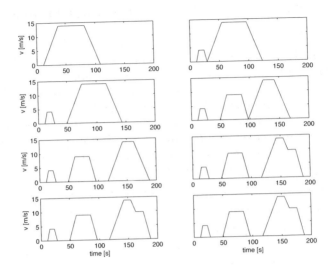

Fig. 8.17. Estimated velocity profile during the ECE cycle. The various plots refer to the different instants (see Table 8.8) at which a new piece of information is available.

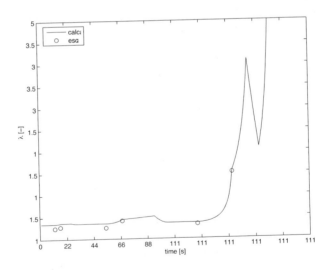

Fig. 8.18. Calculated curve of λ and values estimated with the T-ECMS, for the ECE cycle. The cycle-averaged value of λ is 0.35.

8.5 Case Study 5: Optimal Rendez-Vous Maneuvers

The problem discussed in this section is the fuel-optimal control of a vehicle which approaches a leading car whose speed is known using telemetry. The speed of the leader can be zero (fixed obstacle) or greater than zero (moving obstacle) and can also vary with time. The process ends when the follower has reached the speed of the leader. The distance between the two vehicles at this point can be rigidly specified (e.g. zero), or just forced to be nonnegative. Similarly, the time needed for the process can be specified or unconstrained. In all cases the legal speed limits have to be taken into account.

Two problem settings will be discussed:

- *unspecified rendez-vous time t_f and specified final distance $\delta(t_f)$; and*
- *specified rendez-vous time t_f and unspecified final distance $\delta(t_f)$.*

For both cases, the optimal trajectory will be found with the theory of optimal control, in such a way as to minimize the specific fuel consumption. The work described in this section was first published in [228].

8.5.1 Modeling and Problem Formulation

The two vehicles are assumed to drive at time $t = 0$ with speed $w(0) = w_0$ (leader) and $v(0) = v_0$ (follower). It is assumed that the speed trajectory of the leading vehicle may be predicted to be

$$w(t) = \begin{cases} w_0 + \dfrac{v_1 - w_0}{t_1} \cdot t & t < t_1 \\ v_1 & t \geq t_1 \end{cases} . \tag{8.19}$$

Accordingly, the distance it travels will be given by

$$e(t) = \begin{cases} w_0 \cdot t + \dfrac{v_1 - w_0}{2 \cdot t_1} \cdot t^2 & t < t_1 \\ v_1 \cdot t - \dfrac{v_1 - w_0}{2} \cdot t_1 & t \geq t_1 \end{cases} . \tag{8.20}$$

The velocity v_1 is the actual top speed limitation and may be obtained using GPS-based on-board navigation systems. The time interval t_1 is the leading vehicle's acceleration time which may be estimated using measurements of its previous velocity.

The dynamics of the leader vehicle are completely described by the equations (8.19) and (8.20). The follower's dynamics are given by

$$m\frac{d}{dt}v(t) = F(t) - c_0 - c_2 v^2(t) , \tag{8.21}$$

$$\frac{d}{dt}d(t) = v(t) , \tag{8.22}$$

with the two initial conditions $v(0) = v_0 \geq 0$ and $d(0) = 0$. Accordingly, the distance between the two vehicles is given by

$$\delta(t) = e(t) - d(t) + \delta(0) , \qquad (8.23)$$

where of course the constraint $d(t) \geq 0$ has to be satisfied for all times t by all admissible solutions (the initial distance $\delta(0)$ is assumed to be known as well). In the case analyzed in Sect. 8.5.2, the conditions that have to be met at the final time are

$$v(t_f) = v_1 \quad \text{and} \quad d(t_f) = d_f = e(t_f) + \delta(0) . \qquad (8.24)$$

Constant gear ratios are assumed here, i.e., vehicle and engine/motor speed are assumed to be strictly proportional. The extension to more than one gear ratio is straightforward and can be accomplished using the same ideas as introduced below.

An important point to clarify is the connection between the mass flow $\overset{*}{m}_f(t)$ of the fuel consumed by the engine and the mechanical power $F(t)v(t)$ delivered to the vehicle. Many approaches have been proposed for this relation, most of which are too complicated to be used for practical optimal control problems. A simple yet sufficiently precise description of this relation is given by the Willans formulation as introduced in Sect. 3.1.3. For the purposes of this case study, the following formulation is used

$$F(t) = \frac{e \cdot H_l}{v(t)} \cdot \overset{*}{m}_f(t) - F_0 . \qquad (8.25)$$

Here, the positive constants e, H_l, and F_0 describe the engine's internal efficiency, the fuel's lower heating value, and the engine's mechanical friction and gas exchange losses, respectively. The propulsion force is assumed to be limited by

$$F \in [-F_0, F_{max}] , \qquad (8.26)$$

where the upper limit corresponds to the maximum force (torque) of the engine and the lower limit to its friction force (torque). In general, these parameters are functions of the engine speed, however, since the dependency is not very strong it may be neglected in a first analysis. Notice that $F = -F_0$ implies a fuel mass flow of zero. In this condition the engine's internal friction losses are compensated by the diminishing kinetic energy of the vehicle (fuel cut-off). Several extensions of this problem formulation could be investigated, for instance the effects of irreversible or regenerative braking. However, this would exceed the scope of this case study.

The criterion which has to be minimized is the fuel consumed per distance travelled while satisfying the requirements of the rendez-vous maneuver as described above, i.e.,

$$J = \frac{\int_0^{t_f} \overset{*}{m}_f(t) \, dt}{\int_0^{t_f} v(t) \, dt} = \frac{\int_0^{t_f} (F(t) + F_0) \cdot v(t) \, dt}{e \cdot H_l \cdot d(t_f)} . \qquad (8.27)$$

8.5.2 Optimal Control for a Specified Final Distance

The classical theory of optimal control is applied to the problem described above, with t_f treated as a constant in a first step. The optimal final time may then be found in a second step solving a one-parameter nonlinear programming problem.

Obviously, for a fixed final time t_f and distance $d(t_f)$, minimizing the criterion $J(F)$ given by (8.27) is equivalent to minimizing the criterion

$$\tilde{J} = \int_0^{t_f} F(t) \cdot v(t) \, dt \, , \tag{8.28}$$

because e, H_l, F_0, and $d(t_f)$ are all positive constants and because

$$\int_0^{t_f} F_0 \cdot v(t) \, dt = F_0 \cdot d_f \tag{8.29}$$

is a positive constant for all admissible solutions as well.

The Hamiltonian function associated with this problem is given by

$$H(t) = F(t) \cdot \left(v(t) + \frac{\lambda_1(t)}{m} \right) - \frac{\lambda_1(t)}{m} \cdot \left(c_0 + c_2 \cdot v^2(t) \right) + \lambda_2(t) \cdot v(t) \, , \tag{8.30}$$

and the dynamic behavior of the adjoint state variables is described by

$$\frac{d}{dt}\lambda_1(t) = 2 \cdot c_2 \cdot v(t)\frac{\lambda_1(t)}{m} - F(t) - \lambda_2(t) \, , \tag{8.31}$$

$$\frac{d}{dt}\lambda_2(t) = 0 \, . \tag{8.32}$$

Notice that the Hamiltonian is time invariant and affine in the control signal $F(t)$. Obviously, this Hamiltonian reaches its minimum at the extreme values of the control signal and, according to the Maximum Principle, the optimal control signal will be discontinuous ("bang-bang control") in the regular case

$$F_{opt} = \begin{cases} F_{max} & \sigma(t) > 0 \\ F_\sigma & \sigma(t) \equiv 0 \\ -F_0 & \sigma(t) < 0 \end{cases} , \tag{8.33}$$

where $\sigma(t) = v(t) + \lambda_1(t)/m$. In the singular case ($F_{opt} = F_\sigma$) the optimal force has to be chosen such that the vehicle velocity is constant (see Lemma 1 below). In the regular case, i.e., if the solution does not include singular arcs, there is at most one switching event. This can be explained by analyzing the limiting case where $c_2 \approx 0$ (negligible aerodynamic drag). In this case the time derivative of the switching function $\sigma(t)$ is given by the expression

$$\frac{d}{dt}\sigma(t) = -\frac{c_0 + \lambda_2}{m} .$$

(8.34)

Since $\lambda_2(t) = \lambda_2$ is constant – see (8.32) – the switching function is an affine function of the time t and it can switch its sign at most once. Since the ODEs (8.21) and (8.31) are smooth functions this assertion will be true for sufficiently small values of c_2 as well.

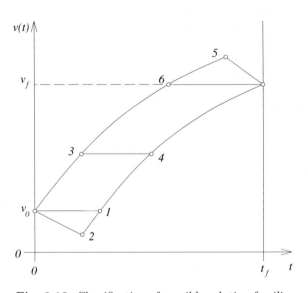

Fig. 8.19. Classification of possible solution families.

Lemma 1: During the singular arc phase the vehicle velocity is constant.
Proof: The singular arc is characterized by the condition $\sigma(t) \equiv 0$ for all $t \in [t_a, t_b]$ and therefore the condition $\dot{\sigma}(t) \equiv 0$ has to be satisfied in this interval as well. From the first condition the equality $v(t) = -\lambda_1(t)/m$ may be derived and from the second condition

$$\frac{c_0 + \lambda_2}{m} = \frac{c_2}{m} \cdot \left(\frac{2 \cdot v(t) \cdot \lambda_1(t)}{m} - v^2(t) \right) .$$

(8.35)

Inserting in this expression the equality $v(t) = -\lambda_1(t)/m$ the following expression is obtained

$$\frac{c_0 + \lambda_2}{c_2} = -\frac{3 \cdot \lambda_1^2(t)}{m^2} .$$

(8.36)

Since the expression on the left-hand side of (8.35) is a constant – again, see (8.32) – during the time interval $t \in [t_a, t_b]$ the adjoint variable $\lambda_1(t)$ has to be constant as well. Using the relation $v(t) = -\lambda_1(t)/m$, which is true for $t \in [t_a, t_b]$, the assertion follows.
The specific value of the singular velocity may be found as

$$v_\sigma = \sqrt{\frac{-c_0 - \lambda_2}{3 \cdot c_2}}. \tag{8.37}$$

Going back to the general problem the following qualitative points are important to be understood:

- With the exception of two particular combinations of v_1 and d_f, the problem is over-determined. The two special cases are indicated in Fig. 8.19 by the curves $v_0 \rightarrow 2 \rightarrow v_f$ and $v_0 \rightarrow 5 \rightarrow v_f$, respectively, and they correspond to the case where d_f is exactly equal to the area under the speed trajectory obtained for F equal to either F_{max} or $-F_0$.
- For values of d_f larger than $d_f(v_0 \rightarrow 5 \rightarrow v_f)$ or smaller than $d_f(v_0 \rightarrow 2 \rightarrow v_f)$, no solution is possible. For values of d_f that are intermediate between the two extreme values, only solutions that include singular arcs are possible, as shown by the example $v_0 \rightarrow 3 \rightarrow 4 \rightarrow v_f$.
- The two values of d_f corresponding to the curves $v_0 \rightarrow 1 \rightarrow v_f$ and $v_0 \rightarrow 6 \rightarrow v_f$ separate two regions in which the control signal changes its sign at both sides of the singular arc, from a region in which the control signal keeps its sign.
- In all cases, the control law is determined by one or two real parameters: either the regular switching time t_s or the singular arc initial and final time $\{t_a, t_b\}$. These one or two degrees of freedom are used to satisfy the two boundary conditions.

As mentioned above, the optimization problem analyzed in this section has no unused degrees of freedom, i.e., the two conditions $v(t_f) = w(t_f)$ and $d(t_f) = d_f$ fully define the unknown control parameters $\{t_a, t_b\}$ after optimal control theory has been used to determine the *qualitative* form of the optimal control signal. Therefore, the problem may be re-stated as a simple parameter optimization procedure.

Moreover, the differential equations (8.21)–(8.22) have to be solved for a piecewise constant input F. The resulting ODEs are separable and a closed-form solutions becomes possible. An example for the case $v_0 \rightarrow 6 \rightarrow v_f$ and $v_0 \rightarrow 5 \rightarrow v_f$ of Fig. 8.19 is presented in the original paper [228]. Analogously, closed-form expressions for $v(t)$ and $d(t)$ may be obtained for all the other situations depicted in Fig. 8.19.

Unfortunately, it is generally not possible to explicitly solve the equations

$$v(t_f) = w(t_f), \tag{8.38}$$

$$d(t_f) = d_f \tag{8.39}$$

using the closed-form expressions for $v(t)$ and $d(t)$. Numerical approaches will have to be used for that, for instance a minimization of the criterion

$$\epsilon = \frac{(v(t_f) - w(t_f))^2}{w^2(t_f)} + \frac{(d(t_f) - d_f)^2}{d_f^2}. \tag{8.40}$$

However, these computations are not very heavy because:

- No differential equations have to be solved numerically, i.e., the velocity and the distance of the follower vehicle are defined in closed form.
- The gradients of the criterion (8.40) with respect to the switching times $\{t_a, t_b\}$ (the only unknown parameters) may be computed explicitly.

Figure 8.20 shows two solutions obtained with the approach described above, for the values $t_1 = 25\,\text{s}$ ($d_f = 1.15\,\text{km}$) and $t_1 = 35\,\text{s}$ ($d_f = 1.05\,\text{km}$), respectively. The other data are identical in both cases: $v_0 = 15\,\text{m/s}$, $w_0 = 10\,\text{m/s}$, $v_1 = 30\,\text{m/s}$, $t_f = 40\,\text{s}$, $\delta_0 = 200\,\text{m}$, and the vehicle parameters were chosen to be $F_{max} = 1900\,\text{N}$, $F_0 = 300\,\text{N}$, $c_0 = 150\,\text{N}$, $c_2 = 0.43\,\text{N}\,\text{s}^2/\text{m}^2$, $m = 1500\,\text{kg}$.

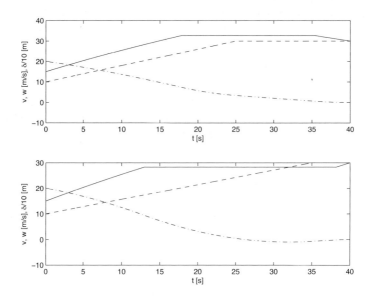

Fig. 8.20. Case a) feasible fuel-optimal trajectory (top) and case b) not feasible fuel-optimal trajectory (bottom). Solid lines: $v(t)$; Dashed lines: $w(t)$; Dashdot lines: $\delta(t)$.

Figure 8.20 also shows that solutions in terms of $v(t)$ and $d(t)$ are not automatically guaranteed to be feasible. The additional constraint $\delta(t) > 0$ $\forall t \in [0, t_f]$ has to be checked a posteriori.

Up to now the final time t_f has been assumed to be constant. If this parameter is varied as well, the fuel consumption decreases for $t_f > t_1$ because the acceleration part becomes less important compared to the cruising part in which the follower vehicle has reached the desired speed v_1. Assuming the leader vehicle to follow the expected trajectory (8.19) for all times, the limiting case $t_f \to \infty$ is characterized by the fuel consumption (expressed in the more familiar $1/100\,\mathrm{km}$ unit)

$$\overset{*}{m}_{f,\infty} = \frac{c_0 + c_2 \cdot v_1^2 + F_0}{e \cdot H_l \cdot \rho_f} , \tag{8.41}$$

with ρ_f being the fuel density.

8.5.3 Optimal Control for an Unspecified Final Distance

When the distance traveled is not imposed, the performance index defined by (8.27) is still valid. However, the distance is not a constant anymore and therefore the simplified formulation (8.28) is no longer correct. Using the plant dynamics (8.21)–(8.22) to express $F(t)$ as a function of $v(t)$ and $\dot{v}(t)$ and using partial integration, the performance index (8.27) can be written as

$$J = \frac{1}{e \cdot H_l} \cdot \left\{ F_0 + c_0 + m \cdot \frac{v^2(t_f) - v_0^2}{2 \cdot d(t_f)} + \frac{c_2}{d(t_f)} \int_0^{t_f} v^3(t)\, dt \right\} . \tag{8.42}$$

The main difference with respect to the case of Sect. 8.5.2 is that (8.42) now includes also an integral term. To transform it into the standard Euler–Lagrange formulation, a third state variable is needed. One possible choice is the normalized mass of fuel consumed m_f, so that the new third state equation will be

$$\frac{d}{dt} m_f(t) = (F(t) + F_0) \cdot v(t) . \tag{8.43}$$

The performance index is now simply

$$J = \frac{m_f(t_f)}{d(t_f)} \tag{8.44}$$

which does not contain any integral term. The resulting Euler–Lagrange equations are

$$\frac{d}{dt} \lambda_1(t) = \frac{2 \cdot c_2 \cdot v(t)}{m} \cdot \lambda_1(t) - \lambda_2(t) - (F(t) + F_0) \cdot \lambda_3(t) , \tag{8.45}$$

$$v(t_f) = w(t_f) , \tag{8.46}$$

$$\frac{d}{dt} \lambda_2(t) = 0 , \tag{8.47}$$

$$\lambda_2(t_f) = -\frac{m_f(t_f)}{d^2(t_f)} , \tag{8.48}$$

$$\frac{d}{dt} \lambda_3(t) = 0 , \tag{8.49}$$

$$\lambda_3(t_f) = \frac{1}{d(t_f)} . \tag{8.50}$$

Obviously, the optimal control law is again discontinuous with its switching time defined by the switching function $\sigma(t)$

$$F = \begin{cases} F_{max} & \sigma(t) < 0 \\ F_\sigma & \sigma(t) \equiv 0 \\ -F_0 & \sigma(t) > 0 \end{cases} , \tag{8.51}$$

where $\sigma(t) = \lambda_1(t)/m + v(t) \cdot \lambda_3$. A sensitivity analysis with respect to the parameter c_2 of the solutions in terms of speed (obtained using parameter optimization procedures) is shown in Fig. 8.21. The following points are worth mentioning:

- From $c_2 = 0$ up to a certain limit value ($c_2 = c_{2,crit} \approx 0.13$), the optimal speed trajectory is qualitatively similar to the case $c_2 = 0$. A "bang-bang" controller is optimal, and the switching time is given by the condition that the switching function σ changes its sign. The switching time increases with an increase of c_2.
- The limit value of $c_{2,crit}$ is smaller than typical values of c_2 for series production vehicles ($c_2 = 0.5 \cdot \rho_a \cdot c_d \cdot A_f > 0.25$). Singular arcs are therefore likely to appear.
- From the limit value of c_2 onward, the switching function reaches at a time t_a the zero value with a zero derivative, i.e., $\sigma(t_a) \rightarrow 0$, $\dot\sigma(t_a) \rightarrow 0$, and tends to remain constant. In this case, a singular arc occurs. The corresponding control signal cannot be determined by the condition $\dot\sigma = 0$, because the first derivative

$$\frac{d}{dt}\sigma(t) = \frac{1}{m}\left(\frac{2 \cdot c_2 \cdot \lambda_1(t)}{m} \cdot v(t) - \lambda_2 - \right.$$
$$\left. - (F_0 + c_0) \cdot \lambda_3 - c_2 \cdot \lambda_3 \cdot v^2(t) \right) \tag{8.52}$$

is independent of the control signal. However, using the second derivative of $\sigma(t)$ the missing information may be found.
- Moreover, $\ddot\sigma$ shows that along the singular arcs the vehicle velocity is constant. In fact, from

$$\frac{d^2}{dt^2}\sigma(t) = \frac{2 \cdot c_2 \cdot v(t) \cdot \dot v(t) \cdot \lambda_3}{m} = 0 \tag{8.53}$$

it follows that $\dot v = 0$ (all other quantities are larger than zero).
- The velocity on the singular arc is given by

$$v_\sigma = \sqrt{-\frac{\lambda_2 + F_0 \cdot \lambda_3 + c_0 \cdot \lambda_3}{3 \cdot c_2 \cdot \lambda_3}} . \tag{8.54}$$

This expression is not directly applicable because both λ_2 and λ_3 depend on the unknown final distance $d(t_f)$. Nevertheless, (8.54) is useful because it shows the influence of c_2 on the optimal solution. In fact, for sufficiently small values of c_2, the singular velocity is larger than the top speed of the regular "bang-bang" solution and no singular arcs will appear.

- As c_2 increases, v_σ decreases until it falls below v_1 for a certain limit value $c_2 = c_{2,sw} \approx 0.25$. In this case, the singular arc separates two periods with $F(t) = F_{max}$.
- The (unrealistic) upper limit of $c_2 \approx 1.80$ corresponds to the situation in which the final speed no longer can be reached, even with constant maximum thrust.
- The optimal trajectories obtained represent the best trade-off between the losses $\left(v_f^2 - v_0^2\right)/(2 \cdot m \cdot d_f)$, which should require a maximum d_f and correspondingly higher speeds, and the term $c_2/d_f \cdot \int v^3(t)\,dt$ which, for sufficiently large values of c_2, requires lower speed values.

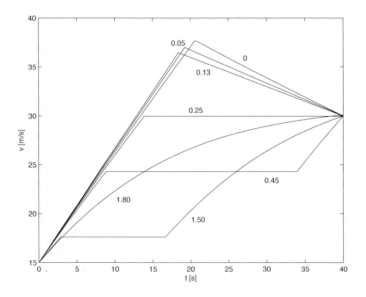

Fig. 8.21. Parameter sweep ($c_2 = 0.0 \rightarrow 1.80$).

The solutions for the two cases $c_2 = 0.20 < c_{2,sw}$ and $c_2 = 0.45 > c_{2,sw}$ are shown for illustration purposes in Fig. 8.22 (the vehicle parameters are the same as those used in Sect. 8.5.2). Table 8.9 shows the parameters of the optimal trajectories as a function of c_2.

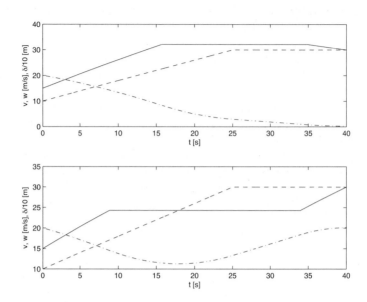

Fig. 8.22. Solid lines: velocity $v(t)$; dashed lines: velocity $w(t)$; dashdot lines: distance $\delta(t)$, for $c_2 < c_{2,sw}$ (top) and for $c_2 > c_{2,sw}$ (bottom).

Table 8.9. Trajectory parameters as a function of c_2.

c_2	t_a (s)	t_b (s)	t_s (s)	v_σ (m/s)	J (J/m)
0	–	–	18.40	–	847.2
0.05	–	–	19.26	–	922.8
0.13	20.54	20.86	–	37.62	1002.0
0.20	15.76	34.96	–	32.13	1069.6
0.25	13.88	39.98	–	30.00	1111.8
0.45	8.88	33.96	–	24.29	1246.6
1.50	2.90	16.66	–	17.59	1843.1
1.80	–	–	–	–	2200.0

8.6 Case Study 6: Fuel Optimal Trajectories of a Racing FCEV

The problem discussed in this section consists of determining the trajectories of an ultra-light, three-wheeler FCEV that minimize the hydrogen consumption over a given race circuit. The latter is characterized by the total length and road slope versus distance covered. Moreover, the particular case analyzed here imposes a lower limit to the average speed in such a way that the time to cover the circuit is also a problem constraint.

The problem is solved using optimal control theory. This requires a mathematical model of the vehicle powertrain that is developed and validated in the first section. Then the optimal control law is analytically determined. Finally, the results presented are compared with those yielded by a conventional PID controller.

The work described in this section was first published in [231].

8.6.1 Modeling

The system considered here consists of: (i) a PEM fuel cell system, (ii) a DC traction motor equipped with a DC/DC converter, (iii) a fixed gear reduction, (iv) a lightweight, three-wheeler car body.

The model derived belongs to the class of dynamic models. Each power converter is represented by a submodel, whose input and output variables are the power factors at the input and output stages of the component. The choice of the input variables and of the output variables is made according to the physical direct causality. Figure 8.23 illustrates the variables exchanged by the various submodels.

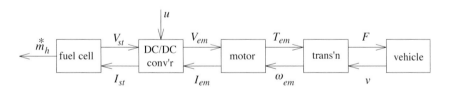

Fig. 8.23. Block diagram of the FCEV.

The fuel-cell submodel describes a fuel-cell stack with all the auxiliary devices that are necessary to supply hydrogen, air, and the coolant flows. A fraction of the current delivered by the fuel cell stack $I_{fc}(t)$ is drawn off to drive such auxiliaries. The power balance of (6.28) is written in terms of currents as

$$I_{fc}(t) = I_{st}(t) + I_{aux}(t) . \tag{8.55}$$

The stack current is related to the fuel cell voltage by means of the static polarization curve. Measurements taken at the fuel cell terminals are shown in

Fig. 8.24. The experimental data have been fitted with a nonlinear polarization model (M1)

$$U_{st}(t) = c_0 + c_1 \cdot \exp\left(-c_2 \cdot I_{fc}(t)\right) - c_3 \cdot I_{fc}(t) \tag{8.56}$$

derived from (6.21)–(6.22) and with a linear model (M2)

$$U_{st}(t) = c_4 - c_5 \cdot I_{fc}(t) \,, \tag{8.57}$$

which is derived from (6.25). In both cases, the effect of the concentration overvoltage has been neglected, as it occurs only at very high current densities.

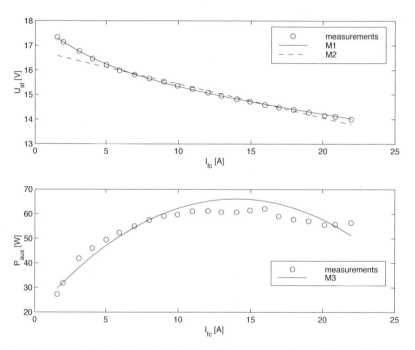

Fig. 8.24. Fuel cell voltage (top) and auxiliary power (bottom) as a function of the stack current. Measured data (circles), fitted with (8.56) (solid), and with (8.57) (dashed).

The power drained by the auxiliaries depends on the mass flow rates of air and hydrogen and on the heat generated by the stack that is removed through the coolant flow. Ultimately, the auxiliary power is a function of the fuel cell current, as shown in Fig. 8.24. The data measured are fitted with (6.71). Since (6.32) shows that the hydrogen mass flow rate is proportional to the fuel cell current, $\overset{*}{m}_h(t) = c_9 \cdot I_{fc}(t)$, (6.71) is rewritten as

$$P_{aux}(t) = c_6 + c_7 \cdot I_{fc}(t) + c_8 \cdot I_{fc}^2(t) \,. \tag{8.58}$$

The motor controller senses the actual motor speed and the accelerator signal $u(t)$ (control input to the system) and determines the motor armature voltage according to the following rule

$$U_{em}(t) = \kappa_{em} \cdot \omega_{em}(t) + K \cdot R_{em} \cdot u(t) , \qquad (8.59)$$

where κ_{em} is the motor torque constant, R_{em} the armature resistance, and K a static accelerator gain ($K = 30$).

The current required at the input stage of the converter is given by a balance of power

$$I_{st}(t) = \frac{U_{em}(t) \cdot I_{em}(t)}{\eta_c \cdot U_{st}(t)} , \qquad (8.60)$$

where η_c is the constant efficiency of the converter.

The motor submodel follows the standard armature representation of DC motors, (4.10)

$$I_{em}(t) = \frac{U_{em}(t) - \kappa_{em} \cdot \omega_{em}(t)}{R_{em}} , \qquad (8.61)$$

$$T_{em}(t) = \kappa_{em} \cdot I_{em}(t) . \qquad (8.62)$$

The transmission consists of a fixed gear placed between the motor and the drive wheel axle. The tractive force and input speed are calculated as

$$F(t) = \eta_t^{\pm 1} \cdot \frac{\gamma \cdot T_{em}(t)}{r_w} , \qquad (8.63)$$

$$\omega_{em}(t) = \frac{\gamma \cdot v(t)}{r_w} , \qquad (8.64)$$

where η_t is the constant transmission efficiency, γ is the constant transmission ratio, and r_w is the wheel radius. The positive sign in (8.63) is valid for $T_{em} > 0$ (traction), the negative sign for $T_{em}(t) < 0$ (braking). The numerical values of the model parameters are listed in Table 8.10.

Combining (8.55)–(8.64) an overall equation is derived, relating the stack current to the acceleration signal and the vehicle speed,

$$I_{fc}(t) = \frac{P_{aux}(t)}{U_{st}(I_{fc}(t))} + \frac{K \cdot u(t)}{\eta_c \cdot U_{st}(I_{fc}(t))} \left(K \cdot R_{em} \cdot u(t) + \kappa_{em} \cdot \frac{\gamma}{r_w} \cdot v(t) \right) . \qquad (8.65)$$

This equation is implicit since $P_{aux}(t)$ and $U_{st}(t)$ are functions of $I_{fc}(t)$ through (8.58) and (8.56) or (8.57). The dependency of the tractive force on the acceleration signal is given by

$$F(t) = \frac{\eta_t \cdot \gamma}{r_w} \cdot \kappa_{em} \cdot K \cdot u(t) . \qquad (8.66)$$

Using (8.66), the vehicle dynamics may be described by[7]

---

[7] An approximation valid for small grade has been introduced, $\sin \alpha \approx \alpha$, $\cos \alpha \approx 1$. Moreover, the vehicle mass m_v is increased by the quantity m_r representing the inertia of the rotational masses.

Table 8.10. Numerical values for the fuel cell vehicle model.

c_0	15.93	γ	20.4
c_1	2.06	r_w	$0.25\,\mathrm{m}$
c_2	0.200	m_v	$115\,\mathrm{kg}$
c_3	0.0876	m_r	1.1
c_4	16.80	ρ_a	$1.2\,\mathrm{kg/m^3}$
c_5	0.137	A_f	$0.3\,\mathrm{m^2}$
c_6	19.89	c_d	0.3
c_7	6.60	c_r	0.002
c_8	-0.236	g	$9.81\,\mathrm{m/s^2}$
c_9	0.208	κ_{em}	0.0168
η_c	0.95	R_{em}	$0.08\,\Omega$

$$\frac{d}{dt}v(t) = h_1 \cdot u(t) - h_2 \cdot v^2(t) - g_0 - g_1 \cdot \alpha(x(t)) , \qquad (8.67)$$

where the new parameters h_1, h_2, g_0, and g_1 are defined as functions of the physical quantities c_r, c_d, A_f, and m_v, and $x(t)$ is the distance travelled

$$\frac{d}{dt}x(t) = v(t) . \qquad (8.68)$$

The vehicle model (8.65)–(8.67) is not suitable for use in optimal control theory due to its highly nonlinear structure. Therefore, a simpler model was sought to relate the hydrogen mass flow rate to the vehicle speed and the acceleration signal. On the basis of (8.65) and (6.32), the following three-parameter structure (M4) was selected

$$\overset{*}{m}_h(t) = b_0 + b_1 \cdot v(t) \cdot u(t) + b_2 \cdot u^2(t) . \qquad (8.69)$$

Notice that the model (8.69) is strictly equivalent to (8.65) if (i) the auxiliary current is approximated by an affine function of the hydrogen mass flow rate, and (ii) the fuel cell voltage is approximated by a constant value.

A comparison of the validated models M1 and M3 and its three-parameter counterpart M4 is shown in Fig. 8.25 in terms of hydrogen mass flow rate as a function of the vehicle speed, with $u(t)$ as a parameter. The comparison shows a good agreement at low acceleration and speed, while higher differences are observed at higher power levels (both higher $u(t)$ and $v(t)$). Therefore, the three-parameter model should be considered to be a valid approximation for low-power operating conditions. The constants b_0, b_1, b_2 are listed in Table 8.11.

8.6.2 Optimal Control

The optimization problem can be stated in mathematical terms as follows: find the control law $u(t)$, $t \in [0, t_f]$ that minimizes the vehicle fuel consumption

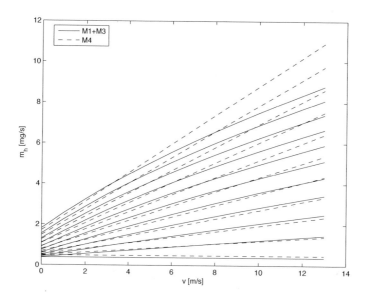

Fig. 8.25. Hydrogen consumption and mass flow rate with the physical model (solid) and the three-parameter model (dashed) as a function of the vehicle speed, for values of u from 0 to 1 regularly spaced with step 0.1.

Table 8.11. Fitted values for the three-parameter model.

b_0	0.475 mg/s
b_1	0.707 mg/m
b_2	1.24 mg/s

$\int_0^{t_f} \overset{*}{m}_h(t)\, dt$, subject to the constraint of a given average speed v_m. The latter condition can be written as $x(t_f) = x_f$ where x_f is the given length of the drive cycle and $t_f = x_f/v_m$. The problem is defined by the choice of v_m and by the particular grade profile $\alpha(x)$.

Using the results of the previous sections, the problem is formally described in the framework of optimal control theory. The system state variables are $v(t)$ and $x(t)$, the system state equations being (8.67) and (8.68). The incremental cost $\overset{*}{m}_h(t)$ is expressed by the three-parameter model of (8.69) as a function of the state and control variables. The Hamiltonian function is thus constructed as

$$H(u, v, x, \lambda_1, \lambda_2) = b_0 + b_1 \cdot v(t) \cdot u(t) + b_2 \cdot u^2(t) + \lambda_1(t) \cdot \{h_1 \cdot u(t) - $$
$$-h_2 \cdot v^2(t) - g_0 - g_1 \cdot \alpha(x(t))\} + \lambda_2(t) \cdot v(t) , \tag{8.70}$$

where $\lambda_1(t)$ and $\lambda_2(t)$ are two Lagrange multipliers.

The resulting Euler–Lagrange equations are

$$\frac{d}{dt}\lambda_1(t) = -\frac{\partial H}{\partial v} = -b_1 \cdot u(t) + 2 \cdot h_2 \cdot \lambda_1(t) \cdot v(t) - \lambda_2(t) \,, \qquad (8.71)$$

$$\frac{d}{dt}\lambda_2(t) = -\frac{\partial H}{\partial x} = g_1 \cdot \lambda_1(t) \cdot \frac{d}{dx}\alpha(x) \,. \qquad (8.72)$$

The condition for u to be optimal is that it minimizes the Hamiltonian. A stationary point of H with respect to u is found as

$$\frac{\partial H}{\partial u} = b_1 \cdot v(t) + 2 \cdot b_2 \cdot u(t) + h_1 \cdot \lambda_1(t) = 0 \,, \qquad (8.73)$$

resulting in

$$u^o(t) = -\frac{h_1 \cdot \lambda_1(t) + b_1 \cdot v(t)}{2 \cdot b_2} \,. \qquad (8.74)$$

This value is optimal, provided that $u^o(t) \in [0,1]$. Otherwise the control variable lies along a constrained arc, i.e., it is either $u^o(t) = 0$ or $u^o(t) = 1$. Notice that since the system is not explicitly time dependent, the Hamiltonian function of (8.70) is constant. This noticeable property can be used to evaluate for instance $\lambda_2(t)$ instead of using (8.72), which involves a derivative of $\alpha(x)$ with respect to x.

The boundary conditions of the optimization problem are split between the initial time $t = 0$ and the terminal time t_f. Two cases will be discussed. The former is representative of a vehicle start scenario, the latter of steady repetitions of the same periodic route. For a start cycle, the boundary conditions are

$$v(0) = 0, \quad x(0) = 0 \,, \qquad (8.75)$$

$$\lambda_1(t_f) = 0, \quad x(t_f) = x_f = v_m \cdot t_f \,, \qquad (8.76)$$

with the second of (8.76) replacing any condition of the Lagrange multiplier $\lambda_2(t_f)$, which is therefore unconstrained.

For a periodic route the boundary conditions are

$$x(0) = 0 \,, \qquad (8.77)$$

$$\lambda_1(t_f) = \lambda_1(0), \quad x(t_f) = x_f = v_m \cdot t_f, \quad v(t_f) = v(0) \,, \qquad (8.78)$$

with the initial value of the speed not assigned but a new constraint imposed at the terminal time.

The system of (8.67), (8.68), (8.71), (8.72), and (8.74) is highly nonlinear and subject to inequality constraints on the control variable. Thus a fully analytical solution cannot be found.

A way of numerically solving the optimal control problem consists of searching the initial values for the Lagrange multipliers and the initial speed, $\lambda_1(0)$, $\lambda_2(0)$, and $v(0)$ which, when applied to the aforementioned system of differential equations, lead to a fulfillment of the terminal boundary conditions. These particular values will be referred to as $\lambda_1^o(0)$, $\lambda_2^o(0)$, and $v^o(0)$ since they generate the optimal control law $u^o(t)$. Notice that in the start-cycle case the initial speed assigned is $v^o(0) = 0$.

The optimization results will be compared with a conventional driving strategy, in which a PID regulator tries to achieve the desired average speed. Such a strategy emulates the behavior of a human driver who tries to follow a target speed. Its mathematical definition for the start-cycle case is

$$u(t) = K_p \cdot (f \cdot v_m - v(t)) + K_i \cdot \int_0^t (f \cdot v_m - v(t)) \, dt \, , \qquad (8.79)$$

where f is a factor, usually very close to unity, which is tuned to achieve an exact match between v_m and the actual average speed. For the periodic route case the reference control law is

$$u(t) = K_1 \cdot \int_0^t (v(0) - v(t)) \, dt + K_2 \cdot \int_0^t (v_m \cdot t_f - x(t)) \, dt \, , \qquad (8.80)$$

where the two integral terms try to keep the terminal speed close to the initial value and the average speed close to the target value, respectively.

8.6.3 Results

As previously stated, the dynamic optimization system is analytically solved through the Euler–Lagrange equations, and thus it is converted to a static parametric optimization with two or three degrees of freedom, namely $\lambda_1^o(0)$, $\lambda_2^o(0)$, and possibly $v^o(0)$.

The optimal controller was developed for a drive cycle having a required average speed of 30 km/h and a known grade profile. The total length of the cycle is 3636 m.

Figure 8.26 shows the optimization results obtained for the start-cycle case. The optimal speed trajectory and the optimal control law are compared with the corresponding traces calculated with the reference PID strategy. The figure also shows the hydrogen mass flow rate and the hydrogen mass consumption as a function of time, as well as the traces of $\lambda_1(t)$ and $\lambda_2(t)$ corresponding to the optimal trajectory.

The PID controller uses the values $K_i = 0$, $K_p = 0.5$, $f = 1.054$. After some trial-and-error tests, these values seemed to provide the best result in terms of fuel consumption, and they guarantee the fulfillment of the constraint over the average speed.

The values found using the initial Lagrange multipliers are $\lambda_1^o(0) = -9.92$, and $\lambda_2^o(0) = -0.305$. The mass of fuel consumed with the PID strategy is 758 mg, with the optimal controller 700 mg. The gain in fuel economy is thus around 7.5%. In both cases the desired average speed is achieved within a

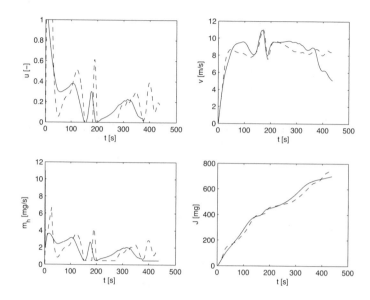

Fig. 8.26. Optimization results: control law, speed trajectory, hydrogen mass flow rate, hydrogen mass consumption as a function of time for the start-cycle case. Solid lines: optimal control; dashed lines: PID control.

tolerance of 0.1%. The optimal trajectory $\lambda_1^o(t)$ converges to zero at the end of the cycle with an even narrower tolerance.

Figure 8.26 clearly shows how both strategies start with the maximum acceleration ($u = 1$) to reach a speed close to the required average value. However, this phase is shorter in the optimal trajectory, thus the optimal speed increases less rapidly. The PID controller is characterized by two long constrained arcs, i.e., time intervals in which the control signal stays along its lower bound $u = 0$. This behavior seems to be avoided by the optimal controller, except for a very short time at the end of the two main downhills (approximately, $t = 150$ s and $t = 190$ s). In the final part of the route, the optimal controller completely releases the accelerator, with the vehicle free to roll. Another difference between the two strategies is that the average value of the optimal control law (0.19) is lower than for the PID controller (0.23).

Figure 8.27 shows the optimization results obtained for the periodic route case. The PID controller uses the values $K_1 = 0.111$, $K_2 = 5.92 \cdot 10^{-6}$. The optimized parameters found are $\lambda_1^o(0) = -20.00$, $\lambda_2^o(0) = -0.25$, $v^o(0) = 8.20$. The mass consumption with the PID strategy is 672 mg, with the optimal controller 613 mg. The gain in fuel economy is thus around 9%.

These traces also show that the optimal control law is much smoother than the PID controller output. The latter consists of several acceleration spikes which try to boost the vehicle speed when it becomes too low. Consequently,

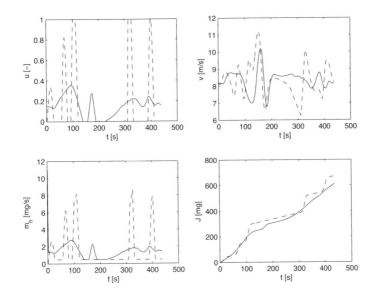

Fig. 8.27. Optimization results: control law, speed trajectory, hydrogen mass flow rate, hydrogen mass consumption as a function of time, periodic route case. Solid lines: optimal control, dashed lines: PID control.

the fuel mass flow rate shows a largely irregular variation as well. However, the average values of the optimal and the PID control laws are almost the same (0.15).

Similarly to vehicle speed, the optimal Lagrange multipliers are both periodic also. The periodic nature of $\lambda_1(t)$ is imposed through the constraint of (8.78), while for $\lambda_2(t)$ it is a consequence of the invariance of the Hamiltonian.

8.7 Case Study 7: Optimal Control of a Series Hybrid Bus

The problem illustrated in this section consists of determining the control law that minimizes the fuel consumption of a series hybrid bus over a specified route, which is assigned in terms of length and altitude profile.

The problem is solved using optimal control theory. This requires a mathematical model of the vehicle powertrain that is developed and validated in the first section. Then the optimal control law is analytically determined. Finally, the results presented are compared with those yielded by a conventional, thermostat-type controller.

The work described in this section was first published in [8].

8.7.1 Modeling and Validation

The system considered here consists of: (i) an electrochemical battery, (ii) an auxiliary power unit (APU) consisting of a natural gas SI engine and an electric generator, (iii) a DC traction motor equipped with a DC/DC converter, (iv) a transmission consisting of a fixed gear and a final drive, and (v) the bus body. The model representation with the QSS toolbox is shown in Fig. 8.28. The relevant model parameters are listed in Table 8.12.

Table 8.12. Relevant data of the system studied in this section.

m_v	4436 kg
c_r	0.015
c_d	0.40
A_f	4.7 m^2
r_w/γ_{gb}	0.0259 m
η_{gb}	0.9

The traction motor is described here by its quasistatic efficiency map. The motor efficiency is provided by the manufacturer as a function of input (electric) power and shaft speed, as shown in Fig. 8.29. From these data, the input power is calculated as a function of speed and torque

$$P_m = f_m (\omega_m, T_m) . \tag{8.81}$$

The battery current is provided by the manufacturer as a function of the power required P_b and the battery depth of discharge $\xi = 1 - q_b$,

$$I_b = f_b (P_b, \xi) . \tag{8.82}$$

The variation of the battery depth of discharge is tabulated as a function of the battery current and power,

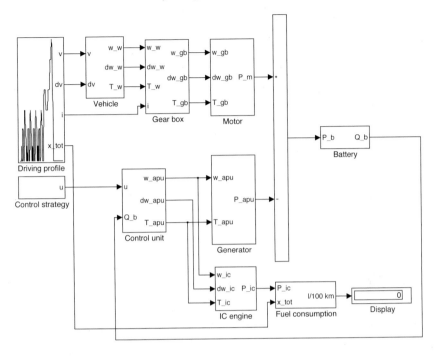

Fig. 8.28. QSS model of the series hybrid bus.

$$\frac{d}{dt}\xi(t) = \begin{cases} \dfrac{P_b(t)}{f_\xi(I_b(t))}, & P_b(t) > 0 \\[3ex] \kappa \cdot P_b(t), & P_b(t) < 0 \end{cases} \tag{8.83}$$

The form of the two functions f_b and f_ξ are shown in Fig. 8.30.

The combination of (8.82) and (8.83) yields the variation of $\xi(t)$ as a function of the battery power. This dependency is fitted with the control-oriented model

$$\frac{d}{dt}\xi(t) = c_{b,0}(\xi) + c_{b,1}(\xi) \cdot P_b(t) + c_{b,2}(\xi) \cdot P_b^2(t) , \tag{8.84}$$

$$c_{b,i}(\xi) = c_{b,i0} + c_{b,i1} \cdot \xi(t), \quad i = 0, 1, 2 , \tag{8.85}$$

whose validation is presented in [8].

The input variable of the APU submodel is the electrical power P_{apu} required at the APU output stage. The operating point of the engine is related only to the power rather than to its single factors, i.e., voltage and current. Thus the operating point is selected in such a way as to maximize the APU efficiency at every power request. The efficiency is a tabulated function of the engine torque and speed,

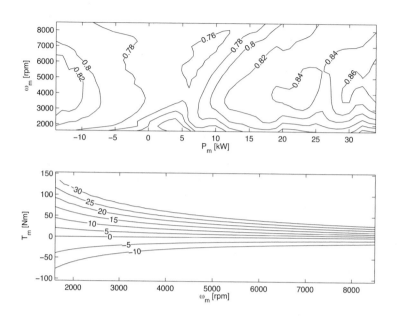

Fig. 8.29. Traction motor efficiency as a function of speed and electric power (top). Motor electric power (kW) as a function of speed and torque (bottom).

$$\eta_{apu} = f_\eta \left(\omega_{apu}, T_{apu} \right) . \tag{8.86}$$

The map f_η provided by the manufacturer is illustrated in Fig. 8.31. The engine speed is controlled as a function of the APU power in such a way as to maximize the efficiency of the APU. A lower limit of 2000 rpm has been introduced to enhance drivability. Combining these two pieces of information the input power, i.e., the chemical power associated with the fuel, is evaluated as a function of the output power. This dependency is shown in Fig. 8.31, together with an affine approximation of the same curve. The agreement between the two curves clearly leads to a control-oriented model of the Willans type [210] for this component,

$$P_f(t) = \frac{P_{apu}(t) + P_{f,0}}{e_{apu}} , \tag{8.87}$$

where $P_{f,0} = 22.5 \text{ kW}$ is the external power loss and $e_{apu} = 0.329$ the internal efficiency. The fuel consumption mass flow rate is $\overset{*}{m}_f(t) = P_f(t)/H_l$, where H_l is the lower heating value of the fuel.

The motor speed $\omega_m(t)$ and the motor power $P_m(t)$ were recorded in a test drive of the bus. The vehicle acceleration is calculated from the speed and filtered. The braking force was not recorded, thus it represents a major source of uncertainty. A relatively flat route was chosen for the test with the aim of isolating the strong influence of grade on the traction power. Although

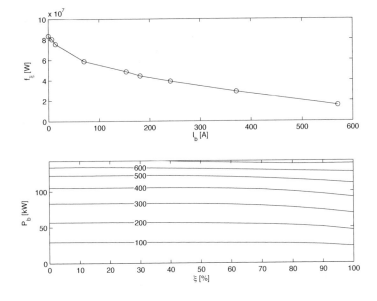

Fig. 8.30. Battery functions f_ξ (top) and f_b (bottom).

always smaller than 2–3%, the road slope has been estimated accurately from altitude data.

The motor power predicted with the system model previously presented is compared with the recorded trace shown in Fig. 8.32. The figure clearly shows the influence of the unknown braking force during heavy decelerations (centered around 38, 72, and 129 s), leading to a substantial underestimation of the total traction power. Nevertheless, during the remainder of the route, the agreement between measured and predicted data is quite satisfactory.

8.7.2 Optimal Control

The quasistatic model derived in Sect. 8.7.1 is characterized by a single state variable – the battery depth of charge (8.84) – and by a single output – the fuel consumption mass flow rate (8.87). The power flows from the two paths are balanced at the DC bus level,

$$P_b(t) + P_{apu}(t) = P_m(t) . \tag{8.88}$$

The control variable $u(t)$ is defined here as the ratio between the power flowing in the battery path and the total power

$$P_b(t) = u(t) \cdot P_m(t), \quad P_{apu}(t) = (1 - u(t)) \cdot P_m(t) . \tag{8.89}$$

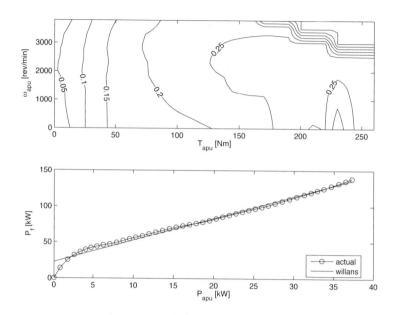

Fig. 8.31. Model of the APU: efficiency as a function of engine speed and torque (top) and fuel power as a function of the APU power (bottom).

The optimization problem can be stated in precise mathematical terms as follows: find the control law $u(t)$, $t \in [0, t_f]$ that minimizes the fuel consumption with a constraint over the battery depth of discharge $\xi(t_f) = \xi(0)$.

Using the results of the previous sections, the problem is formally described in the framework of optimal control theory introduced in Chap. 7. The Hamiltonian function is constructed as

$$H(\xi, u, P_m) = \overset{*}{m}_f(t) + \lambda \cdot \dot{\xi}(t) , \qquad (8.90)$$

where $\lambda(t)$ is a Lagrange multiplier. The corresponding Euler–Lagrange equation is

$$\frac{d}{dt}\lambda(t) = -\frac{\partial H}{\partial \xi} = -\lambda(t) \cdot \frac{\partial \dot{\xi}}{\partial \xi} . \qquad (8.91)$$

The condition for $u(t)$ to be optimal is that it must minimize the Hamiltonian. Notice that since the system is not explicitly time-dependent, the Hamiltonian function is constant with time.

The derivative $\partial \dot{\xi}/\partial \xi$ in (8.91) may be calculated using the parameterization of (8.85). A further simplification consists of neglecting the dependency of $\dot{\xi}$ on ξ which is justified if ξ varies only slightly around its initial level. Since this is actually one of the goals of the supervisory controller, such a simplification will be used in the following, leading to

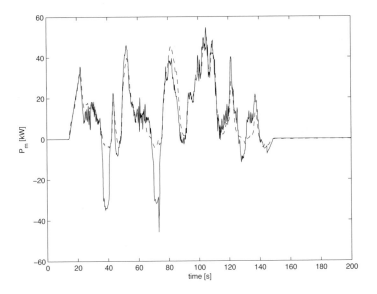

Fig. 8.32. Motor power, recorded (solid) and predicted (dashed).

$$\dot{\lambda}(t) = 0, \quad \lambda(t) = \text{constant} = \lambda^o . \tag{8.92}$$

The value λ^o must be chosen such that the condition $\xi(t_f) = \xi(0)$ is satisfied. Under this assumption, the following sub-optimal control law is found as

$$u^o(t) = \arg\min_u \left\{ \overset{*}{m}_f(u, P_m(t)) + \lambda^o \cdot \dot{\xi}(u, P_m(t)) \right\} . \tag{8.93}$$

This control law corresponds to the minimization of a cost function that only depends on the current driving conditions (via $P_m(t)$) and on λ^o. Therefore, it may be interpreted as a real-time controller, provided that the optimal value of λ^o is known. This in turn depends on the whole drive cycle, which can be reasonably estimated using, for instance, the techniques proposed in [226, 230] for parallel hybrid vehicles.

An alternative way to apply (8.93) is in terms of power flows. The first part of the right-hand term $\overset{*}{m}_f(t)$ can be easily converted into fuel chemical power after multiplication by the lower heating value of the fuel. The term $\dot{\xi}(t)$ can be also converted into electrochemical power after multiplication by the battery open-circuit voltage. This is slightly dependent on the state of charge but, if the latter is kept reasonably constant, this dependency can be neglected as it has been done for $\dot{\xi}(t)$. This means that (8.93) can be rearranged as

$$u^o(t) = \arg\min_u \left\{ P_f(u, P_m(t)) + \lambda^o \cdot P_b(u, P_m(t)) \right\} . \tag{8.94}$$

The search for the optimal values $u^o(t)$ may be done with a simple inspection procedure, i.e., several values of u are tested at each time and the respective values of the cost function are compared to find the minimum. The accuracy of such a method increases with the dimension of the test set and thus with the complexity.

An alternative approach may use the control-oriented model derived in the previous section to calculate the Hamiltonian as an analytical function. After minimization with respect to u the optimal control law then follows. The operating limits of u are given as

$$1 - \frac{P_{apu,max}}{P_m(t)} \leq u(t) \leq 1, \quad P_m(t) > 0 , \tag{8.95}$$

$$1 \leq u(t) \leq 1 - \frac{P_{apu,max}}{P_m(t)}, \quad P_m(t) < 0 , \tag{8.96}$$

since the motor power is always lower than the maximum power available from the battery.

To analytically calculate the optimal control law, various cases should be considered (in the following, $u^-(t) = 1 - P_{apu,max}/P_m(t)$).

For $P_m(t) < 0$ the limits of $u(t)$ given by (8.95)–(8.96) are both positive. Thus the battery power is negative and the Hamiltonian, see (8.83), is calculated as

$$H(u, P_m) = \frac{P_{f,0}}{H_l} + \frac{(1 - u(t)) \cdot P_m(t)}{e_{apu} \cdot H_l} + \lambda^o \cdot \kappa \cdot u(t) \cdot P_m(t) . \tag{8.97}$$

The Hamiltonian varies linearly with the control signal u and it exhibits a minimum for $u = 1$ or $u = u^-(t)$, according to the sign of the derivative of (8.97) with respect to u

$$u^o(t) = \begin{cases} 1, & \lambda^o \cdot \kappa < \dfrac{1}{e_{apu} \cdot H_l} \\[3mm] u^-(t), & \lambda^o \cdot \kappa > \dfrac{1}{e_{apu} \cdot H_l} \end{cases} . \tag{8.98}$$

For $0 < P_m(t) < P_{apu,max}$ the limits of $u(t)$ have opposite signs. In the range $u^-(t) \leq u < 0$ the Hamiltonian varies linearly with u. In $u = 0$ there is a discontinuity, since $H(0^+) > H(0^-)$. For $0 < u < 1$ the Hamiltonian may have a local minimum. The latter is found by setting to zero the derivative of H with respect to u

$$\frac{\partial H}{\partial u} = -\frac{P_m(t)}{e_{apu} \cdot H_l} + \lambda^o \cdot \left(c_{b,1}(\xi(t)) \cdot P_m(t) + 2 \cdot c_{b,2}(\xi(t)) \cdot u^* \cdot P_m^2(t) \right) = 0 , \tag{8.99}$$

which yields

$$u^* = \frac{\dfrac{1}{e_{apu} \cdot H_l \cdot \lambda^o} - c_{b,1}}{2 \cdot P_m(t) \cdot c_{b,2}} \tag{8.100}$$

if $0 < u^* < 1$. Summarizing, the optimal control variable in this case is defined by

$$
u^o(t) = \begin{cases} u^-(t), & \lambda^o \cdot \kappa > \dfrac{1}{e_{apu} \cdot H_l} \\[3mm] \arg\min_u\{H(0^-), H(u^*)\}, & \lambda^o \cdot \kappa < \dfrac{1}{e_{apu} \cdot H_l} \end{cases} . \tag{8.101}
$$

Again, for $P_m(t) > P_{apu,max}$ the limits of u are both positive, thus the battery power is also positive. A possible stationary point for the Hamiltonian is found using (8.100). It is optimal if it respects the constraints on u, otherwise the optimal control trajectory lies on a constrained arc

$$
u^o(t) = \arg\min_u\{H(u^-(t)), H(u^*), H(1)\} . \tag{8.102}
$$

8.7.3 Results

The sub-optimal control law given by (8.93) has been tested along a given route from the city of L'Aquila, Italy to the Faculty of Engineering located on a hill about 950 m high. The road slope $\alpha(x)$ has been calculated from topographic data, while the speed profile $v(t)$ represents a typical mission of a public bus connecting the two localities.

The sub-optimal controller adopted minimizes the Hamiltonian function with a search procedure. The test set of the control variable includes the values $u = -1, -0.5, \ldots, 1$ plus the lower and upper values dictated by the APU operating limits. Moreover, the optimal control law depends on the value of λ. There is only one value of that parameter that fulfills the condition $\xi(t_f) = \xi(0)$ (continuous curve in Fig. 8.33). Higher values of λ tend to excessively penalize the use of the battery as a prime mover, thus the final state of charge is too high (dash-dotted curve in Fig. 8.33). Lower values of λ tend to overfavor the use of the battery, thus the final state of charge is too low (dashed curve in Fig. 8.33).

The control results are compared with those calculated using: (i) the dynamic programming technique, (ii) a thermostatic-type controller (as specified below), and (iii) the procedure described in Sect. 8.7.2 to find the minimum of the Hamiltonian analytically. The results of dynamic programming are practically coincident with those of the sub-optimal controller [8] and thus they do not appear in Fig. 8.34 that shows a comparison of state of charge trajectories and fuel mass consumed. Bellman's dynamic programming algorithm [229] was implemented with a time step of 1 s and a state of charge step size of $5 \cdot 10^{-5}$. The test set of u is the same as for the sub-optimal controller. The thermostatic-type controller uses two limits for the depth of discharge, ξ_{hi} and ξ_{lo}. When $\xi > \xi_{lo}$ the APU output is set to 30 kW. When $\xi < \xi_{hi}$ the APU is turned off. For intermediate levels of SOC the APU status is kept equal to that at the previous time step.

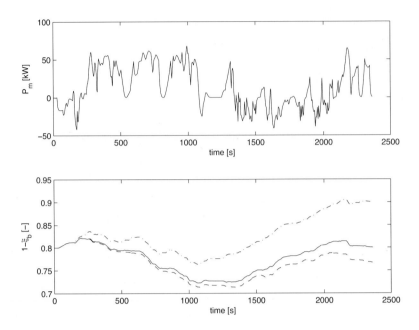

Fig. 8.33. Route L'Aquila–faculty of engineering. Top: electric power required (positive) and provided (negative) by the traction motor. Bottom: trajectory of the battery state of charge (1-ξ) obtained with $\lambda = 7.5 \cdot 10^8$ (dashdot), $\lambda = 6.7 \cdot 10^8$ (solid), and $\lambda = 6 \cdot 10^8$ (dashed).

The improvement in fuel economy obtained with optimal control with respect to a conventional controller is evident from the figures. Due to its inefficient on/off nature, the thermostatic controller exhibits a fuel mass consumption which is 29% higher than the global optimum calculated using the dynamic programming approach.

The figures also confirm that the control laws calculated with the minimization of the Hamiltonian are nearly optimal. Only a difference of less than 1% arises between the fuel mass consumption obtained with dynamic programming and the minimization of the Hamiltonian. The difference slightly increases to 5% if the analytical minimization is considered, mainly because of the approximation errors inherent to the control-oriented model, and despite the fact that the analytical minimization may find optimal values of u which are not included in the test set of the search procedure. However, the analytical minimization has the advantage of a smaller computation time required.

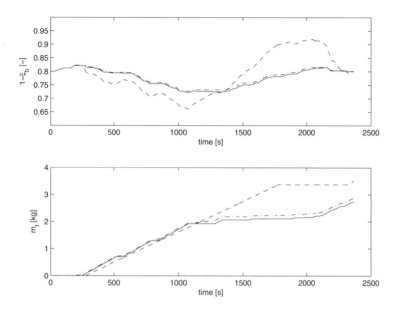

Fig. 8.34. Battery state of charge trajectory (top) and mass of fuel consumed (bottom): sub-optimal controller (solid), sub-optimal controller with analytical minimization (dashdot), on/off thermostat-type controller (dashed line).

8.8 Case Study 8: Hybrid Pneumatic Engine

In this case study a hybrid pneumatic-combustion engine system (HPE) is analyzed, using the concepts introduced in Sect. 5.6. The system is conceived with the aim of enabling energy recuperation using the engine as a pneumatic pump and assisting the conventional engine operation using the air stored in a tank, thus using the engine as a pneumatic motor.

In this section, the potential of the HPE concept in enhancing the fuel economy is analyzed using simple thermodynamic models of the single operation modes. For a description of the system analyzed in this case study refer to Sect. 5.6. The globally optimal supervisory control signals are calculated using a dynamic programming approach. The main constraint is that the tank charge at the start and the end of the drive cycle must be equal.

8.8.1 HPE Modeling

The performance of the system can be predicted using ideal models of the operation modes illustrated in Sect. 5.6. The engine data considered here and listed in Table 8.13 are the same as presented in [111]. The engine friction losses will be added below where the HPE model is combined with a vehicle model.

An ideal thermodynamic analysis of the conventional ICE operation, as described in Sect. 5.6.1, allows calculating the output work per cycle W_o and the input fuel energy per cycle W_i as a function of the design parameters of Table 8.13. The resulting performance curve $W_o = f(W_i)$ is shown in Fig. 8.35a. The curve has been obtained by varying the intake pressure (throttled SI operation) from approximately 0 to 1 bar. The relationship is practically linear with a slope of 0.49 and external losses around 220 J/cycle.

The pneumatic motor operation mode, as introduced in Sect. 5.6.1, is described in terms of W_o and the input energy discharged from the tank per cycle, W_t. Also for this two-stroke mode of operation, the "cycle" to which the quantities W_o and W_t refer to is assumed to be a four-stroke cycle, i.e., W_o and W_t represent twice the amount of work produced during one two-stroke cycle. The performance curves $W_o = f(W_t)$ at various tank pressures p_t are shown in Fig. 8.35b, where each curve has been obtained by varying the charge valve closing CVC only.[8] The curves are quadratic with good approximation in all the cases considered.

Similarly, the pneumatic pump operation, which is introduced in Sect. 5.6.1, is described by W_o and W_t, which both are negative numbers during normal conditions. The performance curves $W_o = f(W_t)$ at various p_t are shown in Fig. 8.36a. Each curve has been obtained by varying the IVC event. The curves are linear with very good approximation, with slopes that increase with p_t.

Complex operating modes, such as the undercharged and the supercharged operation modes, are characterized by all three energy terms W_o, W_i, and

---

[8] In Sect. 5.6 the simultaneous control of both CVC and EVC is illustrated.

Table 8.13. Main parameters of the system analyzed in this section.

V_d	displacement volume	2000 cm^3 (4 × 500 cm^3)
κ	compression ratio	10
p_{in}	intake pressure at wide-open throttle	1 bar
p_{out}	exhaust pressure	1.1 bar
ϑ_{in}	intake temperature	300 K
H_l	fuel lower heating value	42.5 MJ/kg
η_c	combustion efficiency	0.8
ϕ_v	fraction of combustion energy	
	released at constant volume	0.5
ϕ_p	fraction of combustion energy	
	released at constant pressure	0.5
V_t	tank volume	50 dm^3
	tank temperature	300 K
p_t	nominal tank pressure	20 bar
m_v	vehicle curb mass	800 kg
c_r	rolling resistance coefficient	0.00863
c_d	aerodynamic drag coefficient	0.312
A_f	vehicle frontal area	2.06 m^2
ρ_a	air density	1.293 kg/m^3
η_d	transmission efficiency	0.8
k_d	transmission ratio	{13.45, 7.57, 5.01, 3.77, 2.84}
r_w	wheel radius	0.282 m
$z_{\{a,b,c\}}$	friction loss coefficients	{0.776, 0.189, 0.0209 }

W_t. The performance curves $W_o = f(W_i)$ and $W_t = f(W_i)$ at various p_t are shown in Fig. 8.36b. Each curve consists of two branches, on the right (supercharged) and on the left (undercharged) of the point corresponding to the full-load conventional operation. The left branches have been obtained by first varying the intake pressure from a reasonable minimum value up to 1 bar and second the CVC event. Interestingly, the fuel energy and the output work increase monotonically, but the energy charged in the tank first increases and then decreases again. The right branches have been obtained by varying the CVO event. Correspondingly the input energy increases, and so do the input energy from the tank and the output work, the former almost linearly and without significant influence of the pressure tank.

The results presented above are qualitatively confirmed using a more detailed modeling approach, based on zero-dimensional modeling of the combustion process and of the mass and energy exchanges through the valves. For example, performance curves of the ICE conventional operation mode are still linear and almost independent from the speed, with an average slope of 0.43 and external losses of 204 J/cycle. In contrast, pneumatic motor and pump

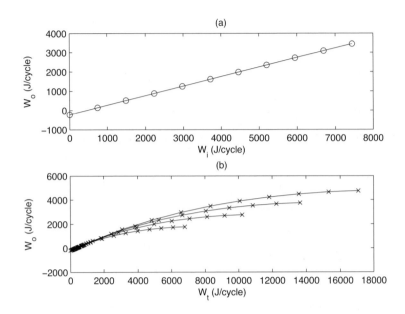

Fig. 8.35. Performance curve for the ICE operation mode (top) and for the pneumatic motor operation mode (bottom). Increasing curves are for tank pressure $p_t = 10, 15, 20, 25$ bar.

operation modes are much more sensitive to speed. Consequently, the single curves of Figs. 8.35b – 8.36a are split in several curves for different speeds.

8.8.2 Driveline Modeling

The energy required at the engine output is calculated at each time step from the vehicle data and the driving profile which defines the vehicle speed $v(t)$ and acceleration $a(t)$. The traction force (negative if braking) is calculated as the sum of the inertia force and the rolling and aerodynamic resistances

$$F(t) = m_v \cdot \frac{d}{dt} v(t) + c_r \cdot m_v \cdot g + \frac{1}{2} \cdot \rho_a \cdot A_f \cdot c_d \cdot v(t)^2 \, , \qquad (8.103)$$

where $m_v, c_r, c_d, A_f, \rho_a$ are the vehicle curb mass, the rolling resistance coefficient, the aerodynamic drag coefficient, the vehicle frontal area and the air density, whose numerical values are listed in Table 8.13.

To evaluate the engine output work per cycle $W_o(t)$, first the engine speed $\omega(t)$ is calculated as

$$\omega(t) = k_d(n(t)) \cdot \frac{v(t)}{r_w} \, , \qquad (8.104)$$

where r_w is the wheel radius and k_d the overall transmission ratio, which depends on the gear number $n(t)$. The latter is evaluated using a simple shifting strategy.

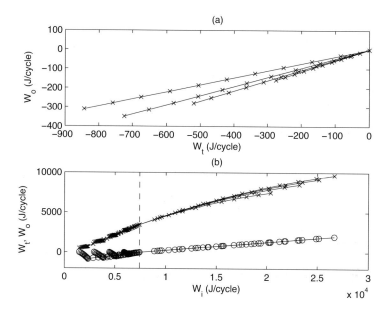

Fig. 8.36. Performance curves for the pneumatic pump operation mode (top) and for the supercharged and undercharged operation modes (bottom). Various curves are for $p_t = 5, 10, 15, 20$ bar. Dashed line represents conventional full-load operation.

The engine output work is here the sum of the work required for traction and the engine friction losses, which are evaluated as a quadratic function of the engine speed

$$
W_o(t) = \frac{4\pi}{\omega(t)} \cdot v(t) \cdot F_t(t) \cdot \eta_d^{-\mathrm{sign}(F_t(t))} +
$$
$$
+ V_d \cdot \left[z_a + z_b \cdot \left(\frac{\pi \cdot \omega(t)}{30} \right) + z_c \cdot \left(\frac{\pi \cdot \omega(t)}{30} \right)^2 \right] , \tag{8.105}
$$

where η_d is the overall transmission efficiency, V_d is the engine displacement volume, and $z_{\{a,b,c\}}$, are loss coefficients which have to be determined experimentally. The engine system must produce at each time step the demanded work $W_o(t)$. The supervisory controller chooses only among the operating modes that are compatible with that value, i.e., that can provide or absorb that work with a suitable selection of the corresponding load control parameter. If $W_o(t)$ is negative, the engine system can absorb only a part of it, while the remaining fraction is intended to be absorbed by a conventional friction braking. For the pneumatic modes, the values shown in Figs. 8.35b – 8.36a are multiplied by two (two-stroke cycles).

8.8.3 Air Tank Modeling

For this case study the air tank is considered as an isothermal system. In fact, in absence of an effective insulation coating, the periods in which the system is subjected only to thermal losses to the environment are much longer than the periods in which the system is in contact via the charge valves with the cylinders at a higher temperature. Thus the tank model is

$$p_t(t) = \frac{\gamma - 1}{V_t} \cdot U_t(t), \quad \dot{U}_t(t) = -W_t(t) \cdot \frac{w(t)}{4\pi} , \tag{8.106}$$

where V_t is the tank volume, U_t its internal energy, and γ is the ratio of specific heats of air.

8.8.4 Optimal Control Strategy

The supervisory controller of an HPE vehicle must switch between the operating modes illustrated in the previous sections. In this section, a technique is presented to calculate a controller that minimizes the fuel consumption over a prescribed drive cycle, while keeping the final value of the pressure tank at the initial value. The control is defined as the sequence of operating modes $u(t)$, where $u = 0, 1, \ldots, 4$ for disengaged, pneumatic motor, pneumatic pump, ICE, and undercharged/supercharged engine operations, respectively.

The control law must minimize the performance index over a cycle of duration t_f

$$J = \int_0^{t_f} W_i(t, u) \cdot \frac{w(t)}{4\pi} dt \tag{8.107}$$

under the constraint that

$$\Delta U = U_t(t_f) - U_t(0) = \int_0^{t_f} -W_t(t, u) \cdot \frac{w(t)}{4\pi} dt = 0 . \tag{8.108}$$

The problem is solved with a dynamic programming technique (see Appendix III), where $U_t(t)$ (proportional to $p_t(t)$ via Eq. 8.106) is the state variable. The Jacobian

$$f(t, u) = -W_t(t, u) \cdot \frac{w(t)}{4\pi}$$

and the arc cost

$$L(t, u) = W_i(t, u) \cdot \frac{w(t)}{4\pi}$$

depend also on the selected operating mode.

The numerical technique adopted in this case study requires a discretization of the time-state space (t, U_t). Here a discretization step of 1 s for time and of 2 kJ for the state variable is used, with the state variable bounded between 25 kJ and 300 kJ, corresponding to 2 bar and 24 bar, respectively. A cost-to-go matrix $J(t, U_t)$ is introduced, which is defined as the cost over the optimal trajectory passing through the point in the time-state space, up to the target point $(t_f, U_t(0))$. With this definition, the overall cost function J must be equal to $J(0, U_t(0))$. The solving algorithm proceeds backwards from time t_f to time $t = 0$. The cost-to-go matrix is evaluated using the recursive rule

$$J(t, U_t) = \min_u \{ J(t + \Delta t, U_t + f(t, u) \cdot \Delta t) + L(t, u) \cdot \Delta t \} . \tag{8.109}$$

The initial condition needed for the recursive algorithm is imposed at time t_f and is set in such a way that all the state-of-charge deviations with respect to $U_t(0)$ are penalized,

$$J(t_f, U_t) = \begin{cases} 0 & \text{for} \quad U_t = U_t(0) \\ \infty & \text{for} \quad U_t \neq U_t(0) \end{cases} . \tag{8.110}$$

At the last step of the algorithm the quantity $J(0, U_t(0))$ is the best approximation of the optimal cost that is compatible with the state discretization used. This result can be checked by running the model forward, using the stored feedback function $C(t, U_t) = \arg J(t, U_t)$ to evaluate the optimal control law as

$$u_{opt}(t) = C(t, U_t(t)) . \tag{8.111}$$

8.8.5 Optimal Control Results

The simulations of [111] on a MVEG-95 test cycle led to an energy consumption of the ICE-based vehicle of 26.4 MJ, while in the hybrid configuration the energy consumption is 22.4 MJ with an air tank pressure of 20 bar at the beginning and at the end of the cycle. This fuel saving, approximately 15%, was estimated using a control strategy based on a set of heuristic rules. In this section the optimal control law described in the previous section is applied in order to estimate the full potential of the HPE.

The simulation of the MVEG–95 cycle with the ideal models of the single operation modes leads to the results shown in Fig. 8.37, which can be summarized as follows. The fuel energy consumption in the conventional ICE operation is 20.8 MJ[9], while with the optimized HPE it is 16.5 MJ, i.e., a fuel saving of about 21%. This fuel saving is due to (i) the 1.3 MJ spent by the conventional propulsion system during idling, which are suppressed in the HPE operation, (ii) the 1.0 MJ energy recuperation during braking, and (iii) the optimization of power flows during traction phases. A comparison of the

---

[9] This value is substantially lower than that declared by the authors of [111], however the mechanical work at the wheels of 2.3 MJ is very similar to that calculated in that paper.

fuel input energy in the conventional and in the hybrid configuration is shown in Fig. 8.38. The fuel consumption of the conventional configuration during idling is evident from that plot. The globally optimal solution is characterized by frequent switches between motor and undercharged operations, which corresponds to several duty cycles similar to those that are typical of hybrid concepts with short-term storage systems (see, for instance, Chap. 5). However, the switching between the two modes involved does not imply onerous operations such as, e.g., clutch engagement/disengagement.

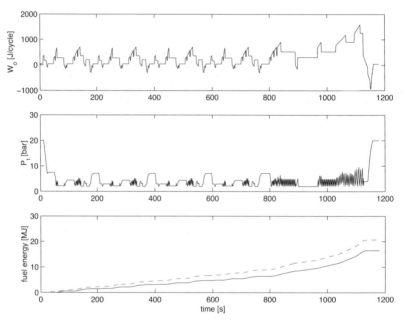

Fig. 8.37. Optimal controller results on the MVEG–95 cycle. Top to down: work required per cycle, tank pressure, fuel energy for the conventional (dashed) and the hybrid (solid) engine system.

The importance of such duty cycles is confirmed by the results obtained for the European urban (ECE–15) test cycle (Fig. 8.39), which show a reduction of about 20% in fuel energy consumption (conventional ICE: 2.25 MJ, HPE: 1.79 MJ). This reduction is explained by the fact that the ECE–15 cycle allows for lower energy recuperation, thus the fuel saving is mainly due to the reason (i) above, i.e., the suppression of the idle fuel consumption. In fact Fig. 8.39 shows that the pump mode is used to recuperate energy only during 14% of the cycle time (the motor operating mode for the 1.5% of time and the undercharged/supercharged mode is never used).

The results obtained with ideal models of the engine system have been confirmed by using more realistic models, both for the MVEG–95 and the

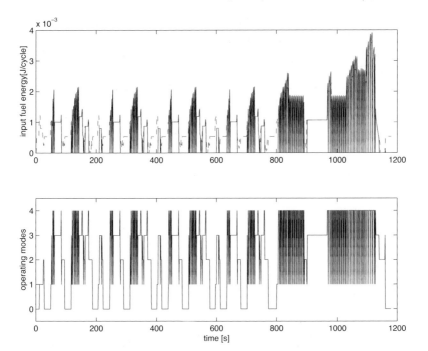

Fig. 8.38. Optimal controller results on the MVEG–95 cycle. Top: fuel energy per cycle as a function of time for the conventional (dashed) and the hybrid (solid) operation. Bottom: operating modes, disengaged ($u = 0$), pneumatic motor ($u = 1$), pneumatic pump ($u = 2$), conventional ($u = 3$), supercharged/undercharged ($u = 4$).

ECE–15 drive cycles. The overall fuel energy calculated for the conventional architecture and for the hybrid configuration optimized by means of dynamic programming are summarized in Table 8.14. The three cases analyzed are referred to: (i) ideal models of all modes, i.e., the results presented above, (ii) zero-dimensional model of the engine mode only, (iii) zero-dimensional models of the engine and the pneumatic motor modes. Since the load control for the pneumatic pump operation and the undercharged/supercharged mode is a critical issue, an effective zero-dimensional modeling approach is not well established for these modes of operation. Thus, they are still modeled as ideal in (ii) and (iii).

With the modeling approach (ii), the fuel consumption calculated for the conventional architecture is higher than in the case (i), as a consequence of an engine efficiency lower than in the ideal case. On the other hand, also the fuel consumption of the HPE increases when estimated with the more realistic models (ii) and (iii). However, the fuel saving in the most realistic case available (iii), being around 23% and 19% for the MVEG–95 and the ECE–15 cycle, respectively, still demonstrates the good potential of the concept.

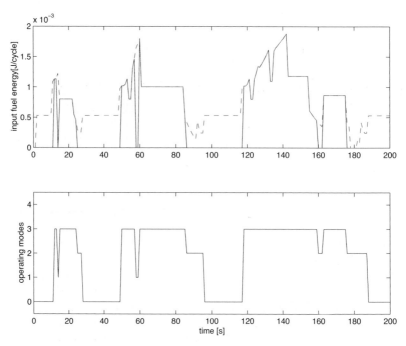

Fig. 8.39. Optimal controller results on the ECE–15 cycle. Top: fuel energy per cycle as a function of time for the conventional (dashed) and the hybrid (solid) operation. Bottom: operating modes, disengaged ($u = 0$), pneumatic motor ($u = 1$), pneumatic pump ($u = 2$), conventional ($u = 3$), supercharged/undercharged ($u = 4$).

Table 8.14. Simulation results over the ECE-15 drive cycle. The three modeling cases (i) to (iii) are described in the text.

case	ECE–15			MVEG–95		
	J_{ICE} (MJ)	J_{HPE} (MJ)	fuel saving	J_{ICE} (MJ)	J_{HPE} (MJ)	fuel saving
(i)	2.25	1.79	-20%	20.8	16.5	-21%
(ii)	2.52	2.02	-20%	23.5	17.1	-27%
(iii)	2.52	2.04	-19%	23.5	18.2	-23%

9

Appendix II – Optimal Control Theory

This appendix briefly summarizes the most useful results of optimal control theory. In the first section static problems are analyzed, i.e., the objective function only includes time-independent control variables. This formulation yields a parameter optimization or nonlinear programming problem for which several closed-form and numerical solution algorithms are known. Several excellent textbooks are available on that subject, for instance [23] and [27].

The second section analyzes dynamic optimal control problems. Starting with a brief repetition of the classical variational calculus theory, the concepts of adjoint (Lagrange) states and Hamiltonian formulations are introduced. To be able to deal with the case of constrained input variables, Pontryagin's minimum principle is briefly introduced. As with the first section, the main objective here is to collect the main facts without any proofs and to introduce the notation. Readers interested to learn more about this field are referred to one of the several available textbooks, for instance [30].

9.1 Parameter Optimization Problems

9.1.1 Problems Without Constraints

The vector $u = [u_1, \ldots, u_m]^T \in \Re^m$ consists of arbitrary parameters and the mapping $L : \Re^m \to \Re$ is a sufficiently differentiable function (the *performance index*) that has to be minimized.[1] Sufficient conditions for a point u^o to be a local minimum are

$$\left.\frac{\partial L(u)}{\partial u}\right|_{u=u^o} = 0, \quad \text{and} \quad \left.\frac{\partial^2 L(u)}{\partial u^2}\right|_{u=u^o} > 0 , \qquad (9.1)$$

i.e., the gradient of the performance index must be zero at the minimum (u^o is a stationary point), and the Hessian matrix of the performance index must

---

[1] A maximization problem can be obtained from a minimization problem by simply multiplying the performance index by -1.

be positive definite (in the neighborhood of u^o, $L(\cdot)$ increases throughout). The condition (9.1) is globally sufficient only for specific cases, for instance if it is known that the function $L(\cdot)$ is globally convex.

Necessary conditions for a local minimum are:

$$\left.\frac{\partial L(u)}{\partial u}\right|_{u=u^o} = 0, \quad \text{and} \quad \left.\frac{\partial^2 L(u)}{\partial u^2}\right|_{u=u^o} \geq 0 . \tag{9.2}$$

To establish whether a minimum exists, additional information is needed.

Example 1 Performance index:

$$L(u) = \frac{1}{2} \cdot u^T \cdot M \cdot u, \quad u = \begin{bmatrix} u_1 \\ u_2 \end{bmatrix}, \quad M = \begin{bmatrix} 1 & 1 \\ 1 & \mu \end{bmatrix}, \quad \mu \in \Re . \tag{9.3}$$

Minimization:

$$\frac{\partial L}{\partial u} = M \cdot u, \quad \frac{\partial^2 L}{\partial u^2} = M , \tag{9.4}$$

that is, if M is non-singular ($\mu \neq 1$) then only one minimum $u^o = [0,0]^T$ exists. For $\mu = 1$ M is semidefinite positive and all the points on the line $\lambda \cdot [1,-1]^T$ are minima. For $\mu > 1$, M is positive definite and u^o is a global minimum. For $\mu < 1$, M is indefinite and u^o is not a minimum, but a saddle point, see Fig. 9.1.

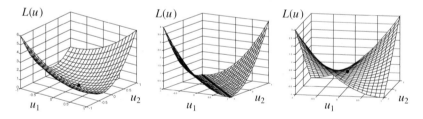

Fig. 9.1. Performance index of Example 1, left $\mu > 1$, center $\mu = 1$, right $\mu < 1$.

Example 2 Performance index:

$$L(u) = (u_1 - u_2^2) \cdot (u_1 - 3 \cdot u_2^2) . \tag{9.5}$$

Minimization:

$$\frac{\partial L}{\partial u} = \begin{bmatrix} 2 \cdot u_1 - 4 \cdot u_2^2 \\ -8 \cdot u_1 \cdot u_2 + 12 \cdot u_2^3 \end{bmatrix} = \begin{bmatrix} 0 \\ 0 \end{bmatrix} \Rightarrow u^o = \begin{bmatrix} 0 \\ 0 \end{bmatrix}, \quad \left.\frac{\partial^2 L}{\partial u^2}\right|_{u=u^o} = \begin{bmatrix} 2 & 0 \\ 0 & 0 \end{bmatrix}, \tag{9.6}$$

that is,[2] $L_u(u) = 0$ has a triple but isolated solution, and this solution is not a local minimum, as shown in Fig. 9.2.

---

[2] L_u denotes the partial derivative of L with respect to u.

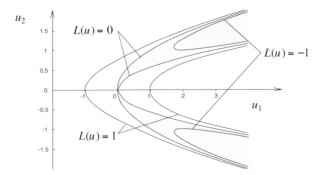

Fig. 9.2. Performance index of Example 2.

9.1.2 Numerical Solution

Only in very rare cases a closed-form solution may be obtained. In the majority of all cases numerical methods are needed. Several different approaches are possible. One class of algorithms utilizes information on the gradients of $L(u)$. These algorithms can be subdivided into:

- semi-analytical methods where the performance index $L(u)$ and its gradient $\partial L/\partial u$ are available in a closed form; and
- fully numerical methods in which only the performance index is known and whose gradient must be approximated through finite differences.

The semi-analytical methods in general converge much faster than the fully numerical methods and they are less sensitive to rounding effects.

The numerical search can be either of first order or of second order. The first-order numerical methods use the idea of the steepest descent such that all first-order algorithms have approximately this structure:

1. guess an initial value for $u(1)$;[3]
2. evaluate the gradient $\Gamma_L(i) = \frac{\partial L}{\partial u}\big|_{u=u(i)}$ either from a known relationship for the gradient or through numerical approximation using finite differences;
3. determine the new iteration point according to the rule $u(i+1) = u(i) - h(i) \cdot \Gamma_L(i)$;
4. check if the variation of the performance index $|L(u(i+1)) - L(u(i))|$ is smaller than a predetermined threshold. In this case, the algorithm ends, otherwise it is repeated starting at point 2.

Crucial for the convergence is the choice of the relaxation factor $h(i)$. If it is chosen too large (see Fig. 9.3), the algorithm may overshoot and even become unstable. One possibility to choose $h(i)$ in an optimal way consists of solving a minimization problem with constraints (see below) along the gradient, i.e.,

---

[3] The argument denotes the iteration index.

$u(2)$ is the point that generates the smallest performance index and at the
same time lies in the direction of the gradient.

Figure 9.3 also shows the problem that arises with narrow and steep "val-
leys": a slow convergence is in this case unavoidable. The choice of the starting
values is very important as well. If, on the basis of simplifying considerations,
a guess of the optimal point is possible, it is often worthwhile making this
additional effort to prevent the search from drifting towards unrealistic solutions.

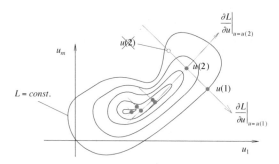

Fig. 9.3. Schematic of the steepest descent algorithm.

Besides first-order methods (which use only the first-order derivatives),
second-order methods are known (Newton-Raphson approaches). They converge faster in the neighborhood of the optimum, but are known to be inefficient far away from it. The basic idea is to approximate the performance
index through a quadratic form

$$L(u) \approx L(u(i)) + \Gamma_L(i) \cdot (u - u(i)) +$$
$$\frac{1}{2} \cdot (u - u(i))^T \cdot \Gamma'_L(i) \cdot (u - u(i)) \tag{9.7}$$

where $\Gamma'_L(i) = \left. \frac{\partial^2 L}{\partial u^2} \right|_{u=u(i)}$, and then to use the solution of the approximate
problem (which is solvable in closed form) as a new iteration value

$$u(i+1) = u(i) - [\Gamma'_L(i)]^{-1} \cdot [\Gamma_L(i)]^T . \tag{9.8}$$

All the algorithms discussed above are available in program libraries or for
instance in Matlab^{TM} (optimization toolbox). In most cases the use of this
software is recommended over any attempts to re-invent the wheel.

9.1.3 Minimization with Equality Constraints

Let $u = [u_1, \ldots, u_m]^T \in \Re^m$ be a vector of arbitrary control variables, $x = [x_1, \ldots, x_n]^T \in \Re^n$ a vector of state variables[4] and $f : \Re^{m+n} \rightarrow \Re^n$, $L : \Re^{m+n} \rightarrow \Re$ sufficiently differentiable functions of these quantities. The optimization problem consists of finding the values u^o, x^o that minimize the performance index L and, at the same time, fulfill the constraint $f(u^o, x^o) = 0$. With the adoption of a new variable $z = [u, x]^T$, the following expressions may be written in a more compact way.

In the case $n = 1$, the solution of this problem can be immediately given with the help of Fig. 9.4. An optimal point z^o is characterized by the fact that both the gradient on an iso-level curve of the performance index and the gradient on the subset defined by the constraint are co-linear,

$$\left. \frac{\partial L}{\partial z} \right|_{z=z^o} + \lambda \cdot \left. \frac{\partial f}{\partial z} \right|_{z=z^o} = 0 , \tag{9.9}$$

where λ is a new arbitrary parameter. These $m + 1$ equations, together with the constraint $f(z^o) = 0$, define the $m + 2$ unknown quantities $z^o = [u^o, x^o]$ and λ^o.

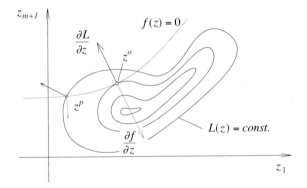

Fig. 9.4. Constrained optimization problem with one constraint.

The solution of the general case can be derived from the fact that in the optimal point any *arbitrary* variation dz that satisfies the constraints may not cause any variation of the performance index, i.e., for

$$df(u, x) = f_u \cdot du + f_x \cdot dx = 0 , \tag{9.10}$$

it has to be true that

---

[4] The distinction between control and state variables is only a matter of convenience.

$$dL(u, x) = L_u \cdot du + L_x \cdot dx = 0 . \qquad (9.11)$$

From (9.10), it follows that the variation dx depends on the variation du,

$$dx = -f_x^{-1} \cdot f_u \cdot du . \qquad (9.12)$$

The $n \times n$ matrix f_x has to be nonsingular in the optimal point, otherwise the problem is not well posed. Inserting (9.12) into (9.11),

$$dL(u, x) = [L_u - L_x \cdot f_x^{-1} \cdot f_u] \cdot du = 0 . \qquad (9.13)$$

Since this has to be valid for any arbitrary variation du, the following necessary condition results

$$L_u - L_x \cdot f_x^{-1} \cdot f_u = 0 . \qquad (9.14)$$

Sufficient conditions for a local minimum have to take into account the second-order variations. Expanding in a Taylor series the performance index around the optimal point x^o, u^o up to the second-order variations,

$$dL \approx [L_x(x^o, u^o), L_u(x^o, u^o)] \cdot \begin{bmatrix} dx \\ du \end{bmatrix} +$$
$$+ \frac{1}{2} \cdot [dx^T, du^T] \cdot \begin{bmatrix} L_{xx}(x^o, u^o) & L_{xu}(x^o, u^o) \\ L_{ux}(x^o, u^o) & L_{uu}(x^o, u^o) \end{bmatrix} \cdot \begin{bmatrix} dx \\ du \end{bmatrix} . \qquad (9.15)$$

The linear term vanishes for (9.11). Moreover, dx cannot be arbitrarily selected, but it has to be related to du in such a way that the constraint $f(x, u) = 0$ is satisfied, i.e., with (9.12). Equation (9.15) then becomes

$$dL \approx \frac{1}{2} \cdot du^T \cdot [-f_u^T \cdot (f_x^{-1})^T, I] \cdot$$
$$\cdot \begin{bmatrix} L_{xx}(x^o, u^o) & L_{xu}(x^o, u^o) \\ L_{ux}(x^o, u^o) & L_{uu}(x^o, u^o) \end{bmatrix} \cdot \begin{bmatrix} -f_x^{-1} \cdot f_u \\ I \end{bmatrix} \cdot du , \qquad (9.16)$$

which has to be fulfilled for arbitrary values of du. The sufficient condition for the point x^o, u^o being optimal is therefore

$$\left. \frac{\partial^2 L}{\partial u^2} \right|_{opt} = (L_{uu} - f_u^T \cdot (f_x^{-1})^T \cdot L_{xu} - $$
$$- L_{ux} \cdot f_x^{-1} \cdot f_u + f_u^T \cdot (f_x^{-1}) \cdot L_{xx} \cdot f_x^{-1} \cdot f_u)\big|_{x^o, u^o} > 0 , \qquad (9.17)$$

where $opt = \{x^o, u^o, f(x, u) = 0\}$. In other words, the Hessian matrix of the performance index in the point x^o, u^o and for variations that satisfy the constraints has to be positive definite. If (9.17) is only semidefinite, x^o, u^o could not be a minimum.

An analogous formulation of the optimization problem, which will be easily extended later, uses a formalism that is based on the approach (9.8). Instead of the original problem, the function

$$H(x, u, \lambda) = L(x, u) + \lambda^T \cdot f(x, u) \tag{9.18}$$

is minimized. The function $H(\cdot)$ is the *Hamiltonian* function of the optimization problem, while $\lambda = [\lambda_1, \ldots, \lambda_n]^T$ are new arbitrary variables that will be referred to as Lagrange multipliers.

In the optimal point $f(x^o, u^o) = 0$ must be zero and therefore in that point $H(x^o, u^o, \lambda^o) = L(x^o, u^o)$. For the formulation (9.18) to be equivalent to the original optimization problem, the variation

$$dH(x, u, \lambda) = H_x(x, u, \lambda) \cdot dx + H_u(x, u, \lambda) \cdot du + f(x, u) \cdot d\lambda \tag{9.19}$$

must vanish if, and only if, dL vanishes. The last addend in (9.19) is in any case identically null (constraint). The new degrees of freedom λ can be defined in such a way that $H_x(x, u, \lambda) = 0$, i.e.,

$$\begin{aligned} H_x(x, u, \lambda) = L_x(x, u) + \lambda^T \cdot f_x(x, u) = 0 \quad &\Rightarrow \\ \Rightarrow \lambda^T(x, u) = -L_x(x, u) \cdot f_x^{-1}(x, u) \; . \end{aligned} \tag{9.20}$$

Inserting λ defined in this way in H_u, the necessary condition (9.14) is obtained. The original optimization problem and the Hamiltonian formulation are therefore equivalent. The necessary condition of the optimization problem can be written as

$$dH(x, u, \lambda) = 0 \tag{9.21}$$

or

$$i) \quad \left. \frac{\partial H}{\partial u} \right|_{u=u^o, x=x^o, \lambda=\lambda^o} = 0 \Rightarrow \text{Condition for optimality}$$

$$ii) \quad \left. \frac{\partial H}{\partial x} \right|_{u=u^o, x=x^o, \lambda=\lambda^o} = 0 \Rightarrow \text{Fulfilled by the choice (9.19) .} \tag{9.22}$$

$$iii) \quad \left. \frac{\partial H}{\partial \lambda} \right|_{u=u^o, x=x^o, \lambda=\lambda^o} = 0 \Rightarrow \text{Constraint}$$

These $m + 2n$ equations are necessary conditions for u^o, x^o, and λ^o representing the solution of the optimization problem. Similarly, also the sufficient condition (9.17) can be derived in the Hamiltonian formulation.

The Lagrange multipliers allow an interesting interpretation, which will be useful in the following,

$$\left. \frac{\partial L}{\partial f} \right|_{u=u^o} = -\lambda^T \; , \tag{9.23}$$

i.e., considering instead of the constraint $f(x, u) = 0$ the slightly modified constraint $f(x, u) - df = 0$ (df being a constant vector), the value of the performance index in the new optimum is approximated as

$$L|_{u=u^o(df)} \approx L|_{u=u^o(0)} - \lambda^T \cdot df \; . \tag{9.24}$$

A further application of the Hamiltonian formalism will be shown in the next section.

9.1.4 Minimization with Inequality Constraints

Let $u = [u_1, \ldots, u_m]^T \in \Re^m$ be a vector of arbitrary variables and the mappings $f : \Re^m \rightarrow \Re^n$, $L : \Re^m \rightarrow \Re$ sufficiently differentiable functions of these quantities. The optimization problem consists of finding that value of u^o which minimizes the performance index L and at the same time satisfies the condition $f(u^o) \leq 0$. Such a problem is often denoted as *nonlinear programming*, in analogy to the concept of *linear programming*, in which both L and f are linear. Often, the number of inequalities n is larger than the number of variables m. Indeed, some of the inequalities in the optimum are not active at all, but it is not known a priori which inequality does not have to be considered. The distinction between control and state variables thus loses its meaning.

If no inequality is active at all, the problem is reduced to the original optimization problem without constraints with the sufficient conditions (9.1). If only one inequality is active, the situation shown in Fig. 9.5 arises.

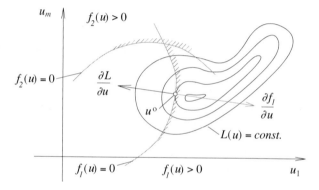

Fig. 9.5. Optimization with inequality constraints: two constraints, with only one active.

This problem is analogous to the situation shown in Fig. 9.4. In particular, the boundary of the inequality constraint becomes a regular constraint, such as the one that was discussed in the last section. Nevertheless, there is a difference: the sign of the Lagrange multiplier is now given, since only positive multipliers are admissible. The sign of the differential dL has to be positive or zero

$$dL|_{u=u^o} = \left.\frac{\partial L}{\partial u}\right|_{u=u^o} \cdot du \geq 0 \qquad (9.25)$$

for admissible variations du, i.e. for

$$df|_{u=u^o} = \left.\frac{\partial f}{\partial u}\right|_{u=u^o} \cdot du \leq 0 . \qquad (9.26)$$

Instead, in the problem shown in Fig. 9.4, the gradients are equally oriented and thus negative Lagrange multipliers are possible.

The situation shown in Fig. 9.5 can be extended to the case of more active constraints. Figure 9.6 shows the basic idea. The gradient of the performance index has to be directed in such a way as to make possible a reduction of the performance index only through a violation of the constraints.

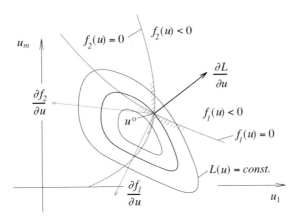

Fig. 9.6. Optimization with inequality constraints: two active constraints.

Accordingly, the gradients of the active constraints at the optimum define the base vectors, whose directions, weighted with negative constants, allow for the evaluation of the gradient of the performance index,

$$\left. \frac{\partial L}{\partial u} \right|_{u=u^o} = -\lambda_1 \cdot \left. \frac{\partial f_1}{\partial u} \right|_{u=u^o} - \lambda_2 \cdot \left. \frac{\partial f_2}{\partial u} \right|_{u=u^o} \cdots - \lambda_\nu \cdot \left. \frac{\partial f_\nu}{\partial u} \right|_{u=u^o} , \qquad (9.27)$$

where $\nu \leq n$ is the number of active constraints.

Introducing the Hamiltonian

$$H(u, \lambda) = L(u) + \lambda^T \cdot f(u) , \qquad (9.28)$$

this condition follows,

$$\frac{\partial H}{\partial u} = \frac{\partial L}{\partial u} + \lambda^T \cdot \frac{\partial f}{\partial u} = 0 , \qquad (9.29)$$

where

$$\begin{matrix} \lambda \geq 0 \ \text{if} \quad f(u) = 0 \ \text{(active constraint)} \\ \lambda = 0 \ \text{if} \quad f(u) < 0 \ \text{(inactive constraint)} \end{matrix} . \qquad (9.30)$$

Thus, in the subspace in which the constraints are not active, a stationary point will be sought in these directions. Where the constraints are active, since the sign of the multipliers is given, a special optimization problem with equality constraints will be solved.

It is possible to imagine "pathological" situations in which the gradients of the constraints are linearly dependent and therefore (9.26) cannot be satisfied. Such situations have to be excluded a priori, see [30].

9.2 Optimal Control

9.2.1 Introduction

The problems discussed in Sect. 9.1 were finite-dimensional, i.e., their solutions could be interpreted as real vectors in a finite-dimensional space. The problems discussed in this section lead instead to infinite-dimensional solutions. For each time point, parameter values are searched in a finite interval, i.e., in a continuum. In the mathematical literature the notion *calculus of variations* is used. In systems theory, the notion of *optimal control* is more common.

The problem formulation is: let $t_a, t_b \in \Re$ be the initial and final time, $u : [t_a, t_b] \to \Re^m$ are arbitrary functions, $x : [t_a, t_b] \to \Re^n$ is the state vector defined by $u(t)$ and the constraints (9.32), and $L : \Re^n \times \Re^m \times [t_a, t_b] \to \Re$, $\varphi : \Re^n \times \Re \to \Re$ are sufficiently differentiable functions. Find the control law $u(t)$ that minimizes the performance index

$$J(u) = \varphi(x(t_b), t_b) + \int_{t_a}^{t_b} L(x(t), u(t), t) \, dt , \qquad (9.31)$$

while the state $x(t)$ satisfies the differential equation

$$\dot{x} = f(x(t), u(t), t), \quad x(t_a) = x_a . \qquad (9.32)$$

Starting from this formulation, many variants can be derived (final time unspecified, final state partially or completely specified, constraints over the state vector or the control vector, etc.), some of which will be discussed below. The next sections follow closely the text of [30] and the same notation is used. Readers interested in a thorough treatment of the subject are referred to that monograph.

9.2.2 Optimal Control for the Basic Problem

Instead of dealing with (9.31) and (9.32) separately and on the basis of the results of Sect. 9.1, it is reasonable to solve a combined optimization problem in which the constraints are integrated in one performance index by means of the Lagrange multipliers,

$$\tilde{J}(u) = \varphi(x(t_b), t_b) + \int_{t_a}^{t_b} \left\{ L(x(t), u(t), t) + \lambda^T(t) \cdot [f(x(t), u(t), t) - \dot{x}(t)] \right\} dt .$$

$$(9.33)$$

Note that the Lagrange multipliers are now functions of time. As in Sect. 9.1, a Hamiltonian formulation is introduced with the definition

$$H(u(t), x(t), \lambda(t), t) = L(x(t), u(t), t) + \lambda^T(t) \cdot f(x(t), u(t), t) . \qquad (9.34)$$

The augmented performance index (9.33) is written as

$$\tilde{J}(u) = \varphi(x(t_b), t_b) + \int_{t_a}^{t_b} H(x(t), u(t), \lambda(t), t)dt - \int_{t_a}^{t_b} \lambda^T(t) \cdot \dot{x}(t)\, dt \ . \quad (9.35)$$

An integration by parts of the third term in (9.35) yields the final equation of the optimality condition

$$\tilde{J}(u) = \varphi(x(t_b), t_b) - \lambda^T(t_b) \cdot x(t_b) + \lambda^T(t_a) \cdot x(t_a) +$$
$$+ \int_{t_a}^{t_b} H(x(t), u(t), \lambda(t), t)dt + \int_{t_a}^{t_b} \dot{\lambda}^T(t) \cdot x(t)\, dt \ . \quad (9.36)$$

Now the control variable $u(t)$ will be varied by $\delta u(t)$. With (9.37) the resulting variation of the augmented performance index is obtained by means of the chain rule

$$\delta \tilde{J}(u) = \left[\left(\frac{\partial \varphi(x, t)}{\partial x} - \lambda^T(t) \right) \cdot \delta x(\delta u) \right]_{t=t_b} + \lambda^T(t_a) \cdot \delta x_a$$
$$+ \int_{t_a}^{t_b} \left[\frac{\partial H(x, u, \lambda, t)}{\partial u} \cdot \delta u + \left(\frac{\partial H(x, u, \lambda, t)}{\partial x} + \dot{\lambda}^T(t) \right) \cdot \delta x(\delta u) \right] dt$$
$$(9.37)$$

In this equation, δx_a is the arbitrary variation of the initial condition, which depends on $\delta u(t)$. Using the constraints (9.32), the variations $\delta x(u)$ are of course related to the variations $\delta u(t)$.

To avoid the calculation of such a dependence, it is possible to use the degrees of freedom $\lambda(t)$, which have not been specified yet. In fact, if the derivative of λ is chosen as

$$\dot{\lambda}(t) = -\left(\frac{\partial H(x, u, \lambda, t)}{\partial x} \right)^T = -\left(\frac{\partial L(x, u, t)}{\partial x} \right)^T - \left(\frac{\partial f(x, u, t)}{\partial x} \right)^T \cdot \lambda(t),$$
$$\lambda(t_b) = \left(\frac{\partial \varphi(x, t)}{\partial x} \right)^T_{t=t_b} \quad (9.38)$$

the variation $\delta \tilde{J}$ will be independent of δx. Equation (9.38) is a system of ordinary differential equations for the Lagrange multipliers. Unfortunately, the final value rather than the initial value of $\lambda(t)$ is specified. Both the systems (9.32) and (9.38) are coupled by the optimality condition for the control vector.

With the choice of (9.38) the variation of the augmented performance index (9.37) becomes

$$\delta \tilde{J} = \lambda^T(t_a) \cdot \delta x_a + \int_{t_a}^{t_b} \left[\frac{\partial H(x, u, \lambda, t)}{\partial u} \cdot \delta u(t) \right] dt \ . \quad (9.39)$$

Two conclusions can be drawn from this equation: (i) if the control vector is kept constant, then $\lambda^T(t_a)$ is the gradient of the performance index with respect to the initial conditions, and (ii) if the initial conditions are kept constant, the optimal control vector has to yield $\delta \tilde{J} = 0$. For arbitrary $\delta u(t)$, the choice

$$\frac{\partial H(x, u, \lambda, t)}{\partial u} = 0, \qquad \forall t \in [t_a, t_b] \tag{9.40}$$

is the only way to accomplish that. Therefore, all the quantities in the equations of the optimization problem defined above are known. Table 9.1 provides a compact summary.

Table 9.1. Sufficient conditions for optimal control (Euler–Lagrange equations).

System	$\dot{x} = f(x(t), u(t), t)$
Performance index	$J(u) = \varphi(x(t_b), t_b) + \int_{t_a}^{t_b} L(x(t), u(t), t)\, dt$
Adjoint system	$\dot{\lambda}(t) = -\left(\dfrac{\partial L(x, u, t)}{\partial x}\right)^T - \left(\dfrac{\partial f(x, u, t)}{\partial x}\right)^T \cdot \lambda(t)$
Optimal control	$\dfrac{\partial L(x, u, t)}{\partial u} + \lambda^T(t) \cdot \dfrac{\partial f(x, u, t)}{\partial u} = 0$
Boundary conditions	$x(t_a) = x_a$ and $\lambda(t_b) = \left(\dfrac{\partial \varphi(x, t)}{\partial x}\right)^T_{t=t_b}$

As already mentioned, this $2n$-dimensional system of coupled differential equations is not easily solved, since n boundary conditions are given at the initial time and the other n values at the final time (two-point boundary problem). If the equations are solvable in closed form, this leads to some implicit relationships, which are sometimes solvable (see the example below). In numerical methods it is necessary to proceed iteratively, since the Lagrange multipliers are not directly related to the system considered, and then no estimation is possible of $\lambda(t_a)$, which strongly complicates the calculation.

Example 3: Optimal "Rendez-Vous" Maneuver A vehicle in the plane has to be controlled to move in a prefixed time t_b from a given initial state as closely as possible to a desired final state $x(t_b) \approx p$ and $v(t_b) \approx w$. The control signal must be optimal in terms of fuel consumption, which is assumed to be proportional to the square of the input $u(t)$. From Table 9.1 the following relationships can be derived:

$$\dot{x}(t) = v(t), \qquad \dot{v}(t) = u(t), \tag{9.41}$$

$$L = \frac{\mu_1}{2} \cdot u^2(t), \qquad \varphi = \frac{\mu_2}{2} \cdot (x(t_b) - p)^2 + \frac{\mu_3}{2} \cdot (v(t_b) - w)^2, \qquad (9.42)$$

$$\dot{\lambda}_1(t) = 0, \qquad \dot{\lambda}_2(t) = -\lambda_1(t), \qquad (9.43)$$

$$u^o(t) = -\frac{1}{\mu_1} \cdot \lambda_2(t), \qquad (9.44)$$

$$x(0) = x_0, \quad v(0) = v_0 \qquad (9.45)$$

$$\lambda_1(t_b) = \mu_2 \cdot (x(t_b) - p), \quad \lambda_2(t_b) = \mu_3 \cdot (v(t_b) - w). \qquad (9.46)$$

The constants $\mu_{\{1,2,3\}}$ are used as tuning parameters with which the relative importance of the otpimization objectives can be qualified. In this simple case, the equations of the system can be solved analytically. For the Lagrange multipliers

$$\lambda_1(t) = \mu_2 \cdot (x(t_b) - p) = c_0(x(t_b)), \qquad (9.47)$$

$$\lambda_2(t) = [\mu_3 \cdot (v(t_b) - w) + \mu_2 \cdot (x(t_b) - p) \cdot t_b] - [\mu_2 \cdot (x(t_b) - p)] \cdot t \qquad (9.48)$$

$$= c_1(x(t_b), v(t_b)) - c_0(x(t_b)) \cdot t.$$

With $\lambda_2(t)$ being linear with time, the optimal control law can be determined and the solution of the system follows

$$v(t) = v_0 - \frac{c_1(x(t_b), v(t_b))}{\mu_1} \cdot t + \frac{1}{2} \cdot \frac{c_0(x(t_b))}{\mu_1} \cdot t^2, \qquad (9.49)$$

$$x(t) = x_0 + v_0 \cdot t - \frac{1}{2} \cdot \frac{c_1(x(t_b), v(t_b))}{\mu_1} \cdot t^2 + \frac{1}{6} \cdot \frac{c_0(x(t_b))}{\mu_1} \cdot t^3. \qquad (9.50)$$

Of course, both of these equations must satisfy the compatibility conditions

$$v(t_b) = v_0 - \frac{c_1(x(t_b), v(t_b))}{\mu_1} \cdot t_b + \frac{1}{2} \cdot \frac{c_0(x(t_b))}{\mu_1} \cdot t_b^2, \qquad (9.51)$$

$$x(t_b) = x_0 + v_0 \cdot t_b - \frac{1}{2} \cdot \frac{c_1(x(t_b), v(t_b))}{\mu_1} \cdot t_b^2 + \frac{1}{6} \cdot \frac{c_0(x(t_b))}{\mu_1} \cdot t_b^3, \qquad (9.52)$$

which yield implicit equations for $x(t_b)$ and $v(t_b)$. After some manipulation, the explicit solution can be found, as

$$v(t_b) = \{12 \cdot \mu_1^2 \cdot v_0 + \mu_2 \cdot \mu_3 \cdot t_b^4 \cdot w + 2 \cdot \mu_1 \cdot t_b \cdot$$
$$\cdot [6 \cdot \mu_3 \cdot w - \mu_2 \cdot t_b \cdot (3 \cdot x_0 + t_b \cdot v_0 - 3 \cdot p)]\} \cdot \qquad (9.53)$$
$$\cdot (12 \cdot \mu_1^2 + 12 \cdot \mu_1 \cdot \mu_3 \cdot t_b + 4 \cdot \mu_1 \cdot \mu_2 \cdot t_b^3 + \mu_2 \cdot \mu_3 \cdot t_b^4)^{-1},$$

$$x(t_b) = \{\mu_2 \cdot \mu_3 \cdot t_b^4 \cdot p + 12 \cdot \mu_1^2 \cdot (t_b \cdot v_0 + x_0) + 2 \cdot \mu_1 \cdot t_b \cdot$$
$$\cdot [2 \cdot \mu_2 \cdot p \cdot t_b^2 + 3 \cdot \mu_3 \cdot (t_b \cdot v_0 + t_b \cdot w + 2 \cdot x_0)]\} \cdot \qquad (9.54)$$
$$\cdot (12 \cdot \mu_1^2 + 12 \cdot \mu_1 \cdot \mu_3 \cdot t_b + 4 \cdot \mu_1 \cdot \mu_2 \cdot t_b^3 + \mu_2 \cdot \mu_3 \cdot t_b^4)^{-1}.$$

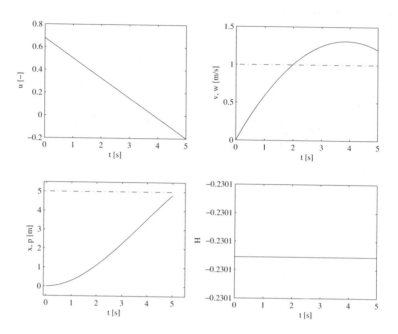

Fig. 9.7. Control variable, state variables, and Hamiltonian of Example 3.

Figure 9.7 shows the system behavior for the case $\mu_1 = \mu_2 = \mu_3 = 1$, $v_0 = x_0 = 0$, $w = 1$, $p = 5$, $t_a = 0$, $t_b = 5$. The behavior computed for the Hamiltonian will be discussed further below. Figure 9.8 shows the sensitivity of the quadratic errors in position and speed

$$\epsilon_p = (x(t_b) - p)^2, \qquad \epsilon_v = (v(t_b) - w)^2 , \qquad (9.55)$$

and of the control effort

$$\epsilon_u = \int_{t_a}^{t_b} u^2(t)\, dt , \qquad (9.56)$$

as well as of the performance index with respect to variations of the weights $\mu_2 \in [0.1, 1.0]$, $\mu_3 \in [0.1, 2.0]$.

9.2.3 First Integral of the Hamiltonian

For time-invariant problems, i.e., when both the performance index (9.31) and the system (9.32) are not explicitly functions of time, it can be shown that along the optimal solution

$$\frac{d}{dt} H(x^o, u^o, \lambda^o) = 0, \qquad \forall t \in [t_a, t_b] , \qquad (9.57)$$

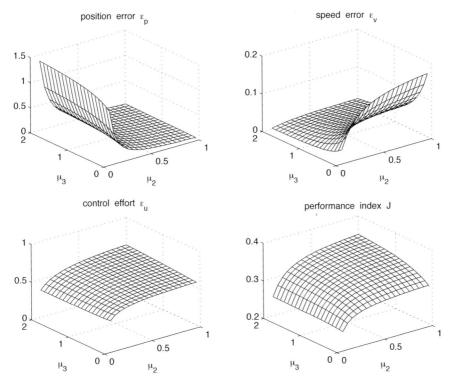

Fig. 9.8. Optimization results of Example 3 for different weights μ_2 and μ_3 ($\mu_1 = 1$).

that is, the Hamiltonian is constant along an optimal solution. This property does not add any new information, but it is sometimes very useful to test the correctness of a solution or of its implementation (see the example above). The proof is easy to derive. The derivative of H is

$$\frac{d}{dt} H(x^o, u^o, \lambda^o) = \frac{\partial H}{\partial t} + \frac{\partial H}{\partial u} \cdot \dot{u} + \frac{\partial H}{\partial x} \cdot \dot{x} + \frac{\partial H}{\partial \lambda} \cdot \dot{\lambda}$$
$$= \frac{\partial H}{\partial t} + H_u \cdot \dot{u} + (H_x + \dot{\lambda}^T) \cdot f \ . \tag{9.58}$$

Along an optimal trajectory $H_u = 0$ and $\dot{\lambda}^T = -H_x$, whence (9.58) can be simplified as

$$\frac{d}{dt} H(x^o, u^o, \lambda^o) = \frac{\partial H}{\partial t} \ . \tag{9.59}$$

For time-invariant systems, the right-hand term of (9.59) must vanish and (9.57) follows therefrom.

9.2.4 Optimal Control with Specified Final State

In the remaining part of this section, three important extensions will be briefly described, without showing the derivation of the results.

In some cases the system is required to be exactly forced to a given point of the state space. In the following it will be assumed that a solution of the problem exists and that $q \leq n$ final states are given. In this situation, it makes no sense to include these states in the end cost $\varphi(x(t_b))$. The corresponding final conditions for $\lambda_j(t_b)$, $j = 1, \ldots, q$ are no longer given in this way, but as $x_j(t_b) = x_{bj}$, $j = 1, \ldots, q$. In other words: the problem is still a two-point boundary problem, but now q initial values of the Lagrange multipliers have to be chosen in such a way as to satisfy the final conditions imposed on the state variables.

Table 9.2. Sufficient conditions for optimal control, specified final states.

System	$\dot{x} = f(x(t), u(t), t)$
Performance index	$J(u) = \varphi(x(t_b), t_b) + \int_{t_a}^{t_b} L(x(t), u(t), t) \, dt$
Adjoint system	$\dot{\lambda}(t) = -\left(\dfrac{\partial L(x, u, t)}{\partial x} \right)^T - \left(\dfrac{\partial f(x, u, t)}{\partial x} \right)^T \cdot \lambda(t)$
Optimal control	$\dfrac{\partial L(x, u, t)}{\partial u} + \lambda^T(t) \cdot \dfrac{\partial f(x, u, t)}{\partial u} = 0$
Boundary conditions	$x(t_a) = x_a, \quad x_j(t_b) = x_{bj}, \quad j = 1, \ldots, q$
	$\lambda(t_b) = \begin{cases} \nu_j \in \mathbb{R} \text{ arbitrary} & j = 1, \ldots, q \\ \left(\dfrac{\partial \varphi(x, t)}{\partial x_j} \right)^T_{t=t_b} & j = q+1, \ldots, n \end{cases}$

Remark: If a state $x_j(t_b)$ is completely irrelevant for the optimization, i.e., it is neither given nor included in $\varphi(x(t_b))$, the variation of the performance index with respect to this variable vanishes

$$\frac{\partial J}{\partial x_j}\bigg|_{t=t_b} = 0 \tag{9.60}$$

and the final values of the corresponding Lagrange multipliers are thus zero, $\lambda_j(t_b) = 0$.

The same is valid when certain initial states $x_j(t_a)$ are not assigned. The corresponding Lagrange multipliers are thus given as $\lambda_j(t_a) = 0$. Also in this case there is no further information. Now, instead of the optimal initial value of the Lagrange multipliers, the corresponding value of the state variable has to be found that still leads to a two-point boundary problem with n unknowns at $t = t_a$.

9.2.5 Optimal Control with Unspecified Final Time

Optimal control problems with unspecified final time and with fully unspecified or partially given final states can be solved in principle through an iterative procedure starting from a given final time. That final time, to which the smallest performance index corresponds, is the solution of the problem. In other words, the problems with unspecified terminal time have an additional degree of freedom.

It is possible to show that using the conditions given in Tables 9.1 and 9.2, this requirement is

$$\left(\frac{\partial \varphi(x,t)}{\partial t} + H(x,u,\lambda,t) \right)_{t=t_b} = 0 . \tag{9.61}$$

For the important case of minimum time problems $(L(x(t), u(t), t) = 1)$, the necessary conditions of Table 9.3 apply.

Table 9.3. Sufficient conditions for minimum time control, fixed final states.

System	$\dot{x} = f(x(t), u(t), t)$
Performance index	$J(u) = \varphi(x(t_b), t_b) + t_b - t_a$
Adjoint system	$\dot{\lambda}(t) = -\left(\dfrac{\partial f(x,u,t)}{\partial x} \right)^T \cdot \lambda(t)$
Optimal control	$\lambda^T(t) \cdot \dfrac{\partial f(x,u,t)}{\partial u} = 0$
Boundary conditions	$x(t_a) = x_a, \quad x_j(t_b) = x_{bj}, \; j = 1, \ldots, q$
	$\lambda(t_b) = \begin{cases} \nu_j \in \mathbb{R} \text{ arbitrary} & j = 1, \ldots, q \\ \left(\dfrac{\partial \varphi(x,t)}{\partial x_j} \right)^T_{t=t_b} & j = q+1, \ldots, n \end{cases}$

9.2.6 Optimal Control with Bounded Inputs

In almost every real application, the control variables of a system are limited. One possible formulation of this fact is to impose[5]

$$F(u, t) \leq 0 . \tag{9.62}$$

The remainder of the problem formulation analyzed below is the same as in Sect. 9.2.

This problem has played a great role in the development of the theory of optimal control. While considering only slight variations in the control signal (u and $\dot{u}$ are limited), it is possible to derive a solution (Euler–Lagrange approach) with analogous considerations to those shown in Sec. 9.2. However, if large variations are allowed (a case that has a great significance in practice), it is necessary to adopt more advanced concepts of optimal control theory. The result of this analysis is well known as *Pontryagin's maximum principle*:

> The Hamiltonian has to be minimized over all the control signals possible.

The key point here is that discontinuities in the control signal $u(t)$ are permitted and that in some cases the minimum is attained at the limits of the range of $u(t)$. The following example will illustrate these ideas.

Example 4: Optimal-Time Rendez-Vous with Limited Control Variable A simple non-trivial problem consists of transferring a material point in the plane from a known rest position (without loss of generality, chosen as the origin of the coordinate system, i.e., $x(t_a) = 0$, $v(t_a) = 0$) to another rest position $x(t_b) = x_b$, $v(t_b) = 0$, with a limited acceleration. After a proper scaling, the problem may be written as

$$\dot{x}(t) = v(t), \qquad \dot{v}(t) = u(t), \qquad |u(t)| \leq 1 , \tag{9.63}$$

$$L = 1 , \tag{9.64}$$

$$\dot{\lambda}_1(t) = 0, \qquad \dot{\lambda}_2(t) = -\lambda_1(t) . \tag{9.65}$$

The Hamiltonian of this problem is linear in the control variable

$$H(t) = 1 + \lambda_1(t) \cdot v(t) + \lambda_2(t) \cdot u(t) , \tag{9.66}$$

thus the minimum $u(t)$ will be reached for a limit value of $u(t)$. According to Pontryagin's minimum principle,

$$u(t) = \begin{cases} +1, \text{ if } \quad \lambda_2(t) < 0 \\ \\ -1, \text{ if } \quad \lambda_2(t) > 0 \end{cases} . \tag{9.67}$$

---

[5] The general case $\tilde{F}(x, u, t) \leq 0$ is clearly more difficult.

The singular case, i.e., when $\lambda_2(t) \equiv 0$ in a finite interval, can be excluded here. In more complex cases, for instance with various input or state limitations, this situation can actually occur.

The optimal control law found in this way is discontinuous[6] and it is clear that, in order to pass from one rest position to another, *at least* one change of sign of the acceleration has to take place. The Lagrange multipliers vary in the following way

$$\lambda_1(t) = c_1$$
$$\lambda_2(t) = -c_1 \cdot t + c_0 \ . \tag{9.68}$$

In accordance with the control law (9.68) *at most* one switch can take place.

The final time t_b has to be minimized depending on the problem formulation, that is, the additional condition

$$\left(\frac{\partial \varphi}{\partial t} + H\right)_{t=t_b} = H(t_b) = 0 \tag{9.69}$$

has to be satisfied. If the problem is time-invariant, $H(t) \equiv 0$, $\forall t \in [0, t_b]$. From (9.66), evaluated at $t = 0$,

$$H(0) = 1 + c_1 \cdot v(0) + (-c_1 \cdot 0 + c_0) \cdot u(0) = 1 + c_0 \cdot u(0) \quad \Rightarrow$$
$$\Rightarrow c_0 = \begin{cases} -1 \text{ if } u(0) > 0 \\ +1 \text{ if } u(0) < 0 \end{cases} \ . \tag{9.70}$$

The initial values of u and λ_2 are also opposite in sign. Inserting the resulting control law in the system equations, e.g., for the case $u(0) = 1$,

$$lllv(t) = t, \qquad x(t) = \tfrac{1}{2} \cdot t^2, \qquad \text{for } t < t_s$$

$$v(t) = 2 \cdot t_s - t, \qquad x(t) = -t_s^2 + 2 \cdot t_s \cdot t - \tfrac{1}{2} \cdot t^2, \qquad \text{for } t > t_s \ . \tag{9.71}$$

The final time and the switch time are

$$t_b = 2 \cdot t_s, \qquad t_s = \sqrt{x(t_b)} \ , \tag{9.72}$$

and the still unknown constant c_1 is

$$c_1 = \frac{-1}{\sqrt{x(t_b)}} \ . \tag{9.73}$$

Figure 9.9 shows the trajectories of the optimally controlled system in the state space (plane v–x, since time no longer appears explicitly). These trajectories are parabolae. Two of them are particularly important, since they lead to the desired final state without any further switches in u.

---

[6] Often referred to as "bang-bang control."

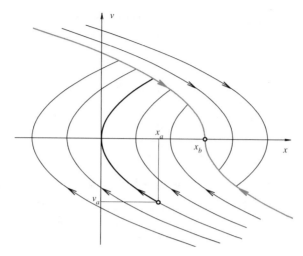

Fig. 9.9. State trajectories of Example 4; dark: regular solutions; gray: switching parabolae.

Incidentally, the same figure shows what happens if instead of a transfer from one rest point to another, any other initial and final states are considered. By choosing one final state, two parabolae are always defined, one with positive $(u(0) > 1)$ and one with negative curvature, which pass through the final point. Only initial points situated on this branch reach the desired final state without any switch. All the other initial conditions lead to one switch.

A further interesting interpretation of Fig. 9.9 results from the implementation of the control law found. Instead of calculating the control variable for each point in advance, the two switching curves can be calculated and then stored in memory. During the normal operation, the controller checks on-line whether the state is in the positive $(u(0) > 1)$ or in the negative semi-plane, in order to determine the correct input. In this way an open-loop control scheme becomes a closed-loop one.

For any arbitrary initial point the total transfer time to the origin is given by

$$t_{total} = v_0 + 2 \cdot \sqrt{\frac{1}{2} \cdot v_0^2 + x_0} \ . \tag{9.74}$$

The *cost-to-go* curves (isolines of transfer time) have the shape shown in Fig. 9.10. This consideration plays an important role in the extension of this approach, towards a formulation as a feed-back solution.

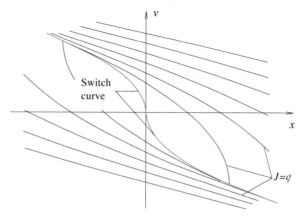

Fig. 9.10. Cost-to-go isolines of Example 4.

10

Appendix III – Dynamic Programming

This appendix summarizes the main concepts of dynamic programming, its implementation, and some automotive applications. Dynamic programming was developed during the 1950's by Richard Bellman [25] and has ever since been used as a tool to design optimal controllers for systems with constraints on the state variables and the control inputs. The first section explains the theory behind the algorithm and its complexity. The second section identifies some of the problems encountered when implementing dynamic programming algorithms. Finally, the third and last section shows an automotive applications of dynamic programming, which illustrates some of the points introduced before. Readers interested in the details of the theory of deterministic and stochastic dynamic programming are referred to the standard text books [25, 27].

10.1 Introduction

For a given system, dynamic programming can be used to find the optimal control input that minimizes a chosen cost function. The benefit of dynamic programming compared to standard optimal control theory is its ability to handle multiple complex constraints on both states and inputs and its low computational burden. The main drawback of the dynamic programming approach is that all disturbances[1] (in the case of deterministic dynamic programming), or at least their stochastic properties (in the case of stochastic dynamic programming) have to be known a priori. Therefore, dynamic programming is often not a useful method for the design of real-time control systems. Only in those cases where the disturbances (their stochastic properties) are known at the outset, deterministic (stochastic) dynamic programming can be used in real-time control applications. Nevertheless, dynamic programming is a very useful tool as it can be used to provide an optimal performance benchmark.

---

[1] All exogenous signals are considered to be "disturbances," e.g. reference signals $r(t_k)$ generated by human drivers are considered to be disturbances as well.

When designing a suboptimal real-time controller, this benchmark can then be used to asses the quality of the suboptimal solution. Moreover, in some cases the optimal non-realizable solution provides insights in how the suboptimal but realizable control system should be designed.

The dynamic programming theory [26, 25, 27] has been used in automotive applications, both using stochastic dynamic programming [153, 141, 139, 127] and using deterministic dynamic programming [226, 86]. Dynamic programming can also be used to optimize the parameters of a power train [240, 241].

In this appendix, the emphasis will be on introducing the main ideas, discussing some implementation issues, and on providing one automotive example illustrating how dynamic programming can be used. The example shown at the end of this chapter deals with the torque split problem in a mild parallel hybrid electric vehicle. In addition to that, two detailed case studies (see Sects. 8.2 and Sects. 8.8) apply dynamic programming ideas to the problem of gear shifting and driving strategy optimization.

10.2 Theory

10.2.1 Introduction

This section provides the basic theoretical concept, which will be used in the other sections. This section is *not* exhaustive and readers interested in a detailed treatment of the subject are referred to the standard textbooks [25, 27]. This section will use the nomenclature introduced in [27].

In dynamic programming problems the following discrete-time dynamic system is considered

$$x_{k+1} = f_k(x_k, u_k, w_k), \qquad k = 0, 1, ..., N - 1. \tag{10.1}$$

The dynamic states $x_k \in X_k \subset \mathbb{R}_\delta^n$, the control inputs $u_k \in U_k \subset \mathbb{R}_\delta^m$, and the disturbances $w_k \in D_k \subset \mathbb{R}_\delta^d$ are discrete variables both in time (index k) and value (thus the subscript $\mathbb{R}_\delta$ of the vector spaces). The control inputs u_k are limited to the subset U_k which can depend on the states x_k, i.e., $u_k \in U_k(x_k)$. As mentioned above, the disturbance w_k must be known in advance for all $k \in [0, N - 1]$.

A specific control sequence (or "policy") is denoted by $\pi = \{\mu_0, \mu_1, ...\mu_{N-1}\}$ and the cost of using π on the problem (10.1) with the initial condition x_0 is defined by

$$J_\pi(x_0) = g_N(x_N) + \sum_{k=0}^{N-1} g_k(x_k, \mu_k(x_k)). \tag{10.2}$$

With these definitions, the optimal trajectory π^o is the trajectory that minimizes J_π

$$J^o(x_0) = \min_{\pi \in \Pi} J_\pi(x_0). \tag{10.3}$$

The principle of optimality [25], as illustrated in Fig. 10.1, provides the main insight in how to solve this problem

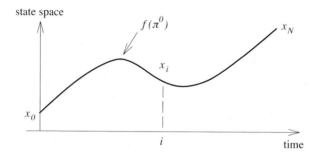

Fig. 10.1. Principle of optimality: assume that $\pi^o(x_0)$ is the optimal policy going from x_0 to x_N and that, when using this policy, a point x_i is reached; then the policy $\pi^o(x_i)$ is the optimal solution of the optimization problem from x_i to x_N.

Principle of Optimality

Let $\pi^o = \{\mu_0^o, \mu_1^o, ...\mu_{N-1}^o\}$ be an optimal policy/trajectory for the basic problem (10.1), and assume that when using π^o a given state x_i is reached at time i. Now consider the optimization problem with initial condition at x_i at time i and cost-to-go from time i to time N defined by

$$E\left\{g_N(x_N) + \sum_{k=i}^{N-1} g_k(x_k, \mu_k(x_k), w_k)\right\}. \tag{10.4}$$

Then the truncated policy $\pi^o(x_i) = \{\mu_i^o, \mu_{i+1}^o, ...\mu_{N-1}^o\}$ is optimal for this new problem.

For every initial state x_0 the optimal cost $J^o(x_0)$ of the deterministic problem (10.1) is equal to $J_0(x_0)$ where J_0 is given by the last step of the following algorithm, which proceeds *backwards* in time from $N - 1$ to 0:

1. end cost calculation step

$$J_N(x_N) = g_N(x_N), \tag{10.5}$$

2. intermediate calculation step

$$J_k(x_k) = \min_{u_k \in U_k(x_k)} \{g_k(x_k, u_k) + J_{k+1}(f_k(x_k, u_k))\}, \tag{10.6}$$

then, if $u_k^o = \mu_k^o(x_k)$ minimizes the right side of (10.6) for each x_k and k, the policy $\pi^o = \{\mu_0^o, ..., \mu_{N-1}^o\}$ is optimal. This algorithm is referred to as deterministic dynamic programming (DDP) since the disturbance has to be known exactly in advance.

Another approach is to consider the same system (10.1), but now to assume that the disturbance $w_k \in D_k$ is random. To be more precise, the random disturbance w_k is assumed to be a Markov process, i.e., its probability distribution at time k does not depend on the previous values of k. However, w_k can depend on the states and control inputs. Of course the probability distribution of the disturbance must be known in advance. The cost criteria

$$J_\pi(x_0) = E_{w_k{}_{k=0,1,\ldots,N-1}} \left\{ g_N(x_N) + \sum_{k=0}^{N-1} g_k(x_k, \mu_k(x_k), w_k) \right\} \qquad (10.7)$$

is the *expected* cost of using the control policy $\pi = \{\mu_0, \mu_1, \ldots \mu_{N-1}\}$ on this stochastic problem with $u_k = \mu(x_k)$ and with the initial condition x_0. The optimal policy π^o is then the policy that minimizes J_π

$$J^o(x_0) = \min_{\pi \in \Pi} J_\pi(x_0). \qquad (10.8)$$

When applying the principle of optimality to the problem (10.1), (10.7)-(10.8) it is possible to create the algorithm:

1. end cost calculation step

$$J_N(x_N) = g_N(x_N), \qquad (10.9)$$

2. intermediate calculation step

$$J_k(x_k) = \min_{u_k \in U_k(x_k)} E_{w_k} \left\{ g_k(x_k, u_k, w_k) + J_{k+1}(f_k(x_k, u_k, w_k)) \right\}, \qquad (10.10)$$

which proceeds backward in time from $N - 1$ to 0. For every initial state x_0, the optimal cost $J^o(x_0)$ of the stochastic problem is equal to $J_0(x_0)$ where J_0 is given by the last step of the algorithm (10.9)-(10.10). The expectation is taken with respect to the probability distribution of w_k, which, in general, depends on x_k and u_k. Furthermore, if $u_k^o = \mu_k^o(x_k)$ minimizes the right side of (10.10) for each x_k and k, the policy $\pi^o = \{\mu_0^o, \ldots, \mu_{N-1}^o\}$ is optimal. The algorithm (10.9)-(10.10) is referred to as the stochastic dynamic programming (SDP) approach.

In some problems a stochastic model of the disturbances can be obtained with relatively low effort a priori and a *real-time* SDP can be developed using the approach described above. Usually, the DDP approach is used to provide a benchmark, which is then used to asses the quality of other suboptimal real-time controllers. However, in those cases where the "disturbance" is known a priori,[2] the DDP approach can be used to design real-time controllers as well.

---

[2] For instance, public transportation systems always follow fixed and well-known routes. The resulting speed and torque profiles, i.e., the "disturbances," can be assumed to be well known a priori.

Adopting a predictive control strategy [90] represents a compromise between the two extreme cases.

In the following sections the term dynamic programming will refer to both the deterministic and the stochastic approach. Which one is meant will be obvious from the context.

10.2.2 Complexity

The objective of the problem (10.1)-(10.3) and (10.7)-(10.8) is to find an optimal solution π^o. The obvious brute force method is to test all possibilities $\pi \in \Pi$. However, this is, in general, not feasible. For instance, to optimize the gear shifting control strategy for a vehicle with 5 gears with a vehicle model sampled with 1 s intervals on a driving cycle lasting 60 s, the number of possible control inputs u_k is 5^{60}. If each evaluation of the model requires 10^{-9} s CPU time, such an approach would require in the order of 10^{25} years. Using the principle of optimality and the dynamic programming algorithm the computation time will be $60 \cdot 5 \cdot 5 \cdot 10^{-9}$ s $= 1.5$ μs. In fact, for this example the DDP algorithm has a number of computations in the order of only $\mathcal{O}(N \cdot p \cdot q)$ where N is the number of time steps, and p and q are the numbers of possible state and input values (value discretization).

In general, the computational burden of all DDP algorithms scales linearly with the problem time N. Unfortunately, all dynamic programming algorithms have a complexity which is exponential in the number of states n and control inputs m

$$\mathcal{O}(N \cdot p^n \cdot q^m). \tag{10.11}$$

This makes the algorithm only suitable for low order systems. The parallel hybrid energy management problem analyzed in this appendix has a long problem duration, but is of low order. Therefore, DDP can be used successfully to derive an optimal control signal.

If the system has a high order (state variables and/or inputs), there are several methods to approximate the optimal solution (approximate dynamic programming) [27]. Obviously, the solution obtained with these methods is not guaranteed to be a global optimum.

10.3 Implementation Issues

There are several issues to consider when implementing the *deterministic* dynamic programming algorithm in general and in particular using Matlab. The following points will be discussed in this section

- grid selection;
- nearest neighbor or interpolation; and
- scalar or set implementation

The grid selection refers to the selection of the discrete state and input spaces for the algorithm. The second issue deals with the problems a discrete state space introduces to the algorithm. The final issue is specific to the Matlab language and refers to the poor efficiency of some of the built in functions. The first two issues are not dependent on the implementation language and are thus relevant in all implementations.

10.3.1 Grid Selection

All dynamic programming algorithms are based on discrete decision processes and therefore a dynamic system with continuous inputs and states has to be approximated by a discrete-value system. This requires the state space to be limited and discretized as a first step. There are several ways of choosing the grid and the number of elements in it. For some problems this is straightforward. For example, the gear shifting case study of Sect. 8.2 has inherently discrete state and input spaces. This is not the case for the parallel hybrid electric vehicle example analyzed below in Sect. 10.3.3, where the input signal (the torque split factor) and the state variable (the battery state of charge) are continuous. A large number of grid elements p and q will give an accurate solution, but will require long computational times and vice versa.

10.3.2 Nearest Neighbor or Interpolation

In all DP algorithms the computation of the last term of (10.6)

$$... + J_{k+1}(f_k(x_k, u_k))$$

poses the following problem: since $x_k \in X_k$ is discretized into a finite set of possible states, the term $J_k(x_k)$ is also only defined for these possible states. A problem arises when the value of the new state, calculated using $f_k(x_k, u_k)$, does *not* match one of the possible states in X_k. Figure 10.2 shows this problem with $X_k = X_{k+1} \forall k$. All possible inputs, u_{1-3}, have to be evaluated when calculating $J_k(x_i)$ and the cost of using respective control input has to be added to the cost-to-go from the resulting state

$$J_k(x_i) = \min \begin{bmatrix} g_k(x_i, u_1) + J_{k+1}(f_k(x_i, u_1)) \\ g_k(x_i, u_2) + J_{k+1}(f_k(x_i, u_2)) \\ g_k(x_i, u_3) + J_{k+1}(f_k(x_i, u_3)) \end{bmatrix} \tag{10.12}$$

The state space X_k, and thus the cost-to-go J_k, is only defined for discrete values of the state x_k. Therefore the term $J_{k+1}(f_k(x_k, u_k))$ in (10.6) must be approximated. There are two main approaches to solve this problem and both have some benefits and some drawbacks. The first, very simple, solution is to take the closest value and evaluate the cost-to-go using this value. Using the example (10.12) illustrated in Fig. 10.2, this method produces the result

$$J_k(x_i) = \min \begin{bmatrix} g_k(x_i, u_1) + J_{k+1}(x^{i+1}) \\ g_k(x_i, u_2) + J_{k+1}(x^i) \\ g_k(x_i, u_3) + J_{k+1}(x^i) \end{bmatrix} \qquad (10.13)$$

The advantage of the nearest neighbor method is its computational speed. Its drawback is a relatively poor accuracy, which can require a very fine grid to achieve the desired precision.

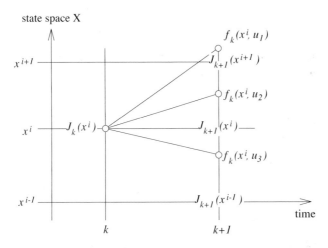

Fig. 10.2. Illustration of the calculation of $J_k(x_i)$. The state variables at time $k+1$ are found starting at the time k with x_i and using the possible inputs u_{1-3}.

The second method to solve the problem is to use an interpolated value of the cost-to-go J_{k+1}. Different interpolation methods have been used, but linear interpolation is often sufficient to provide good accuracy while limiting the extra computational cost. For instance, the approximation of (10.12) when using linear interpolation for the example illustrated in Fig. 10.2 yields the results

$$J_k(x_i) = \min \begin{bmatrix} g_k(x_i, u_1) + (f_k(x_i, u_1) - x^{i+1}) \cdot \frac{J_{k+1}(x^{i+2}) - J_{k+1}(x^{i+1})}{x^{i+2} - x^{i+1}} \\ g_k(x_i, u_2) + (f_k(x_i, u_2) - x^i) \cdot \frac{J_{k+1}(x^{i+1}) - J_{k+1}(x^i)}{x^{i+1} - x^i} \\ g_k(x_i, u_3) + (f_k(x_i, u_3) - x^{i-1}) \cdot \frac{J_{k+1}(x^i) - J_{k+1}(x^{i-1})}{x^i - x^{i-1}} \end{bmatrix}$$

$$(10.14)$$

When implementing the dynamic programming algorithm using interpolation in Matlab there is an advantage of implementing custom interpolation functions since Matlab's interpolation functions `interp1` and `interp2` are not optimized for speed and contain rigorous error checks. This makes custom interpolation functions, optimized for speed, almost a necessity when using interpolation in any dynamic programming algorithms.

Equally Spaced Grids

When using interpolation it can be beneficial to use grids that are equally spaced between the elements. By using such grids the interpolation function's (correctly implemented) computational time does not depend on the size of the grid and this will reduce the computational time substantially. However, if the number of grid points have to be increased in order to make the grid equally spaced, then there is a tradeoff between the amount of time saved and the additional computational time spent for the extra grid points.

Infeasible States or Inputs

A common way to handle infeasible states or inputs is to assign an infinite cost to such states and inputs. However, if an infinite cost is used together with an interpolation scheme, the problem arises that the interpolated value is infinity as well. If this problem is not handled correctly, the number of infeasible states will be artificially increased during the calculations. One solution to this problem is to use a relative large real constant instead of infinity to penalize infeasible states and inputs. This large constant has to be greater than the cost-to-go for any of the states in the solution. Because the maximum cost-to-go is not known in advance, this large constant has to be estimated and some iterations may be necessary.

10.3.3 Scalar or Set Implementation

The way the model (10.1) together with the dynamic programming algorithm is implemented can reduce substantially the computational requirements due to some specific properties of Matlab. A simple implementation of the DDP algorithm will have a form similar to the one shown below.

Standard Pseudo Code (Scalar Implementation)

```
for k = N-1 to 1
    forall indexes i_x in X_grid
        reset Jk
        forall indexes i_u in U_grid
            [x_k+1 g_k] = f(X_grid(i_x),U_grid(i_u),k,...)
            Jk = g_k + J(x_k+1,k+1)
            if Jk < previous Jk
                J_opt = Jk
                u_opt = U_grid(i_u)
            end if
        end forall
        J(i_x,k) = J_opt
        U(i_x,k) = u_opt
    end forall
end for
```

With this standard implementation there is one for-loop for every state and input variable plus one for-loop associated to the time. Unfortunately, Matlab handles for-loops not very efficiently. A much better approach is to utilize Matlab's vector-based algorithms. This requires the for-loops associated with the state and input variables

```
[x_k+1 g_k] = f(X_grid(i_x),U_grid(i_u),k,...)
```

to be replaced by a vector-input vector-output approach

```
[x_k+1 G] = F(X_grid,U_grid,k,...)
```

Such a *set implementation* will have a form similar to the one shown below.

Optimized Pseudo Code (Set Implementation)

```
for k = N-1 to 1
    [x_k+1 G] = F(X_grid,U_grid,k,...)
    Jk = G + V(x_k+1,k+1)
    J(i_x,k) = min Jk
    U(i_x,k) = argmin Jk
end
```

The advantages that can be obtained by using this set implementation are substantial. Fig. 10.3 shows the computation times needed to analyze the example discussed below. Obviously, the speedup factor is so large that it is worth implementing the set function approach, even if this entails some extra programming efforts.

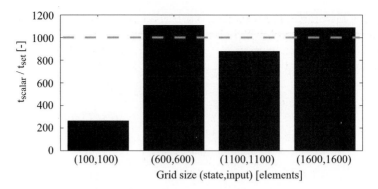

Fig. 10.3. Computation time ratio between scalar implementation and set implementation for different state and input grids for the HEV Example (see Fig. 10.4). The dashed line shows an average (1000 times) obtained with the three largest grids.

Example: Mild Parallel HEV - Torque Split

For the case of a parallel hybrid electric vehicle, this example shows how the optimal torque control strategy can be found using DDP. An overview of the model and its subsystems is shown in Fig. 10.4. The discrete model is very similar to model of the dual clutch gearbox system analyzed in the case study of Sect. 8.2). In addition, the HEV model includes a torque split device, an electric motor, and a battery. The test cycle is again the MVEG-95 and the gear shifting strategy is the standard one (middle plot in Fig. 8.5).

The model has one state x_k, the battery state of charge, and one input u_k, the torque split factor. A time step Δt of 1 s is used. The total torque demanded from the two energy converters is

$$T_{dem} = T_{e0} + T_{m0} + T_g \qquad (10.15)$$

where T_{e0} and T_{m0} are the drag torques of the engine and electric motor. The torque split factor $u \in (-\infty, 1]$ determines the electric motor torque and the internal combustion engine torque according to

$$T_e = (1 - u) \cdot T_{dem} \qquad (10.16)$$

and

$$T_m = u \cdot T_{dem} \qquad (10.17)$$

In the considered HEV example, the motor and engine speed are equal to the gearbox speed on the engine side $\omega_e = \omega_m = \omega_g$. The electric motor model is simply represented by an efficiency map, $\eta_m(\omega_m, T_m)$. The electric power provided to/by the battery is

$$P_m = \frac{T_m \cdot \omega_m}{\eta_m(\omega_m, T_m)} \qquad (10.18)$$

Using the ideas introduced in Sect. 4.4.1, the battery is modeled as a voltage source (open circuit voltage) in series with a resistance (internal resistance). Both the open circuit voltage U_{oc} and the internal resistance R_i depend on the state of charge of the battery. The battery current is

$$I_b = \frac{U_{oc} - \sqrt{U_{oc}^2 - 4 \cdot R_i \cdot P_m}}{2 \cdot R_i} \tag{10.19}$$

The state of charge is calculated using

$$x_{k+1} = \frac{-I_b \, \eta_b(I_b)}{Q_0} \cdot \Delta t + x_k \tag{10.20}$$

where η_b is the battery charging efficiency and Q_0 is the battery capacity.

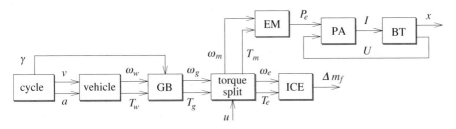

Fig. 10.4. Parallel hybrid electric vehicle model components and the signal flow from the drive cycle through the model.

DDP is used to calculate the optimal torque split factor. A midsize vehicle with a 1.6 l naturally aspirated engine and an electric motor with a maximum power of 23 kW is analyzed. The cost function is defined by

$$g_k(x_k, u_k) = \Delta m_f \tag{10.21}$$

with the final cost

$$g_N(x) = \begin{cases} 0 & x \geq x_0 \\ \infty & x < x_0 \end{cases} \tag{10.22}$$

The following discretization of the state- and input spaces is adopted

$$x \in [0.450, 0.452, \ldots, 0.658, 0.660]$$
$$u \in [-2.0, -1.9, \ldots, 0.9, 1.0] \tag{10.23}$$

The resulting optimal torque-split control strategy, which depends on the state and on the time, is illustrated in Fig. 10.5. This map is then used during a forward simulation of the model to evaluate and collect the different parameters shown in Fig. 10.6.

Figure 10.6 shows that in this example the electric machine is used only for boosting and during braking maneuvers to recuperate energy. The resulting CO_2 emissions for the considered hybrid electric vehicle are 164 g/km.

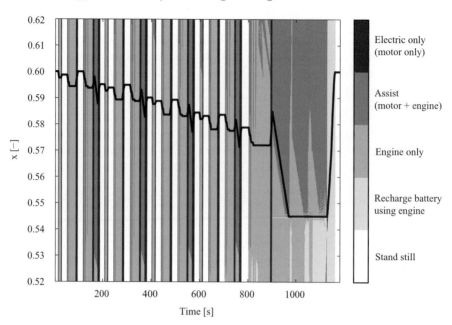

Fig. 10.5. Optimal torque-split control strategy and resulting optimal state of charge trajectory for the special case of $x(0) = \mathrm{SOC}(0) = 0.6$. The input values are grouped into only five classes to better show the main characteristics of the optimal solution.

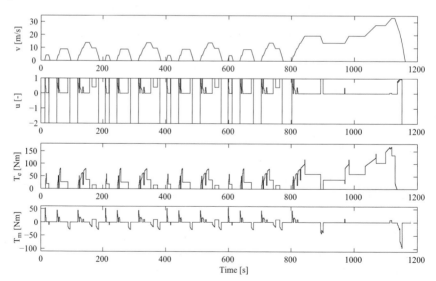

Fig. 10.6. MVEG-95 cycle (top), optimal torque split factor (top, middle), engine torque (bottom, middle), electric motor torque (bottom).

References

1. Ai X, Mohr T, Anderson S (2004) An electro-mechanical infinitely variable speed transmission. SAE Paper 2004-01-0354
2. Amphlett JC, Baumert RM, Mann RF, Peppley BA (1992) System analysis of an integrated methanol steam reformer/PEM fuel cell power generating system. Proc. of the 27th Intersociety Energy Conversion Engineering Conference, San Diego, CA
3. Amphlett JC, Baumert RM, Mann RF, Peppley BA, Roberge PR, Rodrigues A (1994) Parametric modelling of the performance of a 5-kW proton-exchange membrane fuel cell stack. Journal of Power Sources 49:349–356
4. Amphlett JC, Creber KAM, Davis JM, Mann RF, Peppley BA, Stokes DM (1994) Hydrogen production by steam reforming of methanol for polymer electrolyte fuel cells. International Journal of Hydrogen Energy 19(2):131–137
5. Amphlett JC, Mann RF, Peppley BA, Roberge PR, Rodrigues A (1996) A model predicting transient responses of proton exchange membrane fuel cells. Journal of Power Sources 61:183–188
6. Amstutz A, Guzzella L (1999) Fuel cells for transportation – an assessment of its potential for CO_2 reduction. Proc. of the 2nd Conference on Greenhouse Gas Control, Interlaken, Switzerland
7. An F, Vyas A, Anderson J, Santini D (2001) Evaluating commercial and prototype HEVs. SAE Paper 2001-01-0951
8. Anatone M, Cipollone R, Donati A, Sciarretta A (2005) Control-oriented modeling and fuel optimal control of a series hybrid bus. SAE Paper 2005-01-1163
9. Appelbaum J, Weiss R (1982) An electrical model of the lead acid battery. Proc. of the International Telecommunications Energy Conference, Washington, DC
10. Apter R, Präthaler M (2002) Regeneration of power in hybrid vehicles. Proc. of the IEEE 55th Vehicular Technology Conference, Birmingham, AL
11. Arsie I, Pianese C, Rizzo G, Santoro M (2001) A model for the energy management in a parallel hybrid vehicle. Proc. of the 3rd International Conference on Control and Diagnostics in Automotive Applications, Genova, Italy
12. Bach C, Lämmle C, Bill R, Dyntar D, Onder CH, Boulouchos K, Guzzella L, Geering HP (2004) Clean engine vehicle – a natural gas-driven Euro-4/SULEV vehicle with 30% reduced CO2 emissions. SAE Paper 2004-01-0645

13. Back M (2005) Prädiktive Antriebsregelung zum energieoptimalen Betrieb von Hybridfahrzeugen. Diss. Univ. Karlsruhe. online version: http://www.ubka.uni-karlsruhe.de

14. Bailey KE, Cikanek SR, Sureshbabu N (2002) Parallel hybrid electric vehicle torque distribution method. Proc. of the American Control Conference, Anchorage, AK

15. K Bailey KE, Powell BK (1995) A hybrid electric vehicle powertrain dynamic model. Proc. of the American Control Conference, Seattle, WA

16. Bansal D, Rajagopalan S, Choi T, Guezennec YG, Yurkovich S (2004) Pressure and air-fuel ratio control of PEM fuel cell systems for automotive control. Proc. of the IEEE Vehicle Power and Propulsion Symposium, Paris, France

17. Barba G, Glielmo L, Perna V, Vasca F (2001) Current sensorless induction motor observer and control for hybrid electric vehicles. Proc. of the IEEE 32nd Power Electronics Specialists Conference, Vancouver, Canada

18. Barbarisi O, Vasca F, Glielmo L (2006) State of charge Kalman filter estimator for automotive batteries. Control engineering practice 14(3):267–275

19. Barsali S, Ceraolo M, Possenti A (2002) Techniques to control the electricity generation in a series hybrid electrical vehicle. IEEE Transactions on Energy Conversion 17(2):260–266

20. Barsali S, Miulli C, Possenti A (2004) A control strategy to minimize fuel consumption of series hybrid electric vehicles. IEEE Trans. on Energy Conversion 19(1):187–195

21. Barsoukov E, Kim JH, Yoon CO, Lee H (1999) Universal battery parameterization to yield a non-linear equivalent circuit valid for battery simulation at arbitrary load. Journal of Power Sources 83:61–70

22. Baudry P, Neri M, Gueguen M, Lonchampt G (1995) Electro-thermal modelling of polymer lithium batteries for starting period and pulse power. Journal of Power Sources 54:393–396

23. Bazaraa MS, Sherali HD, Shetty CM (1993) Nonlinear programming. Theory and algorithms. John Wiley & Sons, West Sussex, UK

24. Baumann BM, Washington G, Glenn BC, Rizzoni G (2000) Mechatronic design and control of hybrid electric vehicle. IEEE/ASME Transactions on Mechatronics 5(1):58–72

25. Bellman RE (1957) Dynamic programming. Princeton University Press, Princeton, NJ

26. Bellman RE, Lee E (1984) History and development of dynamic programming. IEEE Control Systems Magazine 4(4):24–28

27. Bertsekas DP (2007) Dynamic programming and optimal control. Athena Scientific, Nashua NH

28. Birgerson E (2004) Mathematical modeling of transport phenomena in polymer electrolyte and direct methanol fuel cells. Technical Report, Royal Institute of Technology, Stockholm, Sweden

29. Bowles P, Peng H, Zhang X (2002) Energy management in a parallel hybrid electric vehicle with a continuously variable transmission. Proc. of the American Control Conference, Anchorage, AK

30. Bryson E, Ho YC (1975) Applied Optimal Control. Taylor & Francis, New York, NY

31. Buie L, Fry M, Fussey P, Mitts C (2004) An application of cost based power management control strategies to hybrid fuel cell vehicles. SAE Paper 2004-01-1299

32. Buller S, Thele M, Karden E, De Doncker RW (2003) Impedance-based non-linear dynamic battery modeling for automotive applications. Journal of Power Sources 113:422–430

33. Buntin DL, Howze JW (1995) A switching logic controller for a hybrid electric/ICE vehicle. Proc. of the American Control Conference, Seattle, WA

34. Burnham A, Wang M, Wu Y (2006) Development and Applications of GREET 2.7. Argonne National Laboratory, Report ANL/ESD/06-5, available at http://www.transportation.anl.gov/software/GREET/

35. Butler KL, Ehsani M, Kamath P (1999) A Matlab-based modeling and simulation package for electric and hybrid electric vehicle design. IEEE Transactions on Vehicular Technology 48(6):1770–1778

36. Cai W (2004) Comparison and review of electric machines for integrated starter alternator applications. Proc. of the 39th IEEE Industry Applications Conference, Seattle, WA

37. Candusso D, Rulliere E, Bacha S (2001) Modelling of a fuel cell hybrid power source for a small electric vehicles. Proc. of the 18th Electric Vehicles Symposium, Berlin, Germany

38. Casacca MA, Salameh ZM (1992) Determination of lead–acid battery capacity via mathematical modeling techniques. IEEE Transactions on Energy Conversion 7:442–446

39. Ceraolo M (2000) New dynamical models of lead–acid batteries. IEEE Transactions on Power Systems 15(4):1184–1190

40. Cerruto E, Consoli A, Raciti A, Testa A (1994) Fuzzy logic based efficiency improvement of an urban electric vehicle. Proc. of the 20th International Conference on Industrial Electronics Control and Instrumentation, Bologna, Italy

41. Chan CC, Lo EWC, Weixiang S (2000) The available capacity computation model based on artificial neural network for lead–acid batteries in electric vehicles. Journal of Power Sources 87:201–204

42. Chan CC, Wu J, Zhu GL, Chan TW (1988) Digital simulation of PWM inverter-induction motor drive system for electric vehicles. Proc. of the 14th Conference of Industrial Electronics Society, Singapore

43. Chan HL, Sutanto D (2000) A new battery model for use with battery energy storage systems and electric vehicles power systems. Proc. of the IEEE Power Engineering Society Winter Meeting Conference, Singapore

44. Chan SH, Wang HM (2004) Thermodynamic and kinetic modelling of an autothermal methanol reformer. Journal of Power Sources 126:8–15

45. Chau KT, Wong YS (2002) Overview of power management in hybrid electric vehicles. Energy Conversion and Management 43:1953–1968

46. Chau KT, Wong YS, Chan CC (1999) An overview of energy sources for electric vehicles. Energy Conversion and Management 40:1021–1039

47. Chiasson J, Tolbert L, Lu Y (2002) A library of SIMULINK blocks for real-time control of HEV traction drives. SAE Transactions Journal of Engines 2002:2376-2385

48. Christen T, Carlen MW (2000) Theory of Ragone plots. Journal of Power Sources 91:210–216

49. Christen T, Ohler C (2002) Optimizing energy storage devices using Ragone plots. Journal of Power Sources 110:107–116

50. Chu A, Braatz P, (2002) Comparison of commercial supercapacitors and high-power lithium-ion batteries for power-assist applications in hybrid electric vehicles I. Initial characterization. Journal of Power Sources 112:236–246

51. Cipollone R, Sciarretta A (2006) Analysis of the potential performance of a combined hybrid vehicle with optimal supervisory control. Proc. of the IEEE International Conference on Control Applications, Munich, Germany

52. Cipollone R, Sciarretta A (2001) The quasi-propagatory model: a new approach for describing transient phenomena in engine manifolds. SAE Paper 2001-01-0579

53. College of the Desert (2001) Hydrogen fuel cell engines and related technologies. Technical Report, Energy Technology Training Center, online version: www.eere.energy.gov

54. Conlon B (2005) Comparative analysis of single and combined hybrid electrically variable transmission operating modes. SAE Paper 2005-01-1162.

55. Dahlgren KLM, Marcinkoski J, Holloway D, Levine WS (2002) Powertrain for a hybrid autonomous vehicle. Proc. of the 2nd IFAC Conference on Mechatronic Systems, Berkeley, CA

56. Daihatsu Press Release (2003) World's first hybrid fuel cell minicar approved by Japan's ministry of land, infrastructure and transport. online version: www.daihatsu.com

57. DaimlerChrysler Press Release (2000) ESX3 lowers fuel consumption, emissions and cost. online version: www.daimlerchrysler.com

58. DaimlerChrysler Press Release (2002) Driving test period now starts for the first Citaro bus with fuel cell power system. online version: www.daimlerchrysler.com

59. DaimlerChrysler Press Release (2002) Sodium borohydride: fuel cell vehicle "Natrium" uses clean, safe technology to provide hydrogen on demand. online version: www.daimlerchrysler.com

60. DaimlerChrysler Press Release (2003) DaimlerChrysler initiates "F-Cell" tests in Japan. online version: www.daimlerchrysler.com

61. DaimlerChrysler Press Release (2004) Necar 1 to Necar 5: chemical powerplants on board. online version: www.daimlerchrysler.com

62. Delprat S, Lauber J, Guerra TM, Rimaux J (2004) Control of a parallel hybrid powertrain. IEEE Trans. on Vehicular Technology 53(3):872–881

63. Derdiyok A (2003) A novel speed estimation algorithm for induction machines. Electric Power Systems Research 64:73–80

64. De Vidts P, Delgado J, White RE (1995) Mathematical modeling for the discharge of a metal hydride electrode. Journal of Electrochemical Society 142(12):4006–4013

65. Dietrich P (1999) Gesamtenergetische Bewertung verschiedener Betriebsarten eines Parallel-Hybridantriebes mit Schwungradkomponente und stufenlosem Weitbereichsgetriebe für einen Personenwagen. Dissertation ETH no. 12958, Swiss Federal Institute of Technology, Zurich, Switzerland

66. Dietrich P, Hörler H, Eberle MK (1993) The ETH-hybrid car – a concept to minimize consumption and reduce emissions. Proc. of the 26th ISATA Conference, Aachen, Germany

67. Di Napoli A, Crescimbini F, Solero L, Caricchi F, Giulii Capponi F (2002) Multiple-input DC–DC converter for power-flow management in hybrid vehicles. Proc. of the 37th Industry Applications Conference, Pittsburgh, PA

68. Do K, Guezennec YG, Rizzoni G (2004) Dynamic testing and electrical/thermal model validation of supercapacitors for hybrid electric vehicle powertrains. Proc. of the IEEE Vehicular Power and Propulsion Symposium, Paris, France

69. Dones R, Bauer C, Bolliger R, Burger B, Faist M, Frischknecht R, Heck T, Jungbluth N, Röder A (2004) Sachbilanzen von Energiesystemen: Grundlagen für den ökologischen Vergleich von Energiesystemen und den Einbezug von Energiesystemen in Ökobilanzen für die Schweiz. Final report ecoinvent 2000 no. 6., Paul Scherrer Institut, online version: www.ecoinvent.ch

70. Doyle M, Newmann J, Gozdz AS, Schmutz CN, Tarascon JM (1996) Comparison of modeling predictions with experimental data from plastic lithium ion cells. Journal of Electrochemical Society 143(6):1890–1903

71. Du Pasquier A, Plitz I, Menocal S, Amatucci G (2003) A comparative study of Li-ion battery, supercapacitor and nonaqueous asymmetric hybrid devices for automotive applications. Journal of Power Sources 115:171–178

72. Eaves S, Eaves J (2004) A cost comparison of fuel-cell and battery electric vehicles. Journal of Power Sources 130:208–210

73. EG&G Technical Service, Science Applications International Corporation (2002) Fuel cell handbook. US Department of Energy, National Energy Technology Laboratory, Morgantown WV

74. Ender M (1996) Der Taktbetrieb als teillastverbessernde Massnahme bei Ottomotoren. Dissertation ETH no. 11835, Swiss Federal Institute of Technology, Zurich, Switzerland

75. Ekdunge P (1993) A simplified model of the lead/acid battery. Journal of Power Sources 46:251–262

76. Faggioli E, Rena P, Danel V, Andrieu X, Mallant R, Kahlen H (1999) Supercapacitors for the energy management of electric vehicles. Journal of Power Sources 84:261–269

77. Fan H, Dawson GE, Eastham TR (1993) Model of an electric vehicle induction motor drive system. Proc. of the Canadian Conference on Electrical and Computer Engineering, Vancouver, Canada

78. Fenton J (1998) Handbook of automotive powertrain and chassis design. Professional Engineering Publishing, London, UK

79. Filipi F, Louca L, Daran B, Lin CC, Yildir U, Wu B, Kokkolaras M, Assanis D, Peng H, Papalambros P, Stein J, Szkubiel D, Chapp R (2004) Combined optimisation of design and power management of the hydraulic hybrid propulsion system for the 6 *times* 6 medium truck. International Journal of Heavy Vehicle Systems 11(3/4):371–401

80. Fish S, Savoie T, Vanicek H (2001) Modeling hybrid electric HMMWV power system performance. IEEE Transactions on Magnetics 37(1):480–484

81. Foley DC, Sadegh N, Barth EJ, Vachtsevanos GJ (2001) Model identification and backstepping control of a continuously variable transmission system. Proc. of the American Control Conference, Arlington, VA

82. Ford Press Release (2002) Ford springs surprise with mighty F-350 Tonka. online version: www.ford.com

83. Ford Press Release (2004) Escape Hybrid. online version: media.ford.com

84. Ford Press Release (2004) Focus FCV: combining fuel cell and hybrid technology for the next generation of the automobile. online version: media.ford.com

85. Ford Technical Data (2004) Ford Focus FCV. online version: www.ford.com

86. Haj-Fraj A, Pfeiffer F (2001) Optimal control of gear shift operations in automatic transmissions. Journal of the Franklin Institute 338:371–390

87. Fuchs RD, Hasuda Y, James IB (2002) Full toroidal IVT variator dynamics. SAE paper 2002-01-0586.

88. Gagliardi F, Piccolo A, Vaccaro A, Villacci D (2002) A fuzzy based control unit for the optimal power flow management in parallel hybrid electric vehicles. Proc. of the 19th Electric Vehicles Symposium, Busan, Korea

89. Galler D (2000) Motors. In: Dorf R (ed) The electrical engineering handbook. CRC, Boca Raton, FL

90. Garcia J, Prett J, Morari M (1989) Model predictive control: theory and practice. Automatica, 25: 335–348

91. Gao L, Liu S, Dougal RA (2002) Dynamic lithium-ion battery model for system simulation. IEEE Transactions on Components and Packaging Technologies 25(3):495–505

92. George TM, Bhatia CM, Ahson SI (1988) Implementation of sliding mode control for a chopper-fed separately excited DC motor using a personal computer. Proc. of the 14th Conference of Industrial Electronics Society, Singapore

93. GM Press Release (2003) GM Hy-wire: major step forward in reinventing automobile. online version: www.gm.com

94. Gomadam PM, Weidner JW, Dougal RA, White RE (2002) Mathematical modeling of lithium-ion and nickel battery systems. Journal of Power Sources 110:267–284

95. Gu WB, Yang CY, Li SM, Geng MM, Liaw BY (1999) Modeling discharge and charge characteristics of nickel–metal hydride batteries. Electrochemica Acta 44:4525–4541

96. Guebeli M, Micklem JD, Burrows CR (1993) Maximum transmission efficiency of a steel belt continuously variable transmission. Transactions of the ASME, Journal of Mechanical Design 115:1044–1048

97. Gutmann G (1999) Hybrid electric vehicles and electrochemical storage systems – a technology push–pull couple. Journal of Power Sources 84:275–279

98. Guzzella L (1999) Control oriented modeling of fuel-cell based vehicles. Pres. at NSF Workshop on the Integration of Modeling and Control for Automotive systems, Santa Barbara CA

99. Guzzella L, Amstutz A (1999) CAE tools for quasi-static modeling and optimization of hybrid powertrains. IEEE Transactions on Vehicular Technology 48(6):1762–1769

100. Guzzella L, Onder CH (2004) Introduction to modeling and control of internal combustion engines systems. Springer, Berlin, Germany

101. Guzzella L, Schmid AM (1995) Feedback linearization of spark-ignition engines with continuously variable transmissions. IEEE Transactions on Control Systems Technology 3(1):54–60

102. Guzzella L, Sciarretta A (2003) Global optimization of rendez-vous maneuvers for passenger cars. Proc. of the 5th Stuttgart International Symposium, Stuttgart, Germany

103. Guzzella L, Wenger U, Martin R (2000) IC engine downsizing and pressure wave supercharging for fuel-economy. SAE Paper 2000-01-1019

104. Guzzella L, Wittmer C, Ender M (1996) Optimal operation of drivetrains with SI-engines and flywheels. Proc. of the 13th IFAC World Congress, San Francisco, CA

105. Harding GG (1999) Electric vehicles in the next millennium. Journal of Power Sources 78:220–230

106. Hauer KH (2001) Numerical simulation of two different ultra capacitor hybrid fuel cell vehicles. Proc. of the 18th Electric Vehicle Symposium, Berlin, Germany

107. Hauer KH, Moore RM, Ramaswamy S (2001) A simulation model for an indirect methanol fuel cell vehicle. SAE Paper 2001-01-3083

108. Hawkins L, Murphy B, Zierer J, Hayes R (2002) Shock and vibration testing of an AMB supported energy storage flywheel. Proc. of the 8th International Symposium on Magnetic Bearings, Mito, Japan

109. Hendriksen P, Elst D, Riemersma I, Smokers T, Van den Bosch A, Scheepers M, Van Arkel W, Volkers C (2000) Audi Duo demonstration project – environmental comparison and user survey. Proc. of the 17th Electric Vehicle Symposium, Montreal, Canada

110. Heywood JB (1988) Internal combustion engine fundamentals. McGraw Hill, New York, NY

111. Higelin P, Charlet A, Chamaillard Y (2002) Thermodynamic simulation of a hybrid pneumatic-combustion engine concept. Int. Journal of Applied Thermodynamics 5(1):1-11

112. Hissel D, Péra MC, Francois X, Kauffman JM, Baurens P (2001) Contribution to the modelling of automotive systems powered by polymer electrolyte fuel cells. Proc. of the European Automotive Congress, Bratislava, Slovakia

113. Honda Press Release (2007) The New Honda IMA System is more efficient than ever. online version: world.honda.com

114. Honda Press Release (2001) Honda introduces new fuel cell-powered vehicle, FCX-V4. online version: world.honda.com

115. Hucho WH (1998) Aerodynamics of road vehicles. SAE Publishing, Warrendale, PA

116. Husted H (2003) A comparative study of the production applications of hybrid electric powertrains. SAE Paper 2003-01-2307

117. Ide T, Udagawa A, Kataoka R (1996) Experimental investigation on shift-speed characteristics of a metal V-belt CVT. SAE Paper 9636330

118. Inanc N (2002) A new slinding mode flux and current observer for direct field oriented induction motor drives. Electric Power Systems Research 63:113–118

119. International energy agency, Hybrid and electric vehicle implementing agreement (2000) Annex VII – Overview report 2000, Worldwide developments and activities in the field of hybrid road-vehicle technology. online version: www.ieahev.org

120. Isidori A (2003) Nonlinear control systems. Springer, Berlin, Germany

121. Itagaki K, Teratani T, Kuramochi K, Nakamura S, Tachibana T, Nakao H, Kamijo Y (2002) Development of the Toyota mild-hybrid system (THS-M). SAE Paper 2002-01-0990

122. Iwai N (1999) Analysis of fuel economy and advanced systems of hybrid vehicles. JSAE Review 20:3–11

123. Jeanneret B, Markel T (2004) Adaptive energy management strategy for fuel cell hybrid vehicles. SAE Paper 2004-01-1298

124. Jefferson CM, Ackermann M (1996) A flywheel variator energy storage system. Energy Conversion Management 37(10):1481–1491

125. Jeon SI, Jo ST, Park YI, Lee JM (2002) Multi-mode driving control of a parallel hybrid vehicle using driving pattern recognition. Transactions of the ASME, Journal of Dynamic Systems, Measurement, and Control 124:141–149

126. Jeon SI, Kim KB, Jo ST, Lee JM (2001) Driving simulation of a parallel hybrid electric vehicle using receding horizon control. Proc. of the IEEE International Symposium on Industrial Electronics, Pusan, Korea

127. Johannesson L, Asbogård M, Egardt B (2007) Assessing the potential of predictive control for hybrid vehicle powertrains using stochastic dynamic programming. IEEE Transactions on Intelligent Transportation Systems 8(1):71–83

128. Johnson VH, Wipke KB, Rausen DJ (2000) HEV control strategy for real-time optimization of fuel economy and emissions SAE Paper 2000-01-1543

129. Johnson VH (2002) Battery performance models in ADVISOR. Journal of Power Sources 110:321–329

130. Kahlon GS, Mohan RM, Liu N, Rehman H (1999) A case study of starting power requirement for Visteon integrated starter–alternator system. Proc. of the 18th Digital Avionics Systems Conference, St. Louis, MO

131. Kailath T (1980) Linear systems. Prentice Hall, Englewood Cliffs, NJ

132. Kargul JJ (2004) Affordable advanced technologies. Presented to FACA Meeting, online version: www.epa.gov

133. Kassakian JG, Miller JM, Traub N (2000) Automotive electronics power up. IEEE Spectrum May:34–39

134. Kaushik R, Mawston IG (1989) Discharge characterization of lead/acid batteries. Journal of Power Sources 28:161–169

135. Kelly KJ, Zolot M, Glinsky G, Hieronymus A (2001) Test results and modeling of the Honda Insight using ADVISOR. SAE Paper 2001-01-2537

136. Kempton K, Kubo T (2000) Electric-drive vehicles for peak power in Japan. Energy Policy 28:9–18

137. Kiencke U, Nielsen L (2000) Automotive control systems. Springer, Berlin, Germany

138. Kim KH, Baik IC, Chung SK, Youn MJ (1997) Robust speed control of brushless DC motor using adaptive input–output linearisation technique. IEE Proceedings on Electric Power Applications 144:469–475

139. Kim MJ, Peng H, Lin CC, Stamos E, Tran D (2005) Testing, modeling, and control of a fuel cell hybrid vehicle. Control Engineering Practice 13:41–53

140. Kleimaier A, Schröder D (2002) An approach for the online optimized control of a hybrid powertrain. Proc. of the 7th International Workshop on Advanced Motion Control, Maribor, Slovenia

141. Kolmanovsky I, Siverguina I, Lygoe B (2002) Optimization of powertrain operating policy for feasibility assessment and calibration: Stochastic dynamic programming approach. Proc. of the American Control Conference, Anchorage, AK

142. Koo ES, Lee HD, Sul SK, Kim JS (1998) Torque control strategy for a parallel hybrid vehicle using fuzzy logic. Proc. of the 1998 IEEE Industry Application Conference, St. Louis, MO

143. Koot M, Kessels J, de Jager B, Heemels W, van den Bosch P, Steinbuch M (2005) Energy management strategies for vehicular electric power systems. IEEE Trans. on Vehicular Technology 54(3):1504–1509

144. Kötz R, Baertschi M, Buechi F, Gallay R, Dietrich P (2002) HY.POWER – A fuel cell car boosted with supercapacitors. Proc. of the 12th International Seminar on Double Layer Capacitors and Similar Energy Storage Devices, Deerfield Beach, FL

145. Krause P, Wasynczuk O, Sudhoff S (2002) Analysis of electric machinery. Wiley-IEEE Press, West Sussex, UK

146. Lee JH, Lalk TR, Appleby AJ (1998) Modeling electrochemical performance in large scale proton exchange membrane fuel cell stacks. Jounal of Power Sources 70:258–268

147. Lee CK, Kwok NM (1993) Torque ripple reduction in brushless DC motor velocity control systems using a cascade modified model reference compensator. Proc. of the 24th IEEE Power Electronics Specialists Conference, Seattle, WA

148. Lee HD, Sul SK (1998) Fuzzy-logic-based torque control strategy for parallel-type hybrid electric vehicle. IEEE Transactions on Industrial Electronics 45(4):625–632

149. Lee W, Choi D, Sunwoo M (2002) Modelling and simulation of vehicle electric power system. Journal of Power Sources 109:58–66

150. Liang C, Qingnian W, Youde L, Zhimin M, Ziliang Z, Di L (1999) Study of the electronic control strategy for the power train of hybrid electric vehicle. Proc. of the 1999 IEEE Vehicle Electronics Conference, Changchun, China

151. Lin CC, Kang JM, Grizzle JW, Peng H (2001) Energy management strategy for a parallel hybrid electric truck. Proc. of the American Control Conference, Arlington, VA

152. Lin CC, Peng H, Grizzle JW, Kang JM (2003) Power management strategy for a parallel hybrid electric truck. IEEE Transactions on Control Systems Technology 11(6): 839–849

153. Lin CC, Peng H, Grizzle JW (2004) A stochastic control strategy for hybrid electric vehicle. Proc. of the 2004 American Control Conference, Boston, MA

154. Lin FJ, Wai RJ, Kuo RH, Liu DC (1998) A comparative study of sliding mode and model reference adaptive speed observers for induction motor drive. Electric Power Systems Research 44:163–174

155. Liu S, Stefanopoulou AG (1999) Effects of control structure on performance for an automotive powertrain with a continuously variable transmission. IEEE Transactions on Automatic Control 10(5):701–709

156. Luo FL, Yeo YG (2000) Advanced PM brushless DC motor control and system for electric vehicles. Proc. of the 2000 IEEE Industry Applications Conference, Rome, Italy

157. Maggetto G, Van Mierlo J (2001) Electric vehicles, hybrid vehicles and fuel cell electric vehicles: state of the art and perspectives. Ann Chim Sci Mat 26(4):9–26

158. Maggetto G, Van Mierlo J (2000) Electric and electric hybrid vehicle technology: a survey. IEE Seminar on Electric, Hybrid and Fuel Cell Vehicles

159. Mann RF, Amphlett JC, Hooper MAI, Jensen HM, Peppley BA, Roberge PR (2000) Development and application of a generalized steady-state electrochemical model for a PEM fuel cell. Journal of Power Sources 86:173–180

160. Marino R, Peresada S, Tomei P (2000) On-line stator and rotor resistance estimation for induction motors. IEEE Transactions on Control Systems Technology 8:570–580

161. Martellucci L (1994) Sipre: a family of new hybrid propulsion systems. Proc. of the 27th ISATA Conference, Aachen, Germany

162. Mauracher P, Karden E (1997) Dynamic modelling of lead/acid batteries using impedance spectroscopy for parameter identification. Journal of Power Sources 67:69–84

163. Maxoulis CN, Tsinoglou DN, Koltsakis GC (2004) Modeling of automotive fuel cell operation in driving cycles. Energy Conversion and Management 45:559–573

164. Mayer JS, Wasynczuk O (1989) Analysis and modeling of a single-phase brushless DC motor drive system. IEEE Transactions on Energy Conversion 4:473–479

165. Mazda Press Release (1997) Mazda develops fuel cell electric vehicle, "Demio FCEV". online version: www.mazda.com

166. Mellor PH, Schofield N, Howe D (2000) Flywheel and supercapacitor peak power buffer technologies. Proc. of the IEE Seminar on Electric, Hybrid and Fuel Cell Vehicles, London, UK

167. Miller J (1999) Development of equivalent circuit models for batteries and electrochemical capacitors. Proc. of the 14th IEEE Battery Conference on Applications and Advances, Long Beach, CA

168. Miller J, Everett M (2005) An assessment of ultra-capacitors as the power cache in Toyota THS-II, GM Allison AHS-2 and Ford FHS hybrid propulsion systems. Proc. of the IEEE Applied Power Electronics Conference and Exposition, Austin, TX

169. Miller J, McCleer P, Everett M (2005) Comparative assessment of ultra-capacitors and advanced battery energy storage systems in powersplit electronic-CVT vehicle powertrains. Proc. of the IEEE International Conference on Electric Machines and Drives, San Antonio, TX

170. Morita K (2002) Automotive power sources in the twenty-first century. JSAE Review 24:3–7

171. Murphy BT, Bresie DA, Beno JH (1997) Bearing loads in a vehicular flywheel battery SAE Paper 970213

172. Musardo C, Rizzoni G, Staccia B (2005) A-ECMS: an adaptive algorithm for hybrid electric vehicle energy management. Proc. of the 44th IEEE Conf. on Decision and Control, and the 2005 European Control Conf., Seville, Spain

173. Musser J, Wang CY (2000) Heat transfer in a fuel cell engine. Proc. of the 34th National Heat Transfer Conference, Pittsburgh, PA

174. Nielsen MP, Knudsen Kaer S (2003) Modeling a PEM fuel cell natural gas reformer. Proc. of the 6th International Conference on Efficiency, Costs, Optimization, Simulation and Environmental Impact of Energy Systems, Copenhagen, Denmark

175. Nissan Press Release (2000) Nissan releases Tino Hybrid. online version: www.nissan-global.com

176. Nissan Technical Data (2003) X-Trail FCV – Nissan's fuel cell vehicle of the future limited leasing is being launched. online version: www.nissan-global.com

177. Nitz L, Truckenbrodt A, Epple W (2006) The new two-mode hybrid system from the Global Hybrid Cooperation. Proc. of the 27th International Vienna Motor Symposium, Vienna, Austria.

178. Novotnak RT, Chiasson J, Bodson M (1999) High-performance motion control of an induction motor with magnetic saturation IEEE Transactions on Control Systems Technology 7(3):315–327

179. Ogburn MJ, Nelson DJ, Luttrell W, King B, Postle S, Fahrenkrog R (2000) Systems integration and performance issues in a fuel cell hybrid electric vehicle. SAE Paper 2000-01-0376

180. Ogden J, Steinbuegler M, Kreutz T (1999) A comparison of hydrogen, methanol and gasoline as fuels for fuel cell vehicles: implications for vehicle design and infrastructure development. Journal of Power Sources 79:143–168

181. Opel Press Release (2001) HydroGen3 fuel cell study moves closer to volume production. online version: www.opel.com

182. Paganelli G, Ercole G, Brahma A, Guezennec YG, Rizzoni G (2000) A general formulation for the instantaneous control of the power split in charge-sustaining

hybrid electric vehicles. Proc. of the 5th International Symposium on Advanced Vehicle Control, Ann Arbor, MI

183. Paganelli G, Guerra TM, Delprat S, Santin JJ, Combes E, Delhom M (2000) Simulation and assessment of power control strategies for a parallel hybrid car. Journal of Automobile Engineering 214:705–718

184. Paganelli G, Guezennec Y, Rizzoni G (2002) Optimizing control strategy for hybrid fuel cell vehicle. SAE Paper 2002-01-0102

185. Pathapati PR, Xue X, Tang J (2005) A new dynamic model for predicting transient phenomena in a PEM fuel cell system. Renewable Energy 30:1–22

186. Paxton B, Newman J (1997) Modeling of nickel/metal hydride batteries. Journal of Electrochemical Society 144(11):3818–3831

187. Pell WG, Conway BE, Adams WA, De Oliveira J (1999) Electrochemical efficiency in multiple discharge/recharge cycling of supercapacitors in hybrid EV applications. Journal of Power Sources 80:134–141

188. Pesaran AA (2002) Battery thermal models for hybrid vehicle simulations. Journal of Power Sources 110:377–382

189. Pfiffner R, Guzzella L (2001) Optimal operation of CVT-based powertrains. International Journal of Robust and Nonlinear Control 11(11):1003-1021

190. Pfiffner R, Guzzella L, Onder C (2002) A control-oriented CVT model with nonzero belt mass. Transactions of the ASME, Journal of Dynamic Systems, Measurement, and Control 124:481-484

191. Philips Semiconductors (1996) Philips semiconductor applications handbook, Chap 3. online version: www.semiconductors.philips.com

192. Pillay P, Krishnan R (1998) Modeling of permanent magnet motor drives. IEEE Transactions on Industrial Electronics 35:537–541

193. Piller S, Perrin M, Jossen A (2001) Methods for state-of-charge determination and their applications. Journal of Power Sources 96:113–120

194. Plett GL (2004) Extended Kalman filtering for battery management systems of LiPb-based HEV battery packs. Part 1. Background. Journal of Power Sources 134:252–261

195. Pop V, Bergveld HJ, Notten PHL, Regtien PPL (2005) State-of-the-art of battery state-of-charge determination. Measurement Science and Technology 16:R93–R110

196. Poulikakos D (1994) Conduction and heat transfer. Prentice Hall, Englewood Cliffs, NJ

197. Pourmovahed A, Beachley NH, Fronczak FJ (1992) Modeling of a hydraulic energy regeneration system – Part I: analytical treatment. Transactions of the ASME, Journal of Dynamic Systems, Measurement, and Control 114:155–159

198. Powell BK, Bailey KE, Cikanek SR (1998) Dynamic modeling and control of hybrid electric vehicle powertrain systems. IEEE Control Systems Magazine 18:17–33

199. Powell BK, Pilutti TE (1994) A range extender hybrid electric vehicle dynamic model. Proc. of the 33rd Conference on Decision and Control, Lake Buena Vista, FL

200. PSA Peugeot Citroën Press Release (2006) PSA Peugeot Citroën unveils diesel hybrid technology. online version: www.psa-peugeot-citroen.com/en/

201. PSA Peugeot Citroën Press Release (2004) Fuel savings of 10 to 15% in cities with PSA Peugeot Citroën's new Stop & Start system. online version: www.psa-peugeot-citroen.com/en/

202. Pukrushpan JT, Peng H, Stefanopoulou AG (2004) Control-oriented modeling and analysis for automotive fuel cell systems. Transactions of the ASME, Journal of Dynamic System, Measurement, and Control 126:14–24

203. Pukrushpan JT, Stefanopoulou A, Peng H (2004) Control of fuel cell power systems: principles, modeling, analysis and feedback design. Springer, Berlin, Germany

204. Pukrushpan JT, Stefanopoulou AG, Peng H (2004) Control of fuel cell breathing. IEEE Control System Magazine April:30–46

205. Rabou L (1995) Modelling of a variable-flow methanol reformer for a polymer electrolyte fuel cell. International Journal of Hydrogen Energy 20(10):845–848

206. Rahman KM, Fahimi B, Suresh G, Rajarathnam AV, Ehsani M (2000) Advantages of switched reluctance motor applications to EV and HEV: design and control issues. IEEE Transactions on Industry Applications 36(1):111–121

207. Rahman Z, Butler K, Ehsani M (2002) A comparative study between two parallel hybrid control concetps. SAE Paper 2002-01-0102

208. Rajashekara K (2000) Propulsion system strategies for fuel cell vehicles. SAE Paper 2000-01-0369

209. Rizoluis D, Burl J, Beard J (2001) Control strategies for a series-parallel hybrid electric vehicle. SAE Paper 2001-01-1354

210. Rizzoni G, Guzzella L, Baumann B (1999) Unified modeling of hybrid electric vehicle drivetrains. IEEE/ASME Transactions on Mechatronics 4(3):246–257

211. Rodatz P (2003) Dynamics of the polymer electrolyte fuel cell: experiments and model-based analysis. Dissertation ETH no. 15320, Swiss Federal Institute of Technology, Zurich, Switzerland

212. Rodatz P, Garcia O, Guzzella L, Büchi F, Bärtschi M, Tsukada A, Dietrich P (2003) Performance and operational characteristics of a hybrid vehicle powered by fuel cells and supercapacitors. SAE Paper 2003-01-0418

213. Rodatz P, Paganelli G, Sciarretta A, Guzzella L (2005) Optimal power management of an experimental fuel cell/supercapacitor powered hybrid vehicle. Control Engineering Practice 13(1):41-53

214. Rodatz P, Tsukada A, Mladek M, Guzzella L (2002) Efficiency improvements by pulsed hydrogen supply in PEM fuel cell systems. Proc. of the 15th IFAC World Congress on Automatic Control, Barcelona, Spain

215. Sadler M, Stapleton A, Heath R, Jackson N (2001) Application of modeling techniques to the design and development of fuel cell vehicle systems. SAE Paper 2001-01-0542

216. Society of Automotive Engineers (2006) SAE Standard J1711. online version: http://www.sae.org

217. Salameh ZM, Casacca MA, Lynch WA (1992) A mathematical model for lead–acid batteries. IEEE Transactions on Energy Conversion 7:93–97

218. Salman M, Schouten NJ, Kheir NA (2000) Control strategies for parallel hybrid vehicles. Proc. of the American Control Conference, Chicago, IL

219. Sasaki S (1998) Toyota's newly developed hybrid powertrain. Proc. of the 1998 International Symposium on Power Semiconductor Devices & ICs, Kyoto, Japan

220. Schaefer A, Victor DG (2000) The future mobility of the world population. Transportation Research A 34(3): 171-205

221. Schechter M (1999) New cycles for automobile engines. SAE paper 1999-01-0623

222. Schmidt CL, Skarstad PM (1997) Development of an equivalent-circuit model for the lithium/iodine battery. Journal of Power Sources 65:121–128

223. Schouten NJ, Salman MA, Kheir NA (2003) Energy management strategies for parallel hybrid vehicles using fuzzy logic. Control Engineering Practice 11:171–177

224. Schweighofer B, Raab K, Brasseur G (2002) Modelling of high power automotive batteries by the use of an automated test system. Proc. of the IEEE Instrumentation and Measurement Technology Conference, Anchorage, AK

225. Sciarretta A, Guzzella L (2007) Control of hybrid electric vehicles. Optimal energy-management strategies. IEEE Control Systems Magazine 27(2):60–70

226. Sciarretta A, Back M, Guzzella L (2004) Optimal control of parallel hybrid electric vehicles. IEEE Transactions on Control System Technology 12(3):352–363

227. Sciarretta A, Guzzella L (2003) Rule-based and optimal control strategies for energy management in parallel hybrid vehicles. Proc. of the 6th International Conference on Engines for Automobiles, Capri, Italy

228. Sciarretta A, Guzzella L (2005) Fuel-optimal control of rendezvous maneuvers for passenger cars. at-Automatisierungstechnik 53(6):244–250

229. Sciarretta A, Guzzella L, Onder CH (2003) On the power split control of parallel hybrid vehicles: from global optimization towards real-time control. at-Automatisierungstechnik 51(5):195–203

230. Sciarretta A, Guzzella L, Back M (2004) Real-time optimal control strategy for parallel hybrid vehicles with on-board estimation of the control parameters. Proc. of the IFAC Symposium on Advances in Automotive Control, Salerno, Italy

231. Sciarretta A, Guzzella L, Van Baalen J (2004) Fuel optimal trajectories of a fuel cell vehicle Proc. of the International Mediterranean Modeling Multiconference, Bergeggi, Italy

232. Setlur P, Wagner JR, Dawson DM, Samuels B (2001) Nonlinear control of a continuously variable transmission (CVT) for hybrid vehicle powertrains. Proc. of the American Control Conference, Arlington, VA

233. Shafai E, Simons M, Neff U, Geering HP (1995) Model of a continuously variable transmission. Proc. of the 1st IFAC Workshop on Advances in Automotive Control, Ascona, Switzerland

234. Shepherd CM (1965) Design of primary and secondary cells II. An equation describing battery discharge. Journal of Electrochemical Society 112:252–257

235. Spillane D, O' Sullivan D, Egan MG, Hayes JG (2003) Supervisory control of a HV integrated starter–alternator with ultracapacitor support within the 42 V automotive electrical system. Proc. of the 18th IEEE Applied Power Electronics Conference and Exposition, Miami Beach, FL

236. Steinmauer S, del Re L (2001) Optimal control of dual power sources. Proc. of the 2001 IEEE Int. Conf. on Control Applications, Mexico City, Mexico

237. Stern MO (1988) Energy storage in magnetic fields. Energy 13(2):137–140

238. Stoynov Z, Nishev T, Vacheva V, Stamenova N (1997) Nonstationary analysis and modelling of battery load performance. Journal of Power Sources 64:189–192

239. Sudhoff SD, Krause PC (1990) Average-value model of the brushless DC 120° inverter system. IEEE Transactions on Energy Conversion 5:553–557

240. Sundström O, Stefanopoulou A (2007) Optimum battery size for fuel cell hybrid electric vehicle, part 1. ASME Journal of Fuel Cell Science and Technology 4(2):167–175

241. Sundström O, Stefanopoulou A (2007) Optimum battery size for fuel cell hybrid electric vehicle with transient loading consideration, part 2. ASME Journal of Fuel Cell Science and Technology 4(2):176–184

242. Tai C, Tsao TC, Levin M, Barta G, Schechter M (2003) Using camless valvetrain for air hybrid optimization. SAE paper 2003-01-0038

243. Takano K, Nozaki K, Saito Y, Negishi A, Kato K, Yamaguchi Y (2000) Simulation study of electrical dynamic characteristics of lithium-ion battery. Journal of Power Sources 90:214–223

244. Tamai G, Hoang T, Taylor J, Skaggs C, Downs B (2001) Saturn engine stop–start system with an automatic transmission. SAE Paper 2001-01-0326

245. Tate ED, Boyd SP (2000) Finding ultimate limits of performance for hybrid electric vehicles. SAE Paper 2000-01-3099

246. Thirumalai D, White RE (2000) Steady-state operation of a compressor for a proton exchange membrane fuel cell system. Journal of Applied Electrochemistry 30:551–559

247. Thomas CE, Reardon J, Lomax F, Pinyan J, Kuhn I (2001) Distributed hydrogen fueling systems analysis. Proc. of the 2001 DOE Hydrogen Program Review

248. Thorstensen B (2001) A parametric study of fuel cell system efficiency under full and part load operation. Journal of Power Sources 92:9–16

249. Toyota Environmental Commitment Technical Sheets (2001) Toyota's fuel-cell hybrid vehicles (FCHV). online version: www.toyota.com

250. Toyota Motor Corporation (2003) Toyota Hybrid System THS-II. online version: www.toyota.co.jp

251. Toyota Press Release (2005) Toyota to display fuel cell hybrid and personal mobility concept vehicles at Tokyo Motor Show. online version: www.toyota.co.jp/en/

252. Treffinger P, Gräf M, Goedecke M (2002) Hybridization of fuel cell powered drive trains. Proc. of the VDI-Tagung "Innovative Fahrzeugantriebe," Dresden, Germany

253. T-Raissi A, Banerjee A, Sheinkopf K (1996) Metal hydride storage requirements for transportation applications. Proc. of the 31st Intersociety Energy Conversion Engineering Conference, Washington, DC

254. Tripathy SC (1992) Simulation of flywheel energy storage system for city buses. Energy Conversion Management 33(4):243–250

255. Tzimas E, Filiou C, Peteves SD, Veyret JB (2003) Hydrogen storage: state-of-the art and future perspective. EU Commission, Joint Research Center, Institute for Energy, Petten, The Netherlands

256. US Department of Energy, Office of Hydrogen, Fuel Cells and Infrastructure Technologies (2002) Proc. of the Workshop on Compressed and Liquefied Hydrogen Storage, Southfield, MI (online version: www.eere.energy.gov)

257. Van Druten R (2001) Transmission design of the zero inertia powertrain. Dissertation, TU Eindhoven, The Netherlands

258. Van Mierlo J, Van den Bossche P, Maggetto G (2003) Models of energy sources for EV and HEV: fuel cells, batteries, ultracapacitors, flywheels and engine-generators. Journal of Power Sources 128(1):78–89

259. Vasile I, Higelin P, Charlet A, Chamaillard Y (2006) Downsized engine torque lag compensation by pneumatic hybridization. Proc. of the 13th International Conference on Fluid Flow Technologies, Budapest, Hungary

260. Veenhuizen PA, Bonsen B, Klaassen TWGL, Albers P, Changenet C, Poncy S (2004) Pushbelt CVT efficiency improvement potential of servo-electromechanical actuation and slip control. Proc. of the International Continuously Variable and Hybrid Transmission Congress, Davis, CA

261. Vidyasagar M (1978) Nonlinear systems analysis. Prentice-Hall, Englewood Cliffs, NJ

262. Villeneuve A (2004) Dual mode electric infinitely variable transmission. Proc. of the International Continuously Variable and Hybrid Transmission Congress, Davis, CA

263. Von Raumer T, Dion JM, Dugard L, Thomas JL (1994) Applied nonlinear control of an induction motor using digital signal processing. IEEE Transactions on Control Systems Technology 2:327–335

264. Vroemen BG, Serrarens A, Veldpaus F (2001) Hierarchical control of the zero inertia powertrain. JSAE Review 22:519–526

265. Vroemen BG (2001) Component control for the zero inertia powertrain. Dissertation, TU Eindhoven, The Netherlands

266. Vyas AD, Ng HK, Santini DJ, Anderson JL (1997) Batteries for electric drive vehicles: evaluation of future characteristics and costs through a Delphi study. SAE Paper 970506

267. Yoda S, Ishihara K (1999) The advent of battery based society and the global environment in the 21st century. Journal of Power Sources 81–82:162–169

268. Yoon H, Lee S (2003) An optimized control strategy for parallel hybrid electric vehicle. SAE Paper 2003-01-1329

269. Wallentowitz H, Ludes R (1994) System control application for hybrid vehicles. Proc. of the 3rd IEEE Conference on Control Applications, Glasgow, UK

270. West JGW (1994) DC, induction, reluctance and PM motors for electric vehicles. Power Engineering Journal 8(2):77–88

271. Wiartalla A, Pischinger S, Bornscheuer W, Fieweger K, Ogrzwalla J (2000) Compressor expander units for a fuel cell systems. SAE Paper 2000-01-0380

272. Wipke KB, Cuddy MR, Burch SD (1999) ADVISOR 2.1: a user-friendly advanced powertrain simulation using a combined backward/forward approach. IEEE Transactions on Vehicular Technology 48(6):1751–1761

273. Wipke K, Markel T, Nelson D (2001) Optimizing energy management strategy and degree of hybridization for a hydrogen fuel cell SUV. Proc. of the 18th Electric Vehicles Symposium, Berlin, Germany

274. Wittmer C (1996) Entwurf und Optimierung der Fahrstrategien für ein Personenwagen-Antriebskonzept mit Schwungradkomponente und stufenlosem Zweibereichsgetriebe. Dissertation ETH no. 11672, Swiss Federal Institute of Technology, Zurich, Switzerland

275. Wittmer C, Guzzella L, Dietrich P (1996) Optimized control strategies for a hybrid car with a heavy flywheel. at–Automatisierungstechnik 44(7):331-337

276. Won JS, Langari R, Ehsani M (2005) An energy management and charge sustaining strategy for a parallel hybrid vehicle with CVT. IEEE Trans. on Control Systems Technology 13(2):313–320

277. Wu B, Dougal R, White RE (2001) Resistive companion battery modeling for electric circuit simulations. Journal of Power Sources 93:186–200

278. Wu B, Lin CC, Filipi Z, Peng H, Assanis D (2002) Optimization of power management strategies for a hydraulic hybrid medium truck. Proc. of the 6th International Symposium on Advanced Vehicle Control, Hiroshima, Japan
279. Wyczalek FA (1999) Market mature 1998 hybrid electric vehicles. IEEE AES Systems Magazine March:41–44
280. Wyczalek FA (2001) Hybrid electric vehicles year 2000 status. IEEE AES Systems Magazine March:15–19
281. Xue X, Tang J, Smirnova A, England R, Sammens N (2004) System level lumped-parameter dynamic modeling of PEM fuel cell. Journal of Power Sources 133:188–204
282. Zhang R, Chen Y (2001) Control of hybrid dynamical systems for electric vehicles. Proc. of the American Control Conference, Arlington, VA
283. Zhu Y, Chen Y, Chen Q (2002) Analysis and design of an optimal energy management and control system for hybrid electric vehicles. Proc. of the 19th Electric Vehicles Symposium, Busan, Korea
284. Zhu Y, Chen Y, Tian G, Wu H, Chen Q (2004) A four-step method to design an energy management strategy for hybrid vehicles. Proc. of the 2004 American Control Conference, Boston, MA

Printing: Krips bv, Meppel
Binding: Stürtz, Würzburg